LONDON MATHEMATICAL SOCIETY LECTURE NOTE SERIES

Managing Editor: Professor M. Reid, Mathematics Institute, University of Warwick, Coventry CV4 7AL,
United Kingdom

The titles below are available from booksellers, or from Cambridge University Press at
http://www.cambridge.org/mathematics

London Mathematical Society Lecture Note Series: 411

Moduli Spaces

Edited by

LETICIA BRAMBILA-PAZ
Centro de Investigación en Matemáticas A.C. (CIMAT), Mexico

OSCAR GARCÍA-PRADA
Consejo Superior de Investigaciones Científicas, Madrid

PETER NEWSTEAD
University of Liverpool

RICHARD P. THOMAS
Imperial College London

CAMBRIDGE
UNIVERSITY PRESS

University Printing House, Cambridge CB2 8BS, United Kingdom

Cambridge University Press is part of the University of Cambridge.

It furthers the University's mission by disseminating knowledge in the pursuit of
education, learning and research at the highest international levels of excellence.

www.cambridge.org
Information on this title: www.cambridge.org/9781107636385

First published 2014

A catalogue record for this publication is available from the British Library

ISBN 978-1-107-63638-5 Paperback

Contents

Preface

A programme on Moduli Spaces was held from 4 January to 1 July 2011 at the Isaac Newton Institute for Mathematical Sciences in Cambridge, UK. This volume is based on courses and lectures that took place during this semester and reflects some of the main themes that were covered during the activities.

Moduli spaces play a fundamental role in geometry. They are geometric versions of parameter spaces. That is, they are geometric spaces which parametrise something – each point represents one of the objects being parametrised, such as the solution of a particular equation, or a geometric structure on some other object. In the language of physics, a moduli space models the degrees of freedom of the solutions of some system of equations.

The programme was very successful, with a great deal of activity taking place. There were three main areas of research involved in the programme, namely derived categories, Higgs bundles and character varieties, and vector bundles and coherent systems. Topics that were covered included BPS invariants of 3-folds from derived categories of sheaves, and their motivic and categorified refinements, Hodge polynomials of character varieties, motives of moduli spaces of Higgs bundles and their relation to BPS invariants, Gromov–Witten invariants, notions of stability, Bridgeland stability, stability for pairs, geometric invariant theory constructions, wall-crossing formulae using Kirwan blow-ups, d-manifolds, a motivic version of Göttsche's conjecture, the Hilbert scheme of the moduli space of vector bundles, derived categories of quiver representations, mirror symmetry conjecture, ramified non-abelian Hodge theory correspondence, Hitchin fibration and real forms, parahoric bundles, parabolic Higgs bundles and representations of fundamental groups of punctured surfaces, Higgs bundles on Klein surfaces, Higgs bundles and groups of Hermitian type, Higgs bundles over elliptic curves, geometry of moduli spaces of vortices, coherent systems and geometry of moduli of curves, Brill–Noether loci for fixed determinant, Green's conjecture, Butler's conjecture, etc.

We saw progress on many of these topics in real time; it is fair to say that the state of the art looked very different at the end of the six months than it did in the introductory school at the beginning.

Acknowledgements

We are indebted to the authors of these articles for their outstanding contributions and to the referees for the care with which they have read the articles and the helpful suggestions they have made. We thank the speakers in the School on Moduli Spaces held in January 2011 on which this book is based and all participants in the school and in the more extensive programme on Moduli Spaces held from 4 January to 1 July 2011.

Our most grateful thanks are due to the Isaac Newton Institute for funding and hosting this activity. The staff of the Institute were unfailingly helpful. We also acknowledge Cambridge University Press for their help in publishing the volume.

Leticia Brambila-Paz
Oscar García-Prada
Peter Newstead
Richard Thomas

Contributors

Kai Behrend, *Department of Mathematics, University of British Columbia, Vancouver, BC V6T 1Z2, Canada.* behrend@math.ubc.ca

Wu-Yen Chuang, *Department of Mathematics, National Taiwan University, Taipei, Taiwan.* wychuang@gmail.com

Duiliu-Emanuel Diaconescu, *NHETC, Rutgers University, Piscataway, NJ 08854-0849, USA.* duiliu@physics.rutgers.edu

Guang Pan, *Rutgers University, Piscataway, NJ 08854-0849, USA.* guangpan@gmail.com

Peter B. Gothen, *Centro de Matematica, Universidade de Porto, Porto 4167-007, Portugal.* pbgothen@fc.up.pt

Daniel Huybrechts, *Mathematisches Institut, Universität Bonn, Bonn 53115, Germany.* huybrech@math.uni-bonn.de

Dominic Joyce, *The Mathematical Institute, University of Oxford, Oxford OX2 6GG, UK.* joyce@maths.ox.ac.uk

Rahul Pandharipande, *Departement Mathematik, ETH Zürich, 8092 Zürich, Switzerland.* rahul@math.ethz.ch

Richard P. Thomas, *Department of Mathematics, Imperial College London, London SW7 2AZ, UK.* rpwt@ic.ac.uk

1

Introduction to algebraic stacks

K. Behrend
The University of British Columbia

Abstract

These are lecture notes based on a short course on stacks given at the Isaac Newton Institute in Cambridge in January 2011. They form a self-contained introduction to some of the basic ideas of stack theory.

Contents

Moduli Spaces, eds. L. Brambila-Paz, O. García-Prada, P. Newstead and R. Thomas. Published by Cambridge University Press. © Cambridge University Press 2014.

Introduction

Stacks and algebraic stacks were invented by the Grothendieck school of algebraic geometry in the 1960s. One purpose (see [11]) was to give geometric meaning to higher cohomology classes. The other (see [9] and [2]) was to develop a more general framework for studying moduli problems. It is the latter aspect that interests us in this chapter. Since the 1980s, stacks have become an increasingly important tool in geometry, topology and theoretical physics.

Stack theory examines how mathematical objects can vary in families. For our examples, the mathematical objects will be the triangles, familiar from Euclidean geometry, and closely related concepts. At least to begin with, we will let these vary in continuous families, parametrized by topological spaces.

A surprising number of stacky phenomena can be seen in such simple cases. (In fact, one of the founders of the theory of algebraic stacks, M. Artin, is famously reputed to have said that one need only understand the stack of triangles to understand stacks.)

This chapter is divided into three parts, Sections 1.1, 1.2, and 1.3. Section 1.1 is a very leisurely and elementary introduction to stacks, introducing the main ideas by considering a few elementary examples of topological stacks. The only prerequisites for this section are basic undergraduate courses in abstract algebra (groups and group actions) and topology (topological spaces, covering spaces, the fundamental group).

Section 1.2 introduces the basic formalism of stacks. The prerequisites are the same, although this section is more demanding than the preceding one.

Section 1.3 introduces algebraic stacks, culminating in the Riemann–Roch theorem for stacky curves. The prerequisite here is some basic scheme theory.

We do not cover much of the "algebraic geometry" of algebraic stacks, but we hope that these notes will prepare the reader for the study of more advanced texts, such as [16] or the forthcoming book.[1]

The following outline uses terminology that will be explained in the body of the text.

The first fundamental notion is that of a *symmetry groupoid of a family of objects*. This is introduced first for discrete and then for continuous families of triangles.

In Sections 1.1.1–1.1.3, we consider Euclidean triangles up to similarity (the stack of such triangles is called $\mathfrak{M}$). We define what a fine moduli space is, and show how the symmetries of the isosceles triangles and the equilateral triangle prevent a fine moduli space from existing. We study the coarse moduli space of triangles, and discover that it parametrizes a *modular family*, even though this family is, of course, not universal.

Sections 1.1.4–1.1.6, introduce other examples of moduli problems. In Section 1.1.4, we encounter a fine moduli space (the fine moduli space of scalene triangles); in Section 1.1.5, where we restrict attention to isosceles triangles, we encounter a coarse moduli space supporting several non-isomorphic modular families. Restricting attention entirely to the equilateral triangle, in Section 1.1.6, we come across a coarse moduli space that parametrizes a modular family which is versal, but not universal.

In Section 1.1.7, we finally exhibit an example of a coarse moduli space which does not admit any modular family at all. We start studying *oriented triangles*. We will eventually prefer working with oriented triangles, because they are more closely related to algebraic geometry. The stack of oriented triangles is called $\widetilde{\mathfrak{M}}$.

In Section 1.1.8, we first make a few general and informal remarks about stacks and their role in the study of moduli problems.

The second fundamental concept is that of *versal family*. Versal families replace universal families, where the latter do not exist. Stacks that admit versal families are called *geometric*, which means *topological* in Sections 1.1 and 1.2, but will mean *algebraic* in Section 1.3.

We introduce versal families in Section 1.1.9, and give several examples. We explain how a stack which admits a versal family is essentially equal to the stack of 'generalized moduli maps' (or torsors, in more advanced terminology).

In Section 1.1.10, we start including degenerate triangles in our examinations: triangles whose three vertices are collinear. The main reason we do this

---

[1] Contact Martin Olsson, www.math.berkeley.edu/~molsson.

is to provide examples of compactifications of moduli stacks. There are several different natural ways to compactify the stack of triangles. There is a naïve point of view, which we dismiss rather quickly. We then explain a more interesting and natural, but also more complicated, point of view: in this, the stack of degenerate triangles turns out to be the quotient stack of a bipyramid modulo its symmetries, which form a group of order 12. This stack of degenerate triangles is called $\overline{\mathfrak{M}}$.

We encounter a very useful construction along the way: the construction of a stack by *stackification*, which means first describing families only locally, then constructing a versal family, and then giving the stack as the stack of generalized moduli maps to the universal family (or torsors for the symmetry groupoid of the versal family).

We then consider oriented degenerate triangles and introduce the *Legendre family* of triangles which is parametrized by the Riemann sphere. It exhibits the stack of oriented degenerate triangles as the quotient stack of the Riemann sphere by the action of the dihedral group with six elements. (In particular, it endows the stack of oriented degenerate triangles with the structure of an algebraic, not just topological, stack.) We call this stack $\mathfrak{L}$, and refer to it as the *Legendre compactification* of the stack of oriented triangles $\widetilde{\mathfrak{M}}$.

The Legendre family provides the following illustration of the concept of *generalized moduli map* (or groupoid torsor). We try to characterize, i.e. completely describe, the similarity type of an (oriented, maybe degenerate) triangle, by specifying the complex cross-ratio of its three vertices together with the point at infinity. However, the cross-ratio is not a single-valued invariant, but rather a multi-valued one: the six possible values of the cross-ratio are acted upon by the group S_3. Thus the stack $\mathfrak{L}$ of (oriented, maybe degenerate) triangles is the quotient stack of the Riemann sphere divided by S_3.

In Section 1.1.11, we explain how to relate different versal families for the same stack with one another, and how to recognize two stacks as being essentially the same, by exhibiting a bitorsor for the respective symmetry groupoids of respective versal families. We apply this both ways: we exhibit two different versal families for "non-pinched" triangles and show how a bitorsor intertwines them. Then we construct a bitorsor intertwining two potentially different moduli problems, namely two potentially different ways to treat families containing "pinched" triangles, thus showing that the two moduli problems are equivalent.

In Section 1.1.12, we introduce another compactification of the moduli stack of oriented triangles, which we call the *Weierstrass compactification*, because we construct it from the family of degree 3 polynomials in Weierstrass normal form. We denote this stack by $\mathfrak{W}$. We encounter our first example of a

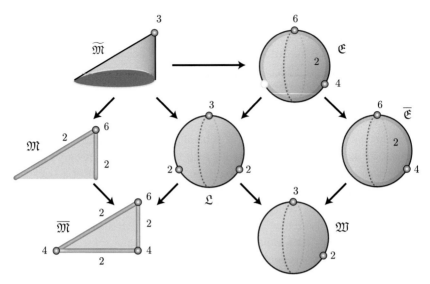

Figure 1.1. Some of the stacks we encounter in these notes, and the morphisms between them. Stacky points (coloured blue) are labeled with the order of their isotropy groups.

non-trivial morphism of stacks, namely the natural morphism $\mathfrak{L} \to \mathfrak{W}$. We also introduce a holomorphic coordinate on the coarse moduli space of oriented triangles known as the j-invariant.

In Section 1.2, we introduce the formalism of stacks. This will allow us to discuss topological stacks in general, without reference to specific objects such as triangles.

In Sections 1.2.1–1.2.4 we discuss the standard notions. We start with categories fibered in groupoids, which formalize what a moduli problem is. Then come the prestacks, which have well-behaved isomorphism spaces, and allow for the general definition of versal family. After a brief discussion of stacks, we define topological stacks to be stacks that admit a versal family. We discuss the basic fact that every topological stack is isomorphic to the stack of torsors for the symmetry groupoid of a versal family. This also formalizes our approach to stackification: start with a prestack, find a versal family, and then replace the given prestack by the stack of torsors for the symmetry groupoid of the versal family.

In Section 1.2.5 we discuss a new idea: symmetry groupoids of versal families should be considered as gluing data for topological stacks, in analogy to atlases for topological manifolds. This also leads to the requirement that the parameter space of a versal family should reflect the local topological structure of a stack faithfully, and, conversely, that a topological stack should locally behave in a manner controlled by the parameter space of a versal family, in order that we can "do geometry" on the stack.

This idea leads to the introduction of *étale versal families*, and the associated stacks, which we call *Deligne–Mumford topological stacks*, in analogy with the algebraic case. We prove a structure theorem that says that every separated Deligne–Mumford topological stack has an open cover by finite group quotient stacks.

This shows that all "well-behaved" moduli problems with discrete symmetry groups are locally described by finite group quotients. Therefore, the seemingly simple examples we start out with in fact turn out to be quite typical of the general case.

We also encounter examples of moduli problems without symmetries, which nevertheless do not admit fine moduli spaces. For sufficiently badly behaved equivalence relations (when the quotient map does not admit local sections), the quotient space is not a fine moduli space.

In Section 1.2.6, we continue our series of examples of moduli problems related to triangles by considering lattices up to homothety. This leads to the stack of elliptic curves, which we call $\mathfrak{E}$, and its compactification, $\overline{\mathfrak{E}}$. We see another example of a morphism of stacks, namely $\mathfrak{E} \to \mathfrak{W}$, which maps a lattice to the triangle of values of the Weierstrass $\wp$-function at the half periods. This is an example of a $\mathbb{Z}_2$-gerbe.

As an illustration of some simple "topology with stacks," we introduce the fundamental group of a topological stack in Section 1.2.7, and compute it for some of our examples.

Section 1.3 is a brief introduction to algebraic stacks. The algebraic theory requires more background than the topological one: we need, for example, the theory of cohomology and base change. We will therefore assume that the reader has a certain familiarity with scheme theory as covered in [15].

We limit our attention to algebraic stacks with affine diagonal. This avoids the need for algebraic spaces as a prerequisite. For many applications, this is not a serious limitation. As typical examples, we discuss the stack of elliptic curves $\mathfrak{E}$ and its compactification $\overline{\mathfrak{E}}$, as well as the stack of vector bundles on a curve.

Our definition of algebraic stack avoids reference to Grothendieck topologies, algebraic spaces, and descent theory. Essentially, a category fibered in groupoids is an algebraic stack if it is equivalent to the stack of torsors for an algebraic groupoid. Sometimes, for example for $\overline{\mathfrak{E}}$, we can verify this condition directly. We discuss a useful theorem, which reduces the verification that a given groupoid fibration is an algebraic stack to the existence of a versal family, with sufficiently well-behaved symmetry groupoid, and the gluing property in the étale topology.

We include a discussion of the coarse moduli space in the algebraic context: the theory is much more involved than in the topological case. We introduce algebraic spaces as algebraic stacks "without stackiness." We sketch the proof that separated Deligne–Mumford stacks admit coarse moduli spaces, which are separated algebraic stacks. As a by-product, we show that separated Deligne–Mumford stacks are locally, in the étale topology of the coarse moduli space, finite group quotients.

We then define what vector bundles and coherent sheaves on stacks are, giving the bundle of modular forms on $\overline{\mathfrak{E}}$ as an example. In a final Section 1.3.6, we study stacky curves, and as an example of some algebraic geometry over stacks we prove the Riemann–Roch theorem for orbifold curves. As an illustration, we compute the well-known dimensions of the spaces of modular forms.

1.1 Topological stacks: triangles

This section is directed at the student of mathematics who has taken an introduction to topology (covering spaces and the fundamental group) and an introduction to abstract algebra (group actions). Most of the formal mathematics has been relegated to exercises, which can be skipped by the reader who lacks the requisite background. The end of these exercises is marked with the symbol "□."

We are interested in two ideas, *symmetry* and *form*, and their role in *classification*.

1.1.1 Families and their symmetry groupoids

Consider a mathematical concept, for example *triangle*, together with a notion of isomorphism, for example *similarity*. This leads to the idea of *symmetry*. Given an object (for example, an isosceles triangle)

a *symmetry* is an isomorphism of the object with itself (for example, the reflection across the "axis of symmetry"). All the symmetries of an object form a *group*, the *symmetry group* of the object. (The symmetry group of our isosceles triangle is {id, refl}.)

To capture the essence of *form*, in particular how form may vary, we consider *families* of objects rather than single objects (for example, the family of four triangles

 (1.1)

consisting of three congruent isosceles triangles and one equilateral triangle).

Definition 1.1. A **symmetry** of a family of objects is an isomorphism of one member of the family with another member of the family.

Example 1.2. The family (1.1) of four triangles has 24 symmetries: there are two symmetries from each of the isosceles triangles to every other (including itself), adding up to 18, plus six symmetries of the equilateral triangle.

If we restrict the family to contain only the latter two isosceles triangles and the equilateral triangle,

the family has 14 symmetries.

Various types of symmetry groupoids
Definition 1.3. The collection of all symmetries of a given family is called the **symmetry groupoid** of the family.

Example 1.4. (Set) The symmetry groupoid of a family of non-isomorphic asymmetric objects

consists of only the trivial symmetries, one for each object. Such a groupoid is essentially the same thing as the set of objects in the family (or, more precisely, the indexing set of the family).

Example 1.5. (Equivalence relation) The symmetry groupoid of a family of asymmetric objects

is *rigid*. From any object to another there is at most one symmetry. A rigid groupoid is essentially the same thing as an equivalence relation on the set of objects (or the indexing set of the family).

Example 1.6. (Group) The symmetry groupoid of a single object

is a group.

Example 1.7. (Family of groups) The symmetry groupoid of a family of non-isomorphic objects

is a family of groups.

Example 1.8. (Transformation groupoid) Consider again the family of triangles (1.1) above, but now rearranged like this:

This figure has dihedral symmetry, and so the dihedral group with six elements, i.e. the symmetric group on three letters S_3, acts on this figure. Each element of S_3 defines four symmetries of the family, because it defines a symmetry originating at each of the four triangles.

For example, the rotation by $\frac{2\pi}{3}$ (or the permutation $1 \longmapsto 3$, $3 \longmapsto 2$, $2 \longmapsto 1$), gives rise to the $\frac{2\pi}{3}$-rotational symmetry of the equilateral triangle in the center of the figure, as well as three isomorphisms, each from one isosceles triangle to another.

The reflection across a vertical line (or the permutation $1 \mapsto 2$, $2 \mapsto 1$, $3 \mapsto 3$) gives rise to reflectional symmetries of the triangles labeled 0 and 3, as well as an isomorphism and its inverse between the two isosceles triangles labeled 1 and 2.

The family (1.1) is in fact so symmetric that every one of its symmetries comes from an element of the dihedral group acting on the figure (1.1).

More formally, let $\mathscr{F} = (\mathscr{F}_i)_{i=0,1,2,3}$ be the family of triangles, and let Γ be its symmetry groupoid. Then we have a bijection

$$\{0, 1, 2, 3\} \times S_3 \longrightarrow \Gamma, \tag{1.2}$$

$$(i, \sigma) \longmapsto \phi_{i,\sigma},$$

where $\phi_{i,\sigma} : \mathscr{F}_i \to \mathscr{F}_{\sigma(i)}$ is the symmetry from the triangle $\mathscr{F}_i$ to the triangle $\mathscr{F}_{\sigma(i)}$ induced by the geometric transformation of the whole figure defined by σ.

The action of S_3 on the figure induces an action on the indexing set $\{0, 1, 2, 3\}$, and the symmetry groupoid is completely described by this group action. The bijection (1.2) is an *isomorphism of groupoids*.

Whenever we have an arbitrary group G acting on a set X, we obtain an associated transformation groupoid $\Gamma = X \times G$.

Example 1.9. Here are three more examples of families of triangles whose symmetry groupoids are transformation groupoids:

In the first, we have added six scalene (i.e. completely asymmetric) triangles to the family (1.1). The indexing set of the family has ten elements, and the group S_3 acts on this set of ten elements, in a way induced by the symmetries of the figure.

In the second case, the family consists of two isosceles and four scalene triangles. The symmetry group of the figure is the dihedral group with four elements (which is isomorphic to $\mathbb{Z}_2 \times \mathbb{Z}_2$).

In the last case, the family consists of four isosceles triangles, and the symmetry group of the figure is the dihedral group with eight elements, D_4.

In each of the three cases, the symmetry groupoid of the family of triangles is equal to the transformation groupoid given by the action of the symmetry group of the figure on the indexing set of the family. Note that, in each case, the number of times a certain triangle appears in the family is equal to the number of elements in the symmetry group of the figure divided by the number of symmetries of the triangle.

Removing triangles breaks the symmetry, and leads to families whose symmetry groupoids are no longer transformation groupoids:

Exercise 1.10. A **groupoid** Γ consists of two sets: the set of objects Γ_0 and the set of arrows Γ_1. Also part of Γ are the *source* and *target* maps $s, t : \Gamma_1 \to \Gamma_0$, as well as the *groupoid operation* $\mu : \Gamma_2 \to \Gamma_1$, where Γ_2 is the set of *composable pairs*, which is the fibered product

$$
\begin{array}{ccc}
\Gamma_2 & \xrightarrow{\;p_2\;} & \Gamma_1 \\
{\scriptstyle p_1}\big\downarrow & & \big\downarrow{\scriptstyle s} \\
\Gamma_1 & \xrightarrow{\;t\;} & \Gamma_0
\end{array}
$$

We write $\mu(\alpha, \beta) = \alpha * \beta$ for composable pairs of arrows $(\alpha, \beta) \in \Gamma_2$, where $t(\alpha) = s(\beta)$:

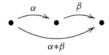

Three properties are required to hold:

(i) (Identities) For every object $x \in \Gamma_0$, there exists an arrow $e_x \in \Gamma_1$, whose source and target are x and such that $e_x * \alpha = \alpha$, for all α with source x, and $\beta * e_x = \beta$, for all β with target x.

(ii) (Inverses) For every arrow $\alpha \in \Gamma_1$, there exists an arrow α^{-1} such that $\alpha * \alpha^{-1} = e_{s(\alpha)}$ and $\alpha^{-1} * \alpha = e_{t(\alpha)}$.

(iii) (Associativity) For every triple (α, β, γ) of composable arrows, we have $(\alpha * \beta) * \gamma = \alpha * (\beta * \gamma)$.

Using the language of categories, we note that a groupoid is nothing but a small category, all of whose arrows are invertible. Often it is more natural to use categorical notation for the groupoid operation: $\beta \circ \alpha = \alpha * \beta$.

The symmetry groupoid of a family parametrized by the set T has $\Gamma_0 = T$.

An **isomorphism of groupoids** consists of two bijections: $\Gamma_0 \to \Gamma_0'$ and $\Gamma_1 \to \Gamma_1'$, compatible with the composition (and hence identities and inverses).

□

1.1.2 Continuous families

So far, we have considered discrete families. More interesting are continuous families. For example, suppose, given a piece of string of length 2 and two pins distanced $\frac{1}{2}$ from each other, we draw part of an ellipse:

(1.3)

We start with a 3:4:5 right triangle, whose sides have lengths $\frac{1}{2}$, $\frac{2}{3}$, and $\frac{5}{6}$, and we end up with a congruent 3:4:5 triangle. This is a family of triangles parametrized by an interval. To make this more explicit, suppose that we are in the plane $\mathbb{R}^2$ and the two pins have coordinates $(-\frac{1}{4}, 0)$ and $(\frac{1}{4}, 0)$. Let us take the interval $[-\frac{1}{4}, \frac{1}{4}]$ as parameter space, and let us denote the family of triangles by $\mathscr{F}$. Then every parameter value $t \in [-\frac{1}{4}, \frac{1}{4}]$ corresponds to a triangle $\mathscr{F}_t$, where $\mathscr{F}_t$ is the triangle subtended by the string when the x-coordinate of the pen point is t:

(1.4)

In (1.4) we see another view of this family, "lying over" the parameter space. The five family members $\mathscr{F}_{-1/4}$, $\mathscr{F}_{-1/8}$, $\mathscr{F}_0$, $\mathscr{F}_{1/8}$, $\mathscr{F}_{1/4}$ are highlighted (but, of course, the mind's eye is supposed to fill in the other family members).

The group with two elements $\mathbb{Z}_2$ acts on diagram (1.3) by reflection across the y-axis. This action induces all symmetries of the family $\mathscr{F}$. The symmetry groupoid of $\mathscr{F}$ is given by the induced action of $\mathbb{Z}_2$ on the parameter space $[-\frac{1}{4}, \frac{1}{4}]$ (where the non-identity element of $\mathbb{Z}_2$ acts by multiplication by -1). The symmetry groupoid of the family (1.4) is the transformation groupoid $[-\frac{1}{4}, \frac{1}{4}] \times \mathbb{Z}_2$.

Gluing families

One essential feature of continuous families is that they can be "glued." The first and last members of the family $\mathscr{F}$ of (1.4) are similar to each other, and we can therefore glue the two endpoints of the parameter interval to obtain a circle, and glue the two corresponding triangles to obtain a family of triangles parametrized by the circle.

First, we bend the second half of the family around:

$\mathscr{F}$:

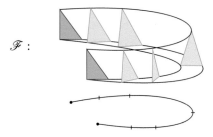

and then we glue the parameter interval and the two end triangles:

$\widetilde{\mathscr{F}}$:

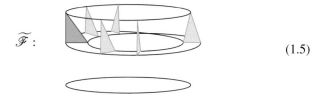

(1.5)

(These figures are not to scale.) The new parameter space is the circle S^1. Let us call the family of triangles we obtain in this way $\widetilde{\mathscr{F}}$ over S^1, or $\widetilde{\mathscr{F}}/S^1$. This family exhibits an interesting feature: the shortest sides of all the triangles in the family put together form a Moebius band. The perimeters of all triangles in the family put together form a Klein bottle. There is no way to label the vertices of the family members in a consistent way with labels A, B, C, say. Equivalently, it is impossible to label the sides of the triangles in the family consistently with labels a, b, c. (On the other hand, such a consistent labeling is always possible if we vary the parameter values inside the circle S^1 only a little, but not too much. We say that *locally* we can label the vertices of the family member triangles consistently.)

Exercise 1.11. As we are studying triangles up to similarity, let us fix the perimeter of the triangles we consider. We can take any value, but for aesthetic reasons we will take the value 2.

Formally, let us then define a continuous family of triangles parametrized by the topological space T to consist of a degree 3 covering map $T' \to T$

and a continuous map $a : T' \to \mathbb{R}_{>0}$. The data $\mathscr{F} = (T', a)$ have to satisfy the triangle inequalities for all $t \in T$. More precisely, for $t \in T$, there are three points of T' lying over t, call them t'_1, t'_2, t'_3, and three positive real numbers, $a(t'_1), a(t'_2), a(t'_3)$. The latter have to satisfy the three triangle inequalities: $a(t'_1) + a(t'_2) > a(t'_3), a(t'_2) + a(t'_3) > a(t'_1), a(t'_3) + a(t'_1) > a(t'_2)$. Moreover, for every $t \in T$, we require $a(t'_1) + a(t'_2) + a(t'_3) = 2$. The triangle $\mathscr{F}_t$ corresponding to the parameter value $t \in T$ is then the triangle whose sides have lengths $a(t'_1), a(t'_2)$, and $a(t'_3)$.

Suppose that $\mathscr{F}/T$ and $\mathscr{G}/T$ are two families of triangles parametrized by the same space T, where $\mathscr{F} = (T', a)$ and $\mathscr{G} = (T'', b)$. Define an **isomorphism** of families of triangles $\phi : \mathscr{F} \to \mathscr{G}$ to consist of a homeomorphism of covering spaces $f : T' \to T''$ (f has to commute with the projections to T) such that $a = b \circ f$.

Prove that the family $\widetilde{\mathscr{F}}$ we constructed in (1.5) is indeed a continuous family of triangles according to this formal definition. Prove that families of triangles can be glued, i.e. that they satisfy the following *gluing axiom*.

Suppose that $T = A \cup B$ is a topological space with closed subsets $A \subset T$, $B \subset T$, that $\mathscr{F}/A$ and $\mathscr{G}/B$ are continuous families of triangles, and that $\phi : \mathscr{F}|_{A \cap B} \to \mathscr{G}|_{A \cap B}$ is an isomorphism. Then there exists, in an essentially unique way, a continuous family of triangles $\mathscr{H}/T$, and isomorphisms $\psi : \mathscr{H}|_A \cong \mathscr{F}$ and $\chi : \mathscr{H}|_B \cong \mathscr{G}$, such that $\phi \circ \psi|_{A \cap B} = \chi|_{A \cap B}$. □

Exercise 1.12. A formulation of the gluing principle which has wider applicability than the one alluded to in Exercise 1.11, is the following.

Suppose that the space T is the union of a family of open subsets $U_i \subset T$, and that over each U_i we have a continuous family of triangles (or other mathematical objects) $\mathscr{F}_i$, parametrized by U_i. Assume that over each intersection $U_{ij} = U_i \cap U_j$ we are given an isomorphism of families $\phi_{ij} : \mathscr{F}_i|_{U_{ij}} \to \mathscr{F}_j|_{U_{ij}}$, and that over all triple overlaps the compatibility condition (the *cocycle* condition) $\phi_{ik}|_{U_{ijk}} = \phi_{jk}|_{U_{ijk}} \circ \phi_{ij}|_{U_{ijk}}$ holds. (The data $(\{\mathscr{F}_i\}, \{\phi_{ij}\})$ are called *gluing data* for a continuous family.)

Then there exists a continuous family $\mathscr{F}$, parametrized by T, together with isomorphisms of families $\phi_i : \mathscr{F}|_{U_i} \to \mathscr{F}_i$, such that over the overlaps we have $\phi_j|_{U_{ij}} = \phi_{ij} \circ \phi_i|_{U_{ij}}$. (The pair $(\mathscr{F}, \{\phi_i\})$ is said to be *obtained by gluing* from the above gluing data.)

The pair $(\mathscr{F}, \{\phi_i\})$ is unique in the following sense: given $(\mathscr{G}, \{\psi_i\})$, solving the same gluing problem, there exists an isomorphism of families $\chi : \mathscr{F} \to \mathscr{G}$ such that on each open U_i we have $\psi_i \circ \chi|_{U_i} = \phi_i$.

If our notion of a continuous family of some type of mathematical object has this gluing property (i.e. for every space T, and for every gluing data over T, the solution exists and is essentially unique in the described way), we say that these types of families *can be glued*.

Prove that families of triangles can be glued.   □

1.1.3 Classification

Our goal is to describe the totality of our mathematical objects as a space: in our example of triangles up to similarity, we would like a space whose points correspond in a one-to-one fashion to similarity classes of triangles. Such a space would be called a *moduli space* of triangles up to similarity, and it would be said to solve the *moduli problem* posed by triangles up to similarity.

In fact, such a space is easily constructed. Every triangle is similar to a triangle of perimeter 2, say, and every triangle of perimeter 2 is given (up to congruence) by the lengths of its sides, which we can label a, b, c, where $a \leq b \leq c$. Thus an example of a space whose points correspond to similarity classes of triangles is

$$M = \{(a, b, c) \in \mathbb{R}^3 \mid a \leq b \leq c, a + b + c = 2, c < a + b\}. \quad (1.6)$$

The space M is a subspace of $\mathbb{R}^3$. Every point (a, b, c) in M defines the triangle whose sides have lengths a, b, and c. Every triangle is similar to one of these. Different points in M give rise to non-similar triangles. This is a pictorial representation of M:

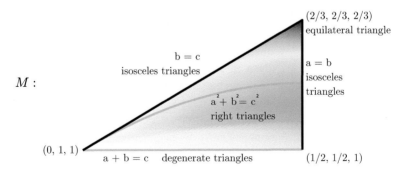

The space M contains two boundary lines of isosceles triangles; it does not contain its third boundary line, where the triangles degenerate. The curve defined by $a^2 + b^2 = c^2$ is indicated, which is the locus of right triangles. Above this curve are the triangles with three acute angles; below this curve are the triangles with one obtuse angle. (The shading corresponds to the size of the angle opposing the side c.)

In the following sketch of M, we mark a few representative points and display the corresponding triangles:

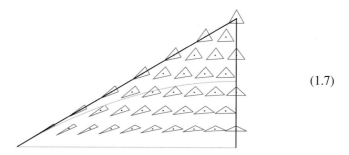

(1.7)

The triangles are displayed with c as base, b as left edge, and a as right edge. Isosceles triangles are highlighted.

This is already quite a satisfying picture: it gives us an overview of all triangles up to similarity. But now we note that it does much more: it also describes *continuous families* of triangles. A path in M

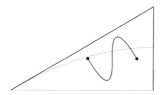

gives rise to a family of triangles:

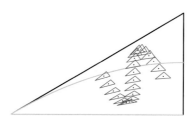

This is a family parametrized by an interval:

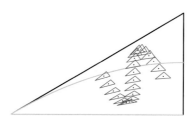

(1.8)

The shape of the triangles in the family is determined completely by the path in M.

Pulling back families

This is an example of *pullback of families*. It is a basic property of families of mathematical objects that they can be pulled back via any map to the parameter space.

The space M itself parametrizes a continuous family of triangles, which is sketched in (1.7) and which we shall denote by $\mathcal{M}$. Our path in M is a continuous map $\gamma : [0, 1] \to M$. Via the path γ, we pull back the family $\mathcal{M}/M$ to obtain a family parametrized by $[0, 1]$, which is denoted by $\gamma^* \mathcal{M}$. This family is defined in such a way that

$$(\gamma^* \mathcal{M})_t = \mathcal{M}_{\gamma(t)}, \qquad \text{for all } t \in [0, 1].$$

The family $\gamma^* \mathcal{M}/[0, 1]$ is displayed in (1.8).

Exercise 1.13. The symmetry groupoid Γ of the family $\mathcal{M}$ is a family of groups, as no distinct family members are isomorphic. As a topological space, Γ looks like this:

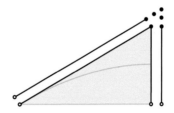

Over the isosceles, but not the equilateral, locus, the fibers of Γ are groups with two elements; over the equilateral locus, the fiber of Γ is isomorphic to S_3.

□

Moduli map of a family

Conversely, every continuous family of triangles $\mathcal{F}/T$, parametrized by a space T, gives rise to a continuous map $T \to M$, the **moduli map** of $\mathcal{F}$. The moduli map takes the point $t \in T$ to the point $(a, b, c) \in M$, where the triangle $\mathcal{F}_t$ has side lengths $a \le b \le c$.

Example 1.14. For example, the moduli map of the family $\mathcal{F}$ of (1.4) is the path in M that starts at the 3:4:5 triangle on the curve of right triangles, follows the line orthogonal to the $b = c$ isosceles edge (this is the line $a = \frac{1}{2}$) until it reaches this edge, and then retraces itself until it comes back to the curve of right triangles:

Example 1.15. The moduli map of the family $\gamma^* \mathcal{M}$ over $[0, 1]$, displayed in (1.8), is, of course, the path $\gamma : [0, 1] \to M$, which gave rise to it.

Exercise 1.16. Prove that the moduli map of a family given by T'/T and $a : T' \to \mathbb{R}_{>0}$ is continuous. Continuity of the moduli map $T \to M$ is a local property, so you can assume that the cover $T' \to T$ is trivial and that the triangle is given by three continuous functions f, g, h on T, representing the lengths of the sides of the triangles in your family. Then $f \leq g \leq h$ defines a closed subspace of T, on which the moduli map is identified with (f, g, h), and is therefore continuous. Other conditions, such as $g \leq f \leq h$, give rise to other moduli maps, such as (g, f, h), which are also continuous. The gluing lemma finishes the proof. □

Fine moduli spaces

One may be tempted to think therefore that M does not classify just triangles, but also *families* of triangles, in the sense that families parametrized by T are in one-to-one correspondence, via their moduli map, with continuous maps $T \to M$. This leads to the following definition.

Definition 1.17. A **fine moduli space** is a space M such that

  (i) the points of M are in one-to-one correspondence with isomorphism classes of the objects we are studying;

 (ii) (technical condition) for every family $\mathscr{F}/T$, the associated moduli map $T \to M$ (which maps the point $t \in T$ to the isomorphism class of the family member $\mathscr{F}_t$) is continuous;

(iii) every continuous map from a space T to M is the moduli map of some family parametrized by T (equivalently, M parametrizes a family $\mathcal{M}$, whose moduli map is the identity id_M);

(iv) if two families have the same moduli map, they are isomorphic families.

Is our space M from (1.6) a fine moduli space for triangles up to similarity? We have constructed M so that (i) would be satisfied. The technical condition (ii) can be checked if one agrees on a formal mathematical definition of a continuous family of triangles (see Exercise 1.16).

We have seen how the continuous map $\gamma : [0, 1] \to M$ gives rise to a family over $[0, 1]$, whose moduli map is γ. We can do the same thing for any map $f : T \to M$, from an arbitrary space T to M. We can use f to pull back the family $\mathcal{M}$ to a family $f^*\mathcal{M}$, and this family has moduli map f. Thus, condition (iii) is satisfied.

Note that applying condition (iii) to $T = M$ and the identity map id_M says that the space M parametrizes a family $\mathcal{M}$, such that $\mathcal{M}_m$ represents the isomorphism class corresponding to m by condition (i), for all $m \in M$. Let us call such a family a **modular family**. Our family of triangles (1.7) is a modular family.

What about condition (iv)?

Consider again the family $\mathcal{F}/I$, where I is the interval $[-\frac{1}{4}, \frac{1}{4}]$, from (1.4).

We will now construct another family $\mathcal{G}/I$, parametrized by the same interval I. The family $\mathcal{G}$ is equal to the family $\mathcal{F}$ over the first half of the interval $[-\frac{1}{4}, 0]$, but then goes back to the starting point rather than continuing on beyond the isosceles triangle in the middle:

For a value $t \in [0, \frac{1}{4}]$, the corresponding triangle is subtended by the string when the pen point has x-coordinate $-t$.

Note that the family $\mathcal{G}$ has the same moduli map as the family $\mathcal{F}$; see Example 1.14.

On the other hand, the two families $\mathcal{F}$ and $\mathcal{G}$ are not isomorphic. Here is another representation of the two families, with $\mathcal{F}$ on the left and $\mathcal{G}$ on the right:

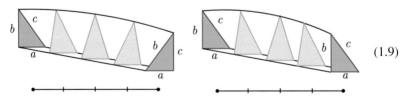

$$(1.9)$$

If we label the sides of the initial triangles a, b, and c in such a way that $a < b < c$, and then label all the following triangles in the respective families in a *continuous* way, then the two end triangles, which are congruent, are labeled differently: one in such a way that $a < b < c$, the other such that $a < c < b$. This shows that the two families are essentially different. We can look at it another way: if we try to construct a continuous isomorphism between the two families, in the first half of the interval, $[-\frac{1}{4}, 0]$, this isomorphism would simply translate a triangle from $\mathscr{F}$ over to the corresponding triangle in $\mathscr{G}$. In the second half of the interval, $[0, \frac{1}{4}]$, an isomorphism would have to translate the triangle from $\mathscr{F}$ over, and then reflect it, to map it onto the corresponding triangle in $\mathscr{G}$. In the middle, at the isosceles triangle, we get two contradicting requirements: continuity from the left requires us to translate the isosceles triangle over; continuity from the right requires us to reflect this isosceles triangle across. There cannot be a continuous isomorphism between the two families $\mathscr{F}$ and $\mathscr{G}$.

We conclude that M is not a fine moduli space because two non-isomorphic families of triangles have the same moduli map. In fact, there does not exist any fine moduli space of triangles. Our two families are pointwise the same, so define the same moduli map to *any* potential fine moduli space.

The family $\mathscr{G}$ is obtained by pulling back the modular family $\mathscr{M}/M$ via the common moduli map of $\mathscr{F}$ and $\mathscr{G}$. The family $\mathscr{F}$ cannot be obtained by pulling back $\mathscr{M}$: any family pulled back from M can be labeled compatibly and continuously with a, b and c, such that $a < b < c$ everywhere, because $\mathscr{M}$ has this property.

Of course, it is easy to see who the culprit is: it is the isosceles triangle, which has a non-trivial symmetry, which allows us to cut the family $\mathscr{F}$ in the middle,

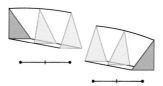

and reassemble it in two different ways. Just gluing it back together, we get $\mathscr{F}$ back:

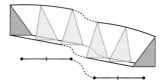

However, flipping the isosceles triangle in the middle while gluing:

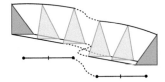

gives us the family $\mathscr{G}$.

Coarse moduli spaces

As property (iv) is violated, M is not a fine moduli space of triangles, but it does satisfy the following definition.

Definition 1.18. A **coarse moduli space** is a space M such that the first two conditions of Definition 1.17 are satisfied, and moreover (technical condition) M carries the finest topology, making condition (ii) true.

Remark 1.19. There is essentially only *one* coarse moduli space for any mathematical notion. It can be constructed as follows: take the set of isomorphism classes of objects under consideration as points of M, and then endow M with the finest topology such that all moduli maps of all continuous families are continuous. This gives a coarse moduli space. Any other coarse moduli space is necessarily homeomorphic to this one. Thus it is customary to speak of *the* coarse moduli space. (This requires the class of objects to be small enough for isomorphism classes to form a set.)

Exercise 1.20. Any space satisfying the first three conditions of Definition 1.17 is a coarse moduli space. (The converse is not true; see Section 1.1.7 for an example.) □

Exercise 1.21. A subset $U \subset M$ of the coarse moduli space is open if and only if it defines an *open condition* on continuous families. This means that, for every continuous family $\mathscr{F}/T$, the set of $t \in T$ such that $[\mathscr{F}_t] \in U$ is open in T. □

Remark 1.22. It is possible for a coarse moduli space to carry several non-isomorphic modular families. (For an example, see Section 1.1.5.)

Remark 1.23. The existence of a fine moduli space implies the existence of pullbacks of families (they correspond to composition of maps).

Remark 1.24. If M is a fine moduli space and $\mathcal{M}$ is a modular family, every continuous family $\mathcal{F}/T$ is the pullback of $\mathcal{M}$ via its moduli map, and therefore $\mathcal{M}$ is called a **universal family**. Any other modular family is isomorphic to the pullback of $\mathcal{M}$ via the identity, in other words isomorphic to $\mathcal{M}$. So there is essentially only one universal family, and one speaks of *the* universal family.

To conclude: the coarse moduli space of triangles up to similarity is isomorphic to the space M from (1.6). It admits a modular family, but no universal family, so there is no fine moduli space of triangles. Moreover, the coarse moduli space of triangles is a 2-dimensional manifold with boundary.

Let us consider a few related classification problems.

1.1.4 Scalene triangles

Here we provide an example of a fine moduli space.

Recall that a triangle is *scalene* if all three sides have different lengths. Scalene triangles are completely asymmetric: they each have a trivial symmetry group. There exists a fine moduli space for scalene triangles. In fact, remove the boundary from M to obtain

$$M' = \{(a, b, c) \in \mathbb{R}^3 \mid a < b < c, a + b + c = 2, c < a + b\}.$$

Let us denote the family parametrized by M' by $\mathcal{M}'/M'$.

We claim that $\mathcal{M}'$ is a universal family for scalene triangles.

To see this, we have to show that every continuous family of scalene triangles $\mathcal{F}/T$ is isomorphic to the pullback of $\mathcal{M}'$ via the moduli map of $\mathcal{F}$.

The key observation is that in a scalene triangle there is never any ambiguity as to which side is the shortest and which side is the longest. So, in a continuous family of scalene triangles we can unambiguously label the sides

a, b, and c, where the shortest side is labeled a and the longest is labeled c. (So it is impossible to construct families of scalene triangles such as $\mathscr{F}$ from (1.4), where the longest side jumps.)

In the family $\mathscr{M}'$, the sides are already labeled in this way. Therefore, in the pullback $f^*\mathscr{M}'$, where $f : T \to M'$ is the moduli map of $\mathscr{F}$, the sides are again labeled in this way.

We can now use this labeling of the sides to define an isomorphism

$$\mathscr{F} \xrightarrow{\sim} f^*\mathscr{M}' , \tag{1.10}$$

by sending the side labeled a to the side labeled a, the side labeled b to the side labeled b, and the side labeled c to the side labeled c. Note that without the canonical labeling of the sides of $\mathscr{F}$ it is impossible to define (1.10).

Exercise 1.25. This defines an isomorphism of families because, for every $t \in T$, the lengths of the sides of $\mathscr{F}_t$ are given by the triple of real numbers $f(t)$ (by definition of the moduli map f), and this triple of real numbers gives the lengths of the sides of $(f^*\mathscr{M}')_t = \mathscr{M}'_{f(t)}$, by the definition of the modular family $\mathscr{M}'$. Prove that (1.10) is a continuous isomorphism of families (i.e. an isomorphism of continuous families). $\qquad\square$

We conclude that scalene triangles admit a fine moduli space (and a universal family), which is a 2-dimensional manifold.

1.1.5 Isosceles triangles

Next we shall demonstrate a coarse moduli space with several modular families.

Let us consider all isosceles triangles. These are classified, up to similarity, by the angle which subtends the two equal sides. This angle can take any value between 0 and π. If $\mathscr{I}/T$ is a continuous family of isosceles triangles, this angle defines a continuous function $T \to (0, \pi)$, the moduli map of $\mathscr{I}$. Of course, the interval $(0, \pi)$ parametrizes a continuous family of triangles:

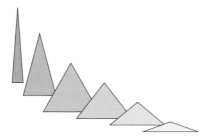

which is a modular family, i.e. the triangle over the point $\gamma \in (0, \pi)$ has angle γ subtending the two equal sides. Therefore, we see that the interval $(0, \pi)$ is a coarse moduli space for isosceles triangles.

Note, however, that there are two further continuous families of isosceles triangles parametrized by $(0, \pi)$, which are modular:

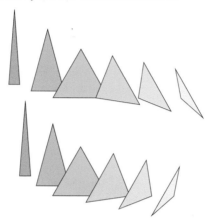

In fact, these latter two families are isomorphic, via reflection across the vertical. But they are not isomorphic to the first family. For the case of the first family, the isosceles angles all fit together continuously (they are always at the top). For the latter two families, the isosceles angle changes position at the equilateral triangle. This is essentially different behavior.

The two essentially different families are competing for the title of "universal family of isosceles triangles." Of course, only one family can carry this title, so there is no universal family. Each of these two families describes one possible way a one-parameter family of isosceles triangles can "pass through" the equilateral triangle.

We conclude that the coarse moduli space of isosceles triangles is a 1-dimensional connected manifold, and it admits two non-isomorphic modular families.

Exercise 1.26. The symmetry groupoids of the two modular families look like this:

$\square$

1.1.6 Equilateral triangles

Let us restrict attention entirely to equilateral triangles. These are, of course, all similar to each other, which seems to indicate that the classification should be quite trivial.

In fact, the one-point space $*$ is, of course, a coarse moduli space for equilateral triangles up to similarity. Let us pick an equilateral triangle, and call it δ. Then the single triangle δ is a continuous family of equilateral triangles parametrized by $*$, and it is a modular family.

Every family pulled back from $\delta/*$ is *trivial*, or *constant*.

On the other hand, there are families of equilateral triangles that are not at all trivial. For example, the following family is parametrized by the circle. It was obtained by taking the trivial family over a closed interval, and gluing the first and last triangle with a $\frac{2\pi}{3}$ twist:

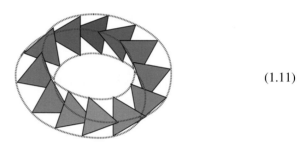

$$(1.11)$$

The dotted line indicates an attempt to label continuously one (and only one) vertex in each triangle with A. This is impossible. Instead, the dotted line defines a degree 3 cyclic cover of the parameter circle.

So $\delta/*$ is not a universal family, and there exists no fine moduli space for equilateral triangles.

The symmetry groupoid of $\delta/*$ is the group S_3, the symmetric group on three letters a, b, c.

Exercise 1.27. In fact, the vertices of any continuous family of equilateral triangles over a topological space T form a degree 3 covering space of T; conversely, every degree 3 covering space of T defines a continuous family of equilateral triangles.   □

1.1.7 Oriented triangles

We obtain an interesting variation on our moduli problem by considering oriented triangles. This means that similarity transformations between triangles are only rotations, translations, and scalings, but not reflections. In this context, the equilateral triangle is the only triangle with non-trivial symmetries; all other isosceles triangles have lost their symmetry. The symmetry group of the oriented equilateral triangle is the cyclic group with three elements.

For now, let us agree that an *oriented triangle* is a triangle with a cyclic ordering of its edges (or vertices). Any isomorphism of triangles has to preserve this cyclic ordering. (There are two ways to order cyclically the edges of a triangle.)

All scalene triangles have two oriented incarnations, which are transformed into each other by a reflection. For example, the two incarnations of the 3:4:5 right triangle are the following:

For the one on the left, the cyclic ordering of the edges according to ascending length is counterclockwise; for the one on the right it is clockwise. There is no oriented similarity transformation of the plane that makes these two triangles equal. On the other hand, isosceles triangles have only one oriented version: if two isosceles triangles are similar, they can be made equal by an oriented similarity transformation, not involving any reflections.

The coarse moduli space for oriented triangles is therefore "twice as big" as the one for unoriented triangles, which we called M. We can construct it by starting with the space

$$\widetilde{M}^{\text{pre}} = \{(a, b, c) \in \mathbb{R}^3 \mid a \leq c, b \leq c, a + b + c = 2, c < a + b\},$$

and gluing together the two boundary lines $b = c$ and $a = c$ of isosceles triangles, as indicated (with the locations of the 3:4:5 and the 4:3:5 triangle marked):

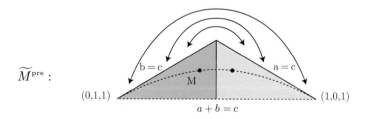

Let us call the resulting space $\widetilde{M}$. Thus, $\widetilde{M}$ can be pictured as the surface of a cone with solid angle $\frac{2\pi}{3}$ steradians at its vertex:

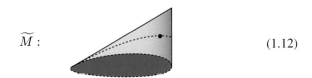

$$\widetilde{M}: \hspace{4cm} (1.12)$$

Every oriented triangle corresponds to a unique point in $\widetilde{M}$: given an oriented triangle, label its sides by a, b, c in such a way that a, b, and c appear in alphabetical order when going around the triangle counterclockwise, and such that the longest side is labeled c. Then rescale the triangle until it has perimeter 2. The resulting side lengths define a unique point in $\widetilde{M}$.

Here is the family of triangles parametrized by $\widetilde{M}^{\mathrm{pre}}$, before gluing:

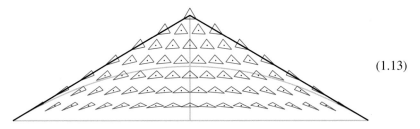

$$(1.13)$$

We can consider the family $\mathscr{F}$ of (1.4), obtained by drawing an ellipse as in (1.3), also as a family of oriented triangles. It is a family starting at the 3:5:4 right triangle and ending at the 3:4:5 right triangle. The corresponding path in $\widetilde{M}^{\mathrm{pre}}$ looks like this:

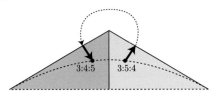

We want the oriented family $\mathscr{F}$ to define a continuous moduli map $[-\frac{1}{4}, \frac{1}{4}] \rightarrow \widetilde{M}$. That is why we have to glue the two boundaries of $\widetilde{M}^{\mathrm{pre}}$ together. We cannot simply remove one of the edges $b = c$ or $a = c$ in defining $\widetilde{M}$, because that would make the moduli map of the oriented family $\mathscr{F}$ discontinuous at the "break point," seen in the sketch.

Exercise 1.28. Formally define a continuous family of oriented triangles parametrized by the topological space T to consist of a *cyclic* degree 3 covering $T' \rightarrow T$ and a continuous map $a : T' \rightarrow T$, satisfying the conditions of Exercise 1.11. (A cyclic cover of degree 3 is a covering space $T' \rightarrow T$, together with a given deck transformation $\sigma : T' \rightarrow T'$, which induces a degree 3 permutation in each fiber of $T' \to T$.) Isomorphisms of families of oriented triangles are defined as in Exercise 1.11, with the additional requirement that the isomorphism of covering spaces has to commute with the respective deck transformations σ.

Prove that $\widetilde{M}$ is a coarse moduli space of oriented triangles. □

Does there exist a modular family over $\widetilde{M}$? Can different families have the same moduli map?

The answer to the second question is "yes." Non-isomorphic families with the same moduli map can be constructed in the same way as before: take a family $\mathscr{F}$, parametrized by $[-\epsilon, \epsilon]$, such that $\mathscr{F}_t$ is non-equilateral for $t \neq 0$ and equilateral for $t = 0$. Then create a new family $\mathscr{F}'$ by gluing $\mathscr{F}|_{[-\epsilon,0]}$ and $\mathscr{F}|_{[0,\epsilon]}$ together at $t = 0$, by using a non-trivial symmetry of the equilateral triangle:

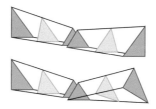

In this sketch, the triangles in the second half of the lower family are obtained by rotating the triangles in the second half of the upper family by $\frac{2\pi}{3}$ clockwise. Then the families are glued together along the central equilateral triangles. The two families have the same moduli map, which is a path in $\widetilde{M}$ connecting the 4:3:5 right triangle with the 3:4:5 right triangle, passing through the "vertex" of $\widetilde{M}$ at the equilateral triangle:

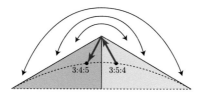

Yet, the two families are essentially different: in one of them, the longest side "jumps"; in the other it does not. Any isomorphism between the two families would consist of a family of translations for $t \in [-\epsilon, 0]$ and a family of translation-rotations for $t \in [0, \epsilon]$. This family of isomorphisms is not continuous at $t = 0$, and so our two families are different as continuous families (as discrete families they would, of course, be isomorphic because they are isomorphic pointwise at each parameter value).

One reason to introduce oriented triangles at this point is that they provide an example where there exists *no* modular family over the coarse moduli space.

To see this, we proceed by contradiction. Suppose that there exists a continuous modular family of oriented triangles $\widetilde{\mathcal{M}}/\widetilde{M}$. Then, very close to the vertex point of $\widetilde{M}$ (i.e. in some, maybe very small, open neighborhood of this point), we can consistently label the vertices of the family $\widetilde{\mathcal{M}}$ in some way. Hence, when we restrict $\widetilde{\mathcal{M}}$ to small enough loops around the vertex of $\widetilde{M}$, these restricted families can also be consistently labeled.

Consider such a loop around the vertex of $\widetilde{M}$:

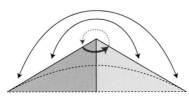

Because this loop avoids all oriented triangles with symmetries, it corresponds to a unique continuous family parametrized by the circle S^1. The family parametrized by the open circle in $\widetilde{M}^{\mathrm{pre}}$ would look something like this:

Of course, if the path in $\widetilde{M}^{\mathrm{pre}}$ is very close to the vertex, the triangles in the family will be very close to equilateral. For clarity, we have depicted the family corresponding to a path further from the vertex. Note how following along a path in (1.13) from the left equilateral edge to the right equilateral edge gives rise to a family as displayed here. The induced loop in $\widetilde{M}$ parametrizes the family obtained by gluing together the two triangles at the end (here implemented by bending the two ends downward):

$$(1.14)$$

Examining this family, we see that it does not admit an unambiguous labeling. Rather, any attempt at such a labeling will run up against a cyclic degree 3 cover of the parameter circle, just like for (1.11).

No matter how close the loop in $\widetilde{M}$ is to the vertex, the corresponding family will always have this feature. So there is no way that any putative modular family $\widetilde{\mathscr{M}}$ could have a consistent vertex labeling, even in a tiny neighborhood of the equilateral vertex. So $\widetilde{\mathscr{M}}$ cannot exist.

Non-equilateral oriented triangles

To look at $\widetilde{M}$ another way, we can flatten out the cone until $\widetilde{M}$ becomes a disc. Alternatively, we can bend the two isosceles edges of $\widetilde{M}^{\text{pre}}$ around, shortening them, and then glue them to get this view of $\widetilde{M}$:

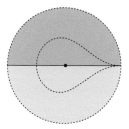

If we follow along with the family parametrized by $\widetilde{M}^{\text{pre}}$, see (1.13), we get

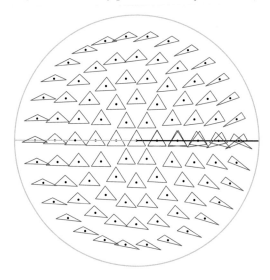

and we see that it is impossible to glue this family together in a consistent way. The equilateral triangle in the middle forces the acute isosceles triangles into this incompatible position.

If we remove the central point and the equilateral triangle, we can rotate all the remaining triangles a little, and then we can glue successfully:

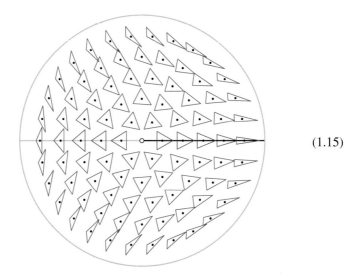

(1.15)

There is no way to put the equilateral triangle back into the center of this picture in a way compatible with the neighboring triangles, because the neighboring triangles exhibit rotation behavior when going around the center in small loops, as we saw in (1.14).

Note that (1.15) is the universal family of non-equilateral oriented triangles.

Recall that the universal family (and therefore *every* family) of scalene unoriented triangles admits a global consistent labeling of vertices. The universal family of non-equilateral oriented triangles does not admit a global labeling as there are families of such triangles which contain twists. Note how, even after removing the symmetric object, the universal family still retains some properties of this object: the universal family contains a twist by the symmetry group of the central object.

1.1.8 Stacks

Stacks are mathematical constructs invented to solve the various problems we encounter when studying moduli problems. Stacks are more general than spaces, but every space is a stack.

There exists no fine moduli space for triangles, but there does exist a fine moduli *stack* of triangles (although in the stack context the word "fine" is usually omitted). The moduli stack of triangles, let us call it $\mathfrak{M}$, parametrizes a universal family $\mathscr{U}/\mathfrak{M}$ of triangles. Every family of triangles $\mathscr{F}/T$ is isomorphic to the pullback of $\mathscr{U}/\mathfrak{M}$ via a continuous map $T \to \mathfrak{M}$, which is *essentially* unique. The word "essential" is key. An important difference between stacks and spaces is that, for a stack such as $\mathfrak{M}$, the continuous maps $T \to \mathfrak{M}$ do not form a set, but rather a groupoid, and, in fact, the groupoid of maps $T \to \mathfrak{M}$ is equivalent to the groupoid of families over T (for all T, in a way compatible with pullbacks of families).

Over the locus of scalene triangles, there is no difference between the coarse moduli space M and the fine moduli stack $\mathfrak{M}$ because, over the scalene locus, M is a fine moduli space. But the isosceles locus consists of so-called "stacky points" of $\mathfrak{M}$. There are two ways a path can pass through a stacky point representing an isosceles triangle, and six ways it can pass through the stacky point representing the equilateral triangle.

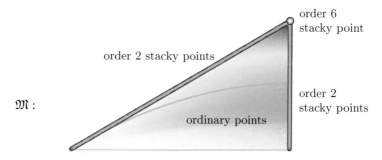

$\mathfrak{M}$:

order 6 stacky point

order 2 stacky points

order 2 stacky points

ordinary points

Let us call the stack of oriented triangles $\widetilde{\mathfrak{M}}$. It has one stacky point of order 3 in the center:

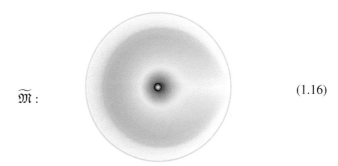

$\widetilde{\mathfrak{M}}$:

(1.16)

The mathematical definition of the notion of stack is a stroke of genius, or a cheap cop-out, depending on your point of view: one simply declares the problem to be its own solution!

The problem we had set ourselves was to describe all continuous families of triangles. We saw that this problem would be solved quite nicely by a universal family, if there was one. But there isn't one. So, instead of trying to single out one family to rule all others, we consider *all* families over *all* parameter spaces to *be* the moduli stack of triangles.

Thus, the notions of *moduli problem* and *stack* become synonymous.

The challenge is then to develop techniques for dealing with such a stack as a geometric object, as if it were a space. For this to be successful, we will need the existence of *versal families*. These fulfill a dual purpose: they allow us to do geometry with the moduli stack, and we can describe all families explicitly in terms of a versal family.

There are two difficulties with this:

(i) the description of the stack of all triangles in terms of a versal family is more complicated than the description in terms of a universal family;

(ii) there are many versal families, and so we also have to study how different versal families relate to each other.

But these problems cannot be avoided if the objects we are studying are symmetric.

1.1.9 Versal families

Consider again the family $\mathscr{F}/I$ from (1.4). The behavior of this family near the isosceles triangle is not modeled anywhere by the modular family $\mathscr{M}$ over the coarse moduli space M. So, to describe all possible families (even locally) we need to enlarge M. Here is a better family. The parameter space is

$$N = \{(a, b, c) \in \mathbb{R}^3 \mid a + b + c = 2, a, b, c < 1\};$$

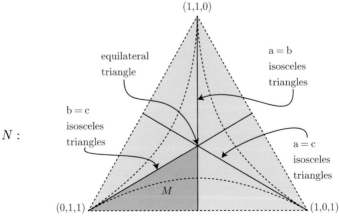

The members of the family $\mathcal{N}/N$ have their vertices labeled A, B, and C in a consistent way, and the side lengths of $\mathcal{N}_{(a,b,c)}$, for $(a, b, c) \in N$, are such that the side opposite vertex A has length a, the side opposite vertex B has length b, and the side opposite C has length c. (It is easy to check that the conditions on a, b, and c imply the three triangle inequalities.)

$\mathcal{N}$: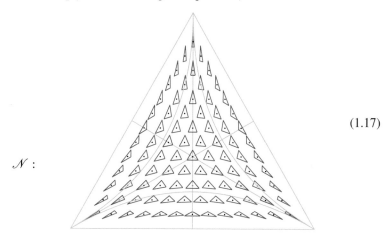

(1.17)

Note that the most symmetric of all triangles, the equilateral one, appears only once in $\mathcal{N}$, the isosceles triangles (except the equilateral one) appear three times each, and the scalene triangles appear six times each. In fact, the number of times a given triangle appears is inversely proportional to the number of its symmetries.

The key feature of the family $\mathcal{N}/N$ is that it models all possible local behaviors of continuous families of triangles, as we shall see in the following.

Let us examine the symmetry groupoid of the family $\mathcal{N}/N$. Note that (1.17) looks very much like a more elaborate (in fact continuous) version of Example 1.8. The symmetry group of (1.17) is the dihedral group with six elements, i.e. S_3. Every element $\sigma \in S_3$ defines a transformation $\sigma : N \to N$ as well as, for every $n \in N$, an isomorphism $\mathcal{N}_n \to \mathcal{N}_{\sigma(n)}$. (In fact, these isomorphisms combine into an isomorphism of families $\mathcal{N} \to \sigma^* \mathcal{N}$.) Every similarity of some triangle $\mathcal{N}_n$ with another triangle $\mathcal{N}_{n'}$ in (1.17) comes about in this way. Therefore, the symmetry groupoid of the family $\mathcal{N}/N$ is equal to the transformation groupoid $N \times S_3$, given by the action of S_3 on the parameter space N. (The geometric action via rotations and reflections on (1.17) induces the permutation action on the components of the points of N.)

Exercise 1.29. Define the *canonical topology* on the symmetry groupoid $\Gamma \rightrightarrows T$ of a continuous family $\mathcal{F}$, parametrized by the space T, to be the finest topology on Γ with the following property: for any space S, and any triple (f, ϕ, g), where $f : S \to T$ and $g : S \to T$ are continuous maps, and $\phi : f^*\mathscr{F} \to g^*\mathscr{F}$ is a continuous isomorphism of families, the induced map $S \to \Gamma$ is continuous. (The induced map $S \to \Gamma$ maps $s \in S$ to the element $\phi_s : \mathscr{F}_{f(s)} \to \mathscr{F}_{g(s)}$ in Γ.)

Define the *tautological isomorphism* over Γ to be the isomorphism $\phi : s^*\mathscr{F} \to t^*\mathscr{F}$ such that, for every $\gamma \in \Gamma$, the isomorphism $\phi_\gamma : \mathscr{F}_{s(\gamma)} \to \mathscr{F}_{t(\gamma)}$ is the isomorphism given by γ itself.

Prove that the canonical topology on the symmetry groupoid of the family $\mathscr{N}/N$ is the product topology on $N \times S_3$. Prove that the tautological isomorphism over Γ is a continuous isomorphism of families (i.e. an isomorphism of continuous families). $\qquad\square$

Exercise 1.30. Prove that N is a fine moduli space of *labeled triangles*. A labeled triangle is a triangle together with a labeling of the edges, with labels a, b, c. A scalene triangle, such as the 3:4:5 right triangle, has six different labelings. An isosceles but not equilateral triangle has three essentially different labelings, but the equilateral triangle has only one. $\qquad\square$

Generalized moduli maps

As individual triangles appear multiple times in $\mathscr{N}$, the moduli map of a family of triangles is multi-valued. For example, a loop of scalene triangles will have a six-valued moduli map, which may look something like this:

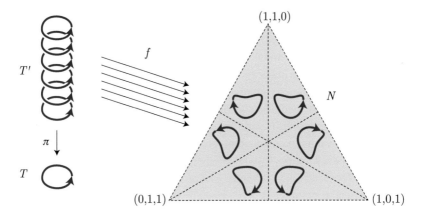

The group S_3 acts on the six image loops in N, and it therefore also acts on T', which is, in this case, a disjoint union of six copies of the parameter space $T = S^1$. Each component of T' corresponds to one way of labeling the triangles in the family.

More interesting is the moduli map of a Moebius family such as $\widetilde{\mathscr{F}}/S^1$ from (1.5):

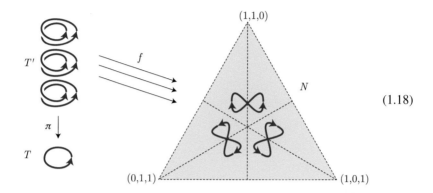

$$(1.18)$$

For this diagram, we have deformed the family a little: in the part before the equilateral triangle, we have made the triangles a little more acute (lengthened the string), and afterwards a little more obtuse (shortened the string). This is to avoid the moduli map collapsing to three lines in N, rather than three figure eights.

The image of the moduli map in N consists of six paths, which are joined head to tail in pairs. Technically, the six-valued moduli map $T \to N$ consists of a degree 6 covering $\pi : T' \to T$ and a continuous map $f : T' \to N$. The space T' can be viewed as the space of all labelings of the triangles in our family: for $t \in T$, the preimage $\pi^{-1}(t)$ consists of the six different ways the vertices of the triangle $\mathscr{F}_t$ can be labeled with the letters A, B, and C. The map $f : T' \to N$ then maps a labeled triangle to the triple of lengths of its sides (because the vertices are labeled, the side lengths form an ordered triple, which is a well-defined point in N).

The covering $\pi : T' \to T$ decomposes into three components: one component consists of those labelings where A is opposite the shortest side (recall that only the shortest side is globally consistent in $\mathscr{F}/T$, the two others swap). Another component consists of labelings where B is opposite the shortest side, and the last component consists of labelings where C is opposite the shortest side. In this way, the three components of T' are labeled with A, B, and C, too.

An important part of the structure of the moduli map $(T'/T, f)$ is the action of the group S_3 on T' and N, and the fact that f respects these actions (f is S_3-*equivariant*). On T' the group S_3 acts by changing the labeling. For example, the permutation which transposes A and B swaps the two components of T' called A and B, and induces the branch swap on the component labeled C. The same permutation acts on N by swapping a and b (it is the reflection across the vertical axis), and exactly mirrors the action on T'.

As another example, consider a family with a $\frac{2\pi}{3}$-twist, such as (1.14). Its moduli map $(T'/T, f)$ might look something like this:

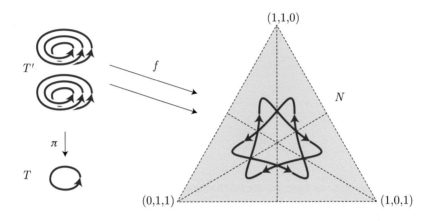

This procedure is general: every continuous family of triangles $\mathscr{F}$ over some parameter space T gives rise to an S_3-covering $T' \to T$, namely the space of labelings of $\mathscr{F}$, and an S_3-equivariant map $f : T' \to N$, given by the triple of the lengths of the labeled sides. The structure $(T'/T, f)$ replaces the concept of moduli map.

Exercise 1.31. Using the formal definition of a continuous family of triangles from Exercise 1.11, we construct the generalized moduli map as follows. Let $T' \to T$ be a degree 3 covering map *without* structure group, and let $a : T' \to \mathbb{R}_{>0}$ be a continuous map. Then define T'' to be the space of all maps $\ell : \{A, B, C\} \to T'$ which are bijections onto a fiber of $T' \to T$. Then $T'' \to T$ is a covering space of degree 6, endowed with a right S_3-action, hence a covering map *with* structure group. The map $f : T'' \to N$ defined by $f(\ell) = \big(a\ell(A), a\ell(B), a\ell(C)\big)$ is S_3-equivariant, and (T'', f) is the generalized moduli map of the continuous family of triangles (T', a). □

Reconstructing a family from its generalized moduli map

One of the problems with the coarse moduli space M was that we were not able to reconstruct a family from its moduli map. This problem we have now solved! Any family of triangles $\mathscr{F}/T$ is, in fact, completely determined by its generalized moduli map $(T'/T, f)$.

Consider, for example, a family $\widetilde{\mathscr{F}}/S^1$ with a Moebius twist, whose moduli map is displayed in (1.18). The way to obtain this family by gluing, as in (1.5), is completely encoded by $(T'/T, f)$. To see this, note that you can pick a closed interval $I \subset T'$, which maps down to $T = S^1$ in a one-to-one fashion, except that the two endpoints t_0, t_1 are glued together by the map π.

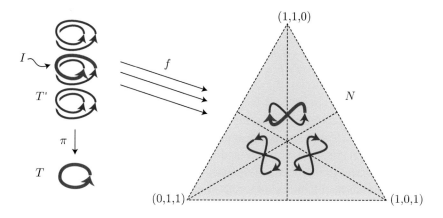

The two points $f(t_0)$ and $f(t_1)$ in N are related by a unique element $\sigma \in S_3$ (in this case, the reflection across the vertical), and this σ defines an isomorphism between the triangles over these two points, which are $\mathscr{N}_{f(t_0)}$ and $\mathscr{N}_{f(t_1)}$. Pull back the family $\mathscr{N}$ to obtain $f^*\mathscr{N}|_I$. This family on I can be glued together to a family over $T = S^1$, using the isomorphism $\sigma : \mathscr{N}_{f(t_0)} \to \mathscr{N}_{f(t_1)}$ of the two end triangles. The family obtained in this way is isomorphic to the family $\widetilde{\mathscr{F}}$ which gave rise to the generalized moduli map $(T'/T, f)$ in the first place.

Exercise 1.32. Using the formal definition of a family of triangles from Exercise 1.11, we construct a family of triangles from a generalized moduli map as follows. Let $T'' \to T$ be a covering map with structure group S_3 and let $f : T'' \to N$ be an S_3-equivariant continuous map. Construct the degree 3 cover *without* structure group $T' \to T$ as $T' = T'' \times_{S_3} \{A, B, C\}$. This notation means that T' is the quotient of $T'' \times \{A, B, C\}$ by the equivalence relation $(\ell\sigma, L) \sim (\ell, \sigma L)$, for all $\sigma \in S_3$, and $(\ell, L) \in T'' \times \{A, B, C\}$ (which amounts to a quotient of $T'' \times \{A, B, C\}$ by an action of S_3). Then define the continuous map $a : T' \to \mathbb{R}_{>0}$ by $a[\ell, L] = \mathrm{pr}_L f(\ell)$. Here, $\mathrm{pr}_L : N \to \mathbb{R}_{>0}$ denotes the projection onto the Lth component, and $[\ell, L] \in T'$ is the equivalence class of $(\ell, L) \in T'' \times \{A, B, C\}$.

Prove that the procedures described here and in Exercise 1.31 are inverses of each other. $\qquad\square$

Versal families: definition

The features of the family $\mathscr{N}/N$, which allow for generalized moduli maps to exist and for us to reconstruct any family from its generalized moduli map, are listed in the following definition:

Definition 1.33. A family $\mathscr{N}/N$ is a **versal family** if it satisfies the following conditions.

(i) Every family $\mathscr{F}/T$ is *locally* induced from $\mathscr{N}/N$ via pullback. This means that, for every $t \in T$, there exists a neighborhood U of t in T and a continuous map $f : U \to N$ such that the restricted family $\mathscr{F}|_U$ is isomorphic to the pullback family $f^*\mathscr{N}$.

(ii) (Technical condition) Endowing the symmetry groupoid $\Gamma \rightrightarrows N$ of the family $\mathscr{N}/N$ with its canonical topology (see Exercise 1.29), the source and target maps $s : \Gamma \to N$ and $t : \Gamma \to N$ are continuous, and the tautological isomorphism of families $\phi : s^*\mathscr{N} \to t^*\mathscr{N}$ is a continuous isomorphism of families (of whatever objects we are studying).

Our family $\mathscr{N}$ of triangles is a versal family. To see that condition (i) is satisfied, it is enough to remark that, in a small enough neighborhood U of any parameter value $t \in T$, the family of triangles $\mathscr{F}$ can be consistently labeled. And once a family over U is labeled, it is completely determined by the three continuous functions $a, b, c : U \to \mathbb{R}_{>0}$ giving the lengths of the three sides. (Even to talk about these three functions, we need labels on the edges of the triangles.) The three functions a, b, c define a continuous map $f : U \to N$, making the family over U isomorphic to $f^*\mathscr{N}$.

Condition (ii) was checked in Exercise 1.29.

Exercise 1.34. Suppose you have a moduli problem satisfying the gluing axiom (see Exercise 1.12), and $\mathscr{N}/N$ is a versal family for this moduli problem, whose symmetry groupoid is a transformation groupoid $N \times G$ for a (discrete) group G.

Associate to a family $\mathscr{F}/T$ a generalized moduli map by endowing the set of isomorphisms

$$T' = \{(n, \phi, t) \mid n \in N, t \in T, \text{ and } \phi : \mathscr{N}_n \to \mathscr{F}_t \text{ is an isomorphism}\}$$

with the structure of a covering space $T' \to T$ and a G-action, and then defining a continuous G-equivariant map $f : T' \to N$ by $f(n, \phi, t) = n$.

Conversely, given a G-cover $T' \to T$ and a G-equivariant map $f : T \to N$, construct a family $\mathscr{F}/T$ whose generalized moduli map is (T', f) as follows:

choose local sections of T'/T, i.e. choose an open covering $T = \bigcup_i U_i$, and continuous sections $s_i : U_i \to T'$, so that $\pi \circ s_i$ is equal to the inclusion $U_i \to T$. Pull back the family $\mathcal{N}$ via each $f \circ s_i$ to a family $\mathcal{F}_i$ over U_i. Over the intersections $U_{ij} = U_i \cap U_j$, you have two sections of π, namely $s_i|_{U_{ij}}$ and $s_j|_{U_{ij}}$, so there is a continuous map $\sigma_{ij} : U_{ij} \to G$, such that $s_i = s_j \sigma_{ij}$, because $T' \to T$ is a G-covering. Then σ_{ij} defines an isomorphism of restricted families $\sigma_{ij} : \mathcal{F}_i|_{U_{ij}} \xrightarrow{\sim} \mathcal{F}_j|_{U_{ij}}$ because G acts by symmetries on the family $\mathcal{N}$. Then $(\{\mathcal{F}_i\}, \{\sigma_{ij}\})$ is gluing data for the family $\mathcal{F}/T$. $\square$

Exercise 1.35. (For families of triangles) Prove that the coarse moduli spaces M and $\widetilde{M}$ are quotient spaces $M = N/S_3$ and $\widetilde{M} = N/\mathbb{Z}_3$.

(For families of arbitrary mathematical objects) Prove that if $\mathcal{N}/N$ is a versal family whose symmetry groupoid is the transformation groupoid $N \times G \rightrightarrows N$, for a (discrete) group G, the quotient space N/G is a coarse moduli space. $\square$

Isosceles triangles

We exhibit versal families for the other previously studied moduli problems related to triangles.

Recall that the coarse moduli space of isosceles triangles, the interval $(0, \pi)$, admitted two distinct modular families. They both exhibit essentially different behavior at the equilateral triangle, so neither of them is versal. To obtain a versal family of isosceles triangles, we can restrict the versal family of triangles $\mathcal{N}/N$ to the isosceles locus in N:

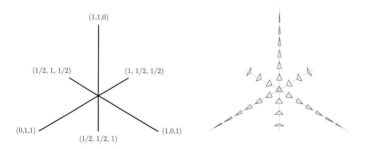

The description of generalized moduli maps is essentially the same as in the case of general triangles.

This versal family of isosceles triangles is not a manifold; it has a singularity at the equilateral triangle point. (In fact, the stack of isosceles triangles is singular.)

Equilateral triangles

The modular family consisting of one equilateral triangle $\delta/*$ is, in fact, a versal family. To see this, first let us recall that a family is constant if it is obtained by pullback from a family with one member, such as $\delta/*$. Then note that if a family of equilateral triangles can be consistently labeled (here indicated by consistent coloring of the vertices)

then it is isomorphic to a constant family:

(In this example, we can imagine the three colored strings being pulled taught, which will rotate the individual equilateral triangles, and rescale them if they have different sizes. In this way, we obtain an isomorphic family.)

So, because every family of equilateral triangles can be locally consistently labeled, every family of equilateral triangles is *locally constant*. Thus $\delta/*$ satisfies the first property required of a versal family.

Exercise 1.36. Complete the proof that $\delta/*$ is a versal family by proving that the canonical topology on the symmetry group S_3 of δ is the discrete topology.

□

For a family of equilateral triangles, the generalized moduli map $(T'/T, f)$ consists only of the S_3-covering T'/T; there is no information contained in $f : T' \to *$ because there is always a unique map to the one-point set from any space.

For a family of equilateral triangles $\mathscr{F}/T$, the associated S_3-cover is the space of all labelings of $\mathscr{F}$. Conversely, an S_3-covering T'/T of a space T encodes gluing data for a family of equilateral triangles because the symmetry group of the equilateral triangle is S_3.

Note that $\delta/*$ is a unique modular family (up to isomorphism), which is nevertheless not universal.

Oriented triangles

The same family $\mathcal{N}/N$ from (1.17) is also a versal family for oriented triangles. Simply declare all triangles in $\mathcal{N}/N$ to have alphabetical cyclic ordering on their edges. This makes $\mathcal{N}/N$ a family of oriented triangles. Any family of oriented triangles can locally be labeled alphabetically, and is therefore locally a pullback from the oriented $\mathcal{N}$.

The difference from the unoriented case is that the symmetry groupoid of the oriented $\mathcal{N}$ consists only of the cyclic subgroup with three elements $\mathbb{Z}_3 \cong A_3 \subset S_3$ acting on N. Therefore, generalized moduli maps of oriented triangles consist of pairs $(T'/T, f)$, where T'/T is a cyclic cover of degree 3, and $f : T \to N$ is an A_3-equivariant continuous map. Examples:

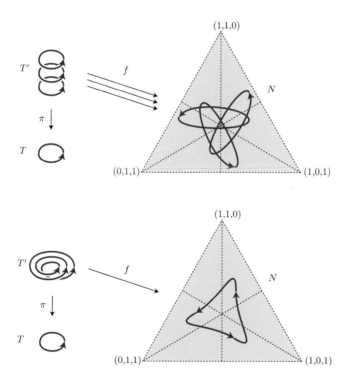

1.1.10 Degenerate triangles

For many reasons, it is nice to have a complete (i.e. compact) moduli space. (Completeness is needed, for example, for the Riemann–Roch theorem, and intersection theory in general.) The coarse moduli spaces M of triangles, and

$\widetilde{M}$ of oriented triangles, are not compact because they do not contain the boundary line of degenerate triangles.

Let us consider degenerate triangles. Things get very interesting now because there are many ways to think of degenerate triangles, all giving rise to different compactifications of the moduli stack $\mathfrak{M}$.

Let us return to our very first continuous family of triangles (1.4), which was obtained by drawing part of an ellipse. When the pen hits the horizontal line through the two pin points, the vertices of the triangle become collinear and the triangle becomes *degenerate*. (The triangle inequality becomes an equality.) Drawing the ellipse to this point or beyond does *not* define a family of triangles.

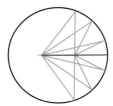

Therefore, we will broaden our point of view to include such degenerate triangles in our moduli problem. We would like to do this in such a way that drawing the complete ellipse leads to a continuous family of degenerate triangles.

On the level of the course moduli space, not much happens: we are just adding the boundary. In fact, one representation of the coarse moduli space of degenerate triangles is

$$\overline{M} = \{(a, b, c) \in \mathbb{R}^3 \mid a \leq b \leq c, a + b + c = 2, c \leq a + b\};$$

$\overline{M}$:

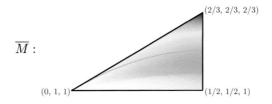

Note that this also adds the triangle $(0, 1, 1)$, which is not only collinear, but also has two of its vertices coinciding. Let us call this the *pinched triangle*. The symmetric degenerate triangle $(\frac{1}{2}, \frac{1}{2}, 1)$ we call the *bisected line segment*. The coarse moduli space $\overline{M}$ is compact.

There are several natural stack structures over this coarse moduli space. They depend on what we mean, exactly, by a continuous family of degenerate triangles. We will cover three approaches.

Lengths-of-sides viewpoint

In this scenario, a degenerate triangle is a set of three real numbers $\{a, b, c\}$, which are not required to be distinct or strictly positive, and which satisfy weak versions of the triangle inequality. More precisely,

(i) $a \geq 0$, $b \geq 0$, $c \geq 0$, but not all three are equal to 0,

(ii) $a + b \geq c$, $a + c \geq b$, and $b + c \geq a$.

We exclude the degenerate triangle whose sides all have length 0 because it is not represented in the coarse moduli space $\overline{M}$ (it cannot be rescaled to have perimeter 2). (It also has infinitely many symmetries, as all rescalings, rotations, and reflections preserve it, and we do not need it to compactify $\mathfrak{M}$. See, however, Exercise 1.59.) Let us denote the stack of degenerate triangles obtained in this way by $\overline{\mathfrak{M}}^{\text{naïve}}$:

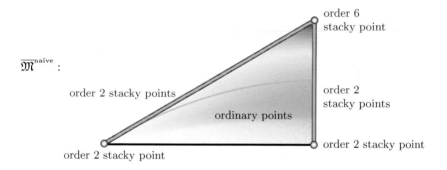

Up to similarity, we have added triangles with $(a, b, c) = (a, 1 - a, 1)$, with $0 \leq a \leq \frac{1}{2}$. None of these have any symmetries, except the two isosceles ones, $(0, 1, 1)$ and $(\frac{1}{2}, \frac{1}{2}, 1)$, which have two symmetries each, namely the swap of the two equal sides.

The closure $\overline{N}$ of N inside $\mathbb{R}^3$ supports a versal family, $\overline{\mathcal{N}}$, whose symmetry groupoid is $\overline{N} \times S_3$, so the behavior of $\overline{\mathfrak{M}}^{\text{naïve}}$ is quite similar to that of $\mathfrak{M}$: families of degenerate triangles are essentially the same thing as S_3-covering spaces together with S_3-equivariant continuous maps to $\overline{N}$.

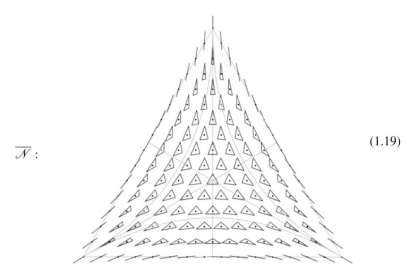

$$(1.19)$$

$\overline{\mathcal{N}}$:

The oriented version is very similar: we only replace S_3 by $\mathbb{Z}_3$. (In this scenario, an oriented triangle is a set of three numbers satisfying weak triangle inequalities with a cyclic ordering on the three numbers.)

This "naïve" point of view on the stack of degenerate triangles has the advantage of being no more complicated than the stack of non-degenerate triangles, and it leads to a "compact" moduli stack. Disadvantages are: the stack is a stacky version of *manifold with boundary*, and it also does not capture correctly the geometric nature of families of triangles.

Let us consider two (unoriented) families obtained by drawing part of the ellipse:

$\mathscr{F}$: $\mathscr{G}$:

Both are parametrized by an interval, start and end at the 3:4:5 right triangle, and have a degenerate triangle in the middle. It is instructive to compare these two families with the two families we considered in (1.9), which had an isosceles triangle in the middle, rather than a degenerate one. (It is of no consequence that this degenerate triangle is also isosceles. Everything would be the same if we shortened or lengthened the string giving rise to our two families a little.)

These two families are completely identical if all we take into account are the lengths of the three sides. If we label their edges consistently, the two labeled families have the same moduli map to $\overline{N}$. The generalized moduli maps of the unlabeled families to $\overline{N}$ are identical. (All this is in stark contrast to the two families from (1.9) crossing the equilateral locus, rather than the degenerate locus.)

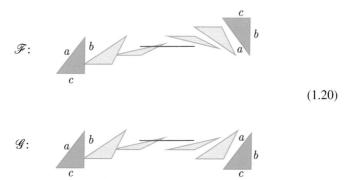

$$(1.20)$$

Yet, the two families are essentially different: an isomorphism between the two families would have to be the identity on the first part of the parameter interval and the reflection across the horizontal for the second half of the parameter interval. So the two families are not continuously isomorphic, if we require isomorphisms to be geometric similarity transformations of the plane the triangles are embedded inside, rather than just permutations and rescalings of the three lengths of the sides.

The difference between the degenerate triangle and the isosceles triangle is that the reflectional symmetry permutes the labels in the latter case, but cannot be detected by labels in the former case. Therefore, the point of view of a triangle as a collection of three numbers giving the lengths of the sides is inadequate for degenerate triangles. This naïve point of view cannot capture the reflectional symmetry of the degenerate triangles, which is present as a *consequence* of the fact that the two families $\mathscr{F}$ and $\mathscr{G}$ of (1.20) are different.

If we consider the two families $\mathscr{F}$ and $\mathscr{G}$ as families of oriented triangles, they are plainly different: $\mathscr{F}$ connects the anticlockwise 3:4:5 triangle with the clockwise one, whereas $\mathscr{G}$ has the anticlockwise one on both ends. But for these statements to hold true, we need "orientation" to be a structure on the ambient plane, not any structure defined on the unordered triple of the sides of the triangles.

We therefore have two notions of *orientation* on a degenerate triangle: a cyclic ordering on the sides (or vertices), or an orientation on the ambient plane. We call the latter an *embedded orientation*. In the case of non-degenerate triangles, these two kinds of orientations determine each other. In the case of degenerate triangles, neither determines the other. (An orientation on the plane can be specified by a cyclic ordering of the edges of an embedded non-degenerate triangle, or by a parametrized circle in the plane.)

Embedded viewpoint

We will now describe the second compactification of the stack of triangles, $\overline{\mathfrak{M}}$, which we call the stack of (degenerate) *embedded* triangles, if it needs emphasizing. We will do this by agreeing on what such families of triangles look like *locally* and exhibiting a versal family. Then the theory of generalized moduli maps gives an explicit description of all continuous families *globally*.

To tell the difference between the two families $\mathscr{F}$ and $\mathscr{G}$ from (1.9), it was sufficient to label the sides of the families consistently. Labels on the sides are not sufficient to tell the difference between the two families $\mathscr{F}$ and $\mathscr{G}$ from (1.20). Therefore, we introduce extra structure: this extra information is an orientation on the plane the triangles are contained in. Once we fix such an orientation (in addition to the labels), it is easy to tell the difference between the two families. For $\mathscr{G}$, the alphabetical orientation of the labels on the sides of the triangle and the given orientation on the plane disagree everywhere. For $\mathscr{F}$, the alphabetical orientation on the edges disagrees up to the degenerate triangle, but agrees afterwards.

We obtain a versal family by considering triangles with labeling and orientation on the ambient plane. As there are two orientations on the plane (counterclockwise and clockwise), this means taking two copies of $\overline{N}$, which we shall call $\overline{N}^+$ and $\overline{N}^-$, and gluing them together along their boundary. The result, which we shall denote $\overline{N}^\pm$, is here sketched as the surface of a bipyramid ($\overline{N}^+$ is at the front, $\overline{N}^-$ at the back):

$$\overline{N}^\pm :$$ 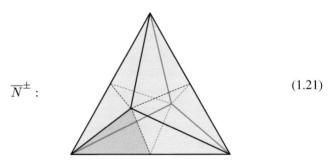 (1.21)

We agree that over $\overline{N}^+$ the plane is oriented in such a way that the alphabetical order on the triangles is counterclockwise, and, over $\overline{N}^-$, clockwise.

Consider the family $\mathscr{F}$ with marking and orientation as in (1.20). Its moduli map to $\overline{N}^\pm$ is displayed on the left-hand side in the following sketch. The first half is in $\overline{N}^-$, the second half in $\overline{N}^+$. (The corresponding moduli map for $\mathscr{G}$ would coincide with the one for $\mathscr{F}$ for the first half of the parameter interval, and then would retrace itself for the second half of the parameter interval.) On the right-hand side, we have displayed the moduli map of a (marked and oriented) family obtained by drawing the complete ellipse.

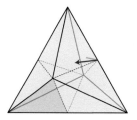 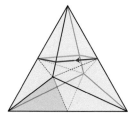

We see how the requirement that these moduli maps be continuous tells us how to glue the two copies of $\overline{N}$ together.

To construct a versal family $\overline{\mathscr{N}}^{\pm}$ parametrized by the bipyramid $\overline{N}^{\pm}$, we glue together the two families of marked and oriented triangles $\overline{\mathscr{N}}^{+}$ parametrized by $\overline{N}^{+}$, and $\overline{\mathscr{N}}^{-}$, parametrized by $\overline{N}^{-}$. Here $\overline{\mathscr{N}}^{+}$ is the labeled family $\overline{\mathscr{N}}$ of (1.19), endowed with the embedded orientation making the labels counterclockwise, and $\overline{\mathscr{N}}^{-}$ is the labeled family $\overline{\mathscr{N}}$ endowed with the clockwise-embedded orientation. When gluing, we use the unique isomorphism of labeled embedded-oriented triangles which exists over the locus of degenerate triangles. (If we think of both $\overline{\mathscr{N}}^{+}$ and $\overline{\mathscr{N}}^{-}$ as identical copies of $\overline{\mathscr{N}}$, this amounts to gluing the two copies of $\overline{\mathscr{N}}$ using the reflectional symmetry of the degenerate triangles, which swaps the ambient orientation, but preserves the labels.) See Figure 1.2 for an attempt at depicting $\overline{\mathscr{N}}^{\pm}$.

Exercise 1.37. (More advanced) The family $\overline{\mathscr{N}}^{\pm}$ is a universal family of labeled degenerate embedded-oriented triangles. (Define a family of labeled degenerate embedded-oriented triangles, parametrized by the topological space T, to consist of a complex line bundle $\mathscr{L}/T$, together with three sections $A, B, C \in \Gamma(T, \mathscr{L})$, such that $A + B + C = 0$, and no more than two sections ever agree anywhere in T. Isomorphisms consist of isomorphisms of line bundles preserving the three sections.) $\square$

Exercise 1.38. The symmetry groupoid of the family $\overline{\mathscr{N}}^{\pm}$ is the transformation groupoid given by the group of symmetries of the bipyramid, which consists of the identity, two rotations by $\frac{2\pi}{3}$, three rotations by π, four reflections, and two rotation-reflections. This group of symmetries of the bipyramid is isomorphic to $S_3 \times \mathbb{Z}_2$, where S_3 acts as before, preserving $\overline{N}^{+}$ and $\overline{N}^{-}$, and $\mathbb{Z}_2$ acts via the reflection across the common base of the two pyramids making up the bipyramid. $\square$

Exercise 1.39. Retaining of the family $\overline{\mathscr{N}}^{\pm}$ only its orientation circle bundle, we obtain a Hopf fibration. More precisely, for every family member of $\overline{\mathscr{N}}^{\pm}$ construct a circle whose center is at the centroid of the triangle, and whose

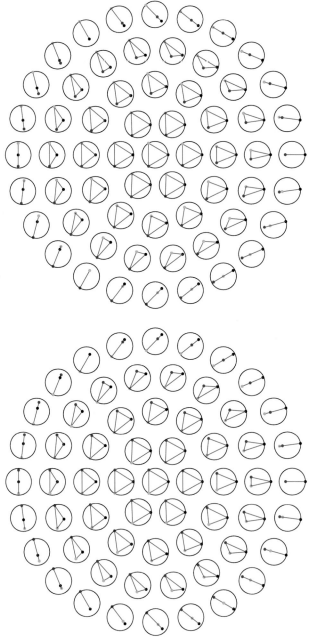

Figure 1.2. The Hopf fibration over $\overline{N}^{\pm}$.

circumference is equal to the perimeter of the triangle (e.g. 2, in our conventions). Consider these circles together with the action of the circle group S^1: it acts via rotations. The embedded orientations of the triangles tell us in which direction S^1 acts on these circles. The union of all these circles forms a topological space $P \to \overline{N}^{\pm}$, together with an action of the group S^1 on P, in such a way that $\overline{N}^{\pm}$ is the quotient of P by this S^1-action. In other words, we have constructed a *principal homogeneous S^1-bundle* P over $\overline{N}^{\pm}$.

We have indicated this circle bundle in Figure 1.2. Above is the family $\overline{\mathscr{N}}^{+}$, below the family $\overline{\mathscr{N}}^{-}$. Both are displayed such that the orientation on the circles is counterclockwise. The labels are indicated by different colours on the vertices of the family members. The two halves of $\overline{N}^{\pm}$ are displayed as discs, rather than triangles. When gluing together these two half families, the circles on the boundary complete a rotation by 2π. Another way to say this is that gluing data for the principal bundle P is given by a map $S^1 \to S^1$, which has winding number ± 1.

This shows that $P \to \overline{N}^{\pm}$ is homeomorphic to the Hopf fibration $S^3 \to S^2$. Since the Hopf fibration is topologically non-trivial, there is no way to embed all the triangles in $\overline{\mathscr{N}}^{\pm}$ into the plane in a compatible way, just as it is impossible to label the Moebius family consistently (1.5).   □

The following is a qualitative sketch of a small neighborhood of the pinched triangle in the family $\overline{\mathscr{N}}^{\pm}$. Note how a line of isosceles triangles and a line of degenerate triangles intersect at the pinched triangle.

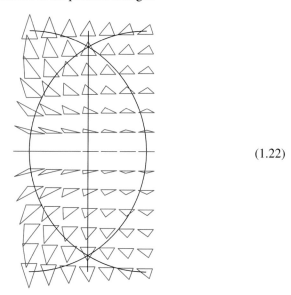

(1.22)

We claim that $\overline{\mathscr{N}}^{\pm}$ is a versal family of degenerate triangles. To prove this, all we need is an understanding of what a family of degenerate triangles looks like locally. Thus, let us agree that, locally, every family of degenerate triangles can be labeled consistently and embedded into one fixed oriented plane, which we will take to be $\mathbb{C}$ (recall that the complex number plane is oriented).

Definition 1.40. If $\mathscr{F}$ is a **continuous family of degenerate embedded triangles** parametrized by the topological space T, then, for every point $t_0 \in T$, there exists an open neighborhood $t_0 \in U \subset T$ of t_0 and three continuous functions $A, B, C : U \to \mathbb{C}$, determining the restriction $\mathscr{F}|_U$. At every point of U, no more than two of the three functions A, B, C are allowed to agree. For $t \in U$, the points $A(t), B(t), C(t) \in \mathbb{C}$ are the three vertices of the triangle $\mathscr{F}_t$.

If two such families are given over all of T by $(A, B, C) : T \to \mathbb{C}$ and $(A', B', C') : T \to \mathbb{C}$, then they are **isomorphic** if, after relabeling of the three functions, there exist continuous functions $R : T \to \mathbb{C}$ and $S : T \to \mathbb{C}^*$, such that $(A', B', C') = (SA + R, SB + R, SC + R)$, or $(A', B', C') = (\overline{SA + R}, \overline{SB + R}, \overline{SC + R})$.

Exercise 1.41. Deduce from the result of Exercise 1.37 that $\overline{\mathscr{N}}^{\pm}$ is a versal family of degenerate embedded triangles. An approach avoiding Exercise 1.37 follows.　□

To prove that $\overline{\mathscr{N}}^{\pm}$ is a versal family, assume that a family of (degenerate, embedded) triangles is given over the whole parameter space T by three functions $A, B, C : T \to \mathbb{C}$. We extract from A, B, C the functions $a, b, c : T \to \mathbb{R}_{\geq 0}$ and $\epsilon : T \to \{+1, 0, -1\}$, where $a = |B - C|$, $b = |C - A|$, $c = |A - B|$, and $\epsilon = 0$ where A, B, C are collinear, $\epsilon = +1$, where A, B, C are in counterclockwise position, and $\epsilon = -1$, where A, B, C are in clockwise position. We define a map

$$f : T \longrightarrow \overline{N}^{\pm},$$

$$t \longmapsto \begin{cases} \frac{2}{a+b+c}\big(a(t), b(t), c(t)\big) \in \overline{N}^{+} & \text{if } \epsilon(t) \geq 0 \\ \frac{2}{a+b+c}\big(a(t), b(t), c(t)\big) \in \overline{N}^{-} & \text{if } \epsilon(t) \leq 0 . \end{cases}$$

The map f is continuous by the pasting lemma. We have to prove that the family given by $A, B, C : T \to \mathbb{C}$ is, at least in a neighborhood of a base point $t_0 \in T$, isomorphic to the one given by $f^* \overline{\mathscr{N}}^{\pm}$.

Let us explain how $f^* \overline{\mathscr{N}}^{\pm}$ gives rise, at least locally, to a family of embedded-oriented triangles given by three continuous functions, say A', B', C', to $\mathbb{C}$. In the case of $f^* \overline{\mathscr{N}}^{\pm}$, the family of triangles is described by three continuous functions $a, b, c : T \to [0, 1]$ and two closed subsets

T^+ and T^-, where $T^+ \cap T^- = \{a = 1\} \cup \{b = 1\} \cup \{c = 1\}$. To this data, we associate A', B', C' as follows: over the locus where $c \neq 0$, define $A'(t) = 0$, $B'(t) = c(t)$, and $C'(t)$ in such a way that $|C'(t)| = b(t)$, $|C'(t) - B'(t)| = a(t)$, and $C'(t)$ is in the upper half plane if $t \in T^+$ and in the lower half plane if $t \in T^-$. Similar constructions can be made over the locus where $a \neq 0$, or $b \neq 0$, respectively.

Now, it is not hard to see that, after relabeling, if necessary, the functions A', B', C' are related to the functions A, B, C by a similarity transformation of $\mathbb{C}$, which depends continuously on $t \in T$. This proves that, indeed, all families of triangles can locally be described as pullbacks from $\overline{\mathscr{N}}^{\pm}$.

Exercise 1.42. Prove that the technical condition on the symmetry groupoid of $\overline{\mathscr{N}}^{\pm}$ is satisfied, i.e. verify that, with our current local description of families of triangles, a continuous isomorphism $\phi : (A, B, C) \to (A', B', C')$ gives rise to a continuous map $T \to \overline{N}^{\pm} \times S_3 \times \mathbb{Z}_2$, and that the tautological isomorphism is continuous. □

As we now have a versal family, and we know its symmetry groupoid, we know what families of degenerate triangles look like globally. The relevant result was stated in Exercise 1.34.

As the symmetry group $S_3 \times \mathbb{Z}_2$ is a product, an $S_3 \times \mathbb{Z}_2$-cover is the same thing as a pair consisting of an S_3-cover and a $\mathbb{Z}_2$-cover. Thus we have:

Proposition 1.43. *Globally, a family of degenerate embedded triangles over T is a pair (T', f), (T'', g), where $T' \to T$ is an S_3-cover and $T'' \to T$ is a degree 2 cover, and $f : T' \to \overline{N}^{\pm}$ and $g : T'' \to \overline{N}^{\pm}$ are equivariant maps. The S_3-cover T' gives all labelings on the family, and the degree 2 cover T'' gives all embedded orientations.*

The stack of embedded triangles, which we shall call $\overline{\mathfrak{M}}$, looks like this:

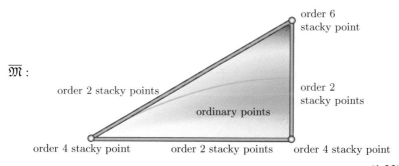

$$\text{(1.23)}$$

As embedded triangles, the pinched triangle, as well as the bisected line segment, has an order 4 symmetry group. In the case of the bisected line segment,
this group consists of the rotation by π and two reflections. In the case of
the pinched triangle, there is the reflectional symmetry, the swap of the two
coinciding points, and the composition of these two symmetries.

Complex viewpoint

There is a canonical family of degenerate embedded triangles parametrized by
the complex plane $\mathbb{C}$. It is customary to write the parameter as $\lambda \in \mathbb{C}$, in honor
of Legendre. Figure 1.3 shows a picture of the λ-plane. The family of triangles
is given by the three functions $A(\lambda) = 0$, $B(\lambda) = 1$, and $C(\lambda) = \lambda$. The twelve
locations of the 3:4:5 triangle have been marked. The locus of acute triangles
is shaded. The family is shown in the lower part of Figure 1.3. As it is labeled
and oriented, it induces a map $f : \mathbb{C} \to \overline{N}^{\pm}$, which is given by

$$f : \mathbb{C} \longrightarrow \overline{N}^{\pm},$$

$$\lambda \longmapsto \begin{cases} \left(\frac{2|\lambda-1|}{1+|\lambda|+|\lambda-1|}, \frac{2|\lambda|}{1+|\lambda|+|\lambda-1|}, \frac{2}{1+|\lambda|+|\lambda-1|} \right) \in \overline{N}^{+} & \text{if } \operatorname{Im}\lambda \geq 0 \\ \left(\frac{2|\lambda-1|}{1+|\lambda|+|\lambda-1|}, \frac{2|\lambda|}{1+|\lambda|+|\lambda-1|}, \frac{2}{1+|\lambda|+|\lambda-1|} \right) \in \overline{N}^{-} & \text{if } \operatorname{Im}\lambda \leq 0. \end{cases}$$

This is a homeomorphism onto the complement of the point $(1, 1, 0)$ in $\overline{N}^{\pm}$,
so it identifies $\overline{N}^{\pm}$ with the one-point compactification $\mathbb{C} \cup \{\infty\}$ of $\mathbb{C}$.

We also see that this family parametrized by $\mathbb{C}$ extends to a family
parametrized by the Riemann sphere $\mathbb{C} \cup \{\infty\}$, by pulling back the family
$\overline{\mathcal{N}}^{\pm}$ via the homeomorphism $\mathbb{C} \cup \{\infty\} \to \overline{N}^{\pm}$.

So $\mathbb{C} \cup \{\infty\}$ is just as good a versal parameter space as $\overline{N}^{\pm}$. In fact, it is
better: it has the structure of a Riemann surface! (It endows the stack $\mathfrak{M}$ with
the structure of *real analytic stack*.)

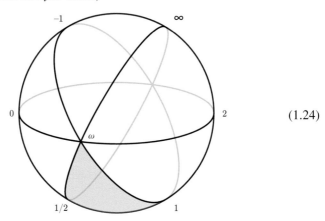

$$\tag{1.24}$$

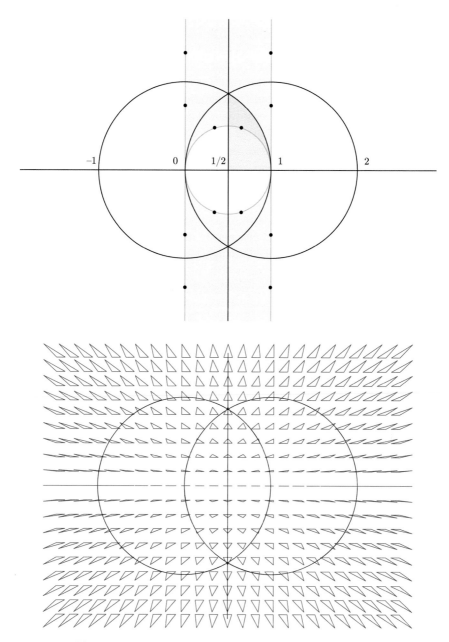

Figure 1.3. The λ-plane (top) and the versal family over it (bottom).

From this viewpoint, the action of S_3 is given by the reflections in the three great circles passing through the sixth root of unity ω, and rotations by multiples of $\frac{2\pi}{3}$ around the axis through ω. The group $\mathbb{Z}_2$ acts by reflection across the great circle representing the real values of λ.

Note how Figure 1.3 exhibits, in a neighborhood of the locus where λ is real, the behavior of the family $\overline{\mathscr{N}}^{\pm}$ near where it has been glued. At the center of Figure 1.3, for example, is a bisected line segment; close by, isosceles triangles, as well as degenerate triangles, appear twice, whereas scalene triangles, appear four times, reflecting the fact that the bisected line segment has four symmetries.

Oriented degenerate triangles

An oriented (degenerate, embedded) triangle is a set of three points in the plane (or rather an unordered triple of points in the plane), no more than two coinciding, together with an orientation on the plane, i.e. a notion of counterclockwise and clockwise. Of course, for non-collinear triangles, such an orientation is the same thing as a cyclic ordering on the vertices, but we have seen that for degenerate triangles the current notion is superior.

Isomorphisms, or similarity transformations, are now required to preserve the orientation, i.e. they are not allowed to be reflections or glide reflections.

Of course, a continuous family of oriented triangles can, at least locally, be endowed with a labeling, if we do not make any requirements about the alphabetical ordering of the labeling agreeing with the counterclockwise orientation on the ambient plane. In other words, just as families of unoriented triangles could be described locally by three complex-valued functions on the parameter space, the same is true for oriented families. The difference manifests only when discussing under what circumstances families are considered isomorphic. (In fact, we can use Definition 1.40 verbatim, adding only the word "oriented," and deleting the last "or" involving complex conjugation.)

Therefore, $\overline{N}^{\pm}$ is a versal parameter space, and $\overline{\mathscr{N}}^{\pm}$ is a versal family. The symmetry groupoid of $\overline{\mathscr{N}}^{\pm}$ as a family of *oriented* triangles is a subgroupoid of the symmetry groupoid of $\overline{\mathscr{N}}^{\pm}$ as a family of triangles. In fact, it is the transformation groupoid of the subgroup of *oriented* symmetries of the bipyramid. This group consists of the rotations by $\frac{2\pi}{3}$ about the axis through the two pyramid vertices, and the three rotations about the axes through the vertices of the common base of the two pyramids. This group is, again, isomorphic to S_3, although it is a subgroup of $S_3 \times \mathbb{Z}_2$ in a different way than the copy of S_3 which acts on $\overline{N}^+$ and $\overline{N}^-$.

In the Legendre picture, where we have identified $\overline{N}^{\pm}$ with the Riemann sphere $\mathbb{C} \cup \{\infty\}$, this symmetry groupoid acts by the six linear fractional transformations

$$\lambda \longmapsto \quad \lambda, \quad \frac{1}{\lambda}, \quad 1 - \lambda, \quad \frac{\lambda}{\lambda - 1}, \quad \frac{1}{1 - \lambda}, \quad \frac{\lambda - 1}{\lambda}.$$

$$(1.25)$$

We will write Λ for the Riemann sphere, endowed with the action of S_3 by these six transformations.

Exercise 1.44. Deduce from Exercise 1.34 that a family of degenerate embedded-oriented triangles parametrized by the space T is given by an S_3-covering space $T' \to T$ and an S_3-equivariant map $T' \to \Lambda = \mathbb{C} \cup \{\infty\}$. □

Exercise 1.45. Prove that, for every family of oriented triangles $\mathscr{F}/T$, the corresponding generalized moduli map to $\Lambda = \mathbb{C} \cup \{\infty\}$ is given by the cross-ratio of the three vertices and ∞. More precisely, the six-valued function $T' \to \mathbb{C} \cup \{\infty\}$ is given by the six cross-ratios of the vertices and ∞. Moreover, the action by S_3 on T' is compatible with the action of S_3 on the six cross-ratios. □

For example, consider a family parametrized by the figure 8, such as the following: over the left loop of the figure 8, this family contains a degree 3 cyclic cover, such as the family (1.14). Over the right loop of the figure 8, the family contains a degree 2 cover (a Moebius band).

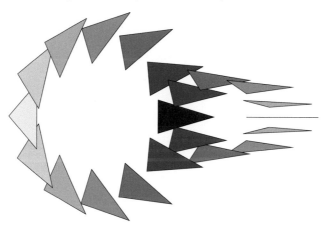

The generalized moduli map of such a family will look like this:

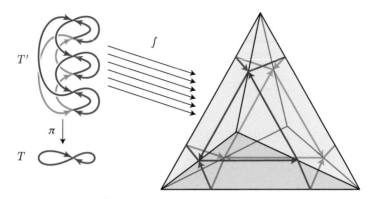

In fact, the image of the moduli map is the Cayley graph of the group S_3. The two cycles of length 3 cover the left loop of the figure 8, and the three cycles of length 2 cover the right loop of the figure 8.

Let us denote the stack of degenerate embedded-oriented triangles by $\mathfrak{L}$. In the sketch of $\mathfrak{L}$ that follows, the front represents triangles oriented according to ascending length of sides, the back represents triangles oriented according to descending length of sides. Identifying front and back, making the sphere flat, we get the picture of $\overline{\mathfrak{M}}$ from (1.23).

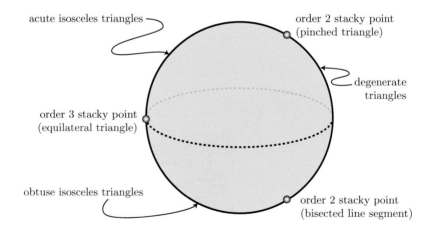

1.1.11 Change of versal family

There can be only one universal family, but there are many versal families. To see examples, let us restrict our discussion to *oriented* triangles. So far, we have seen one versal family, with parameter space $\mathbb{C} \cup \{\infty\} \cong \overline{N}^{\pm}$, and symmetry groupoid given the action of S_3 on this parameter space, by oriented symmetries of a bipyramid.

Oriented triangles by projecting equilateral triangles

Let us construct another versal family of oriented degenerate triangles. Fix a sphere S positioned on a horizontal plane, such that the south pole of the sphere coincides with the origin of the plane. The parameter space will be

$$E = \{\text{great circle equilateral triangles on } S\}.$$

(The vertices or edges of the elements of E are *not* labeled.)

Exercise 1.46. Prove that E is a 3-dimensional manifold. (In fact, it is the quotient of the 3-dimensional Lie group SO_3 of 3×3 orthogonal matrices with determinant 1, by a discrete subgroup isomorphic to S_3.) $\square$

The family of degenerate triangles parametrized by E, which we shall call $\mathscr{E}$, is given by stereographic projection from the north pole of S onto the plane. In other words, $\mathscr{E}_t = p(t)$, for all $t \in E$, where p is the stereographic projection. In the following sketch, a great-circle triangle is displayed, along with its stereographic projection into the plane. Thus, this sketch displays just one family member of $\mathscr{E}$ (as the parameter space is 3-dimensional, it is hard to sketch the entire family).

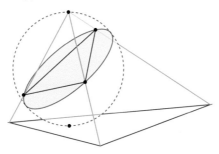

Note that this never produces any triangles whose vertices coincide; instead it produces triangles with a vertex at ∞ (when one of the vertices of the great circle equilateral triangle is at the North pole). Thus, this family takes a different point of view of the "pinched triangles," from the one parametrized by $\mathbb{C} \cup \{\infty\} = \overline{N}^{\pm}$.

The group of oriented similarity transformations, which consists of translations, rotations, and scalings, is isomorphic to the semi-direct product $\mathbb{C}^+ \rtimes \mathbb{C}^*$. The subgroup of translations, $\mathbb{C}^+$, is a normal subgroup, and the subgroup of scaling-rotations with center the origin, is isomorphic to $\mathbb{C}^*$. The conjugation action of the scaling-rotations on the translations is the multiplication action of $\mathbb{C}^*$ on $\mathbb{C}^+$.

Using the stereographic projection, we can translate any geometric state ment about the plane into a statement about the sphere, and conversely. It is, in fact, more convenient to work on the sphere rather than the plane, because we can describe everything we need in terms of the group of con formal symmetries of the sphere. Conformal transformations of the sphere are transformations that preserve angles as well as orientation.

The group of orientation- and angle-preserving (i.e. conformal) transfor mations of the sphere is known as $PSL_2(\mathbb{C})$. If (A, B, C) and (A', B', C') are ordered triples of distinct points on the sphere, there exists a unique $P \in PSL_2(\mathbb{C})$ such that $PA = A'$, $PB = B'$, and $PC = C'$. The oriented similarity transformations of the plane $\mathbb{C}^+ \rtimes \mathbb{C}^*$ correspond via stereographic projection to the subgroup of $PSL_2(\mathbb{C})$ fixing the north pole ∞. The sub group of $PSL_2(\mathbb{C})$ which fixes lengths, as well as angles and orientation, is the group of rotations of the sphere, $SO_3 \subset PSL_2(\mathbb{C})$. The intersection of these two groups is the group of rotations about the axis through the north and south poles. This group is isomorphic to the circle group S^1.

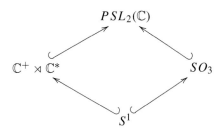

To explain why the family $\mathscr{E}$ is versal, let us work on the sphere S rather than the plane. Let us agree, therefore, that a family of degenerate triangles is locally given by three continuous functions $A, B, C : T \to S$, no two of which agree anywhere in T. Isomorphisms are locally given by relabelings and by continuously varying elements of the group $\mathbb{C}^+ \times \mathbb{C}^* \subset PSL_2(\mathbb{C})$.

To prove that every such family is induced from $\mathscr{E}$, locally, we will prove that, given such a family $(A, B, C)/T$, we can apply a continuous family of elements of $\mathbb{C}^+ \rtimes \mathbb{C}^*$, making it a family of great circle equilateral triangles.

Let e_0 be a fixed great-circle equilateral triangle, which we may as well assume to be labeled, as it is fixed. First, there is a unique continuous map $P : T \to PSL_2(\mathbb{C})$ such that $P(t)e_0 = \big(A(t), B(t), C(t)\big)$ for all $t \in T$ (as labeled families of triangles). This follows from the fact that A, B, C never coincide, and the fact that $PSL_2(\mathbb{C})$ acts simply transitively on the set of ordered triples of distinct points in S. Secondly, there exists a continuous family of rotations $R(t)$ such that $R\infty = P^{-1}\infty$ (at least locally in T). Then $R^{-1}P^{-1}(A, B, C) = R^{-1}e_0$, and $R^{-1}P^{-1}\infty = \infty$. Because of the latter property, $R^{-1}P^{-1}$ is a continuous family of elements of $\mathbb{C}^+ \rtimes \mathbb{C}^*$, and, of course $R^{-1}e_0$ is a continuous family of great-circle equilateral triangles. This proves what we needed.

To determine the symmetry groupoid of the family $\mathscr{E}/E$, assume that $\{A, B, C\} \subset \mathbb{C} \cup \{\infty\}$ is a member of the family $\mathscr{E}$; in other words, a great-circle equilateral triangle.

Let $P \in PSL_2(\mathbb{C})$ such that $P\infty = \infty$ is an arbitrary oriented similarity transformation. Assume that $P\{A, B, C\}$ is another member of $\mathscr{E}$, i.e. another great-circle equilateral triangle. As the rotation group $SO_3 \subset PGL_2(\mathbb{C})$ acts transitively on the great-circle equilateral triangles, there exists a rotation $R \in SO_3$ such that $R\{A, B, C\} = P\{A, B, C\}$. Hence $R^{-1}P$ is in the stabilizer subgroup of $\{A, B, C\}$ inside $PSL_2(\mathbb{C})$. This stabilizer subgroup consists of six rotations, and is isomorphic to S_3. As $R^{-1}P$ is a rotation, it follows that P itself is a rotation. As P fixes the north–south axis, it is in the subgroup $S^1 \subset SO_3 \subset PSL_2(\mathbb{C})$.

Thus, every symmetry of the family $\mathscr{E}$ is induced by an element of S^1, and we see that the symmetry groupoid of the family $\mathscr{E}$ is the transformation groupoid $E \times S^1$, where S^1 acts by rotations about the north–south axis of S.

The following exercise finishes the proof of the versality of $\mathscr{E}$ by proving that the technical conditions on the symmetry groupoid are satisfied.

Exercise 1.47. Prove that with the manifold structure on E from Exercise 1.46, and with the local notions of continuous family and continuous isomorphism described above, the canonical topology on the symmetry groupoid of the family $\mathscr{E}/E$ is the product topology $E \times S^1$.

Prove that the tautological isomorphism of families over the symmetry groupoid is continuous. □

Exercise 1.48. Explain why we cannot define a global notion of family of triangles to be given by a degree 3 cover $T' \to T$ together with a continuous map $T' \to S$. □

Now that we have a versal family, and we know its symmetry groupoid, we can describe all families, globally, in terms of generalized moduli maps. We need a generalization of Exercise 1.34 because the group in our symmetry groupoid is not discrete. Therefore, we have to consider *principal bundles* instead of G-covering spaces.

Definition 1.49. Let G be a topological group and let T be a topological space. A **principal homogeneous bundle** over T with structure group G is given by a topological space T', endowed with a continuous map $\pi : T' \to T$, and a continuous action by the topological group G. The following two conditions are required to hold:

(i) $\pi(tg) = \pi(t)$, for all $t \in T'$ and $g \in G$,

(ii) (local triviality) for every $t \in T$, there exists a neighborhood U of t in T and a continuous section $\sigma : U \to T'$ of $T' \to T$ over U, such that the induced map $U \times G \to \pi^{-1}(U)$, given by $(u, g) \mapsto \sigma(u)g$, is a homeomorphism.

Instead of "principal homogeneous bundle with structure group G," we often say "principal G-bundle," or simply "G-bundle." The terminology "G-torsor" is also common.

Exercise 1.50. Using the concept of principal bundle, generalize both the statement and the results of Exercise 1.34 to the case of a versal family whose symmetry groupoid is a transformation groupoid $T \times G$ with a general topological group G. $\square$

In the case of $G = S^1$, we call principal S^1-bundles **circle bundles**. Thus, in the present case, a generalized moduli map is given by a circle bundle $T' \to T$, together with an S^1-equivariant continuous map $T' \to E$.

Given a family of degenerate triangles $\mathscr{F}/T$, we should think of the circle bundle $T' \to T$ as the space of all ways the members of $\mathscr{F}$ can be obtained by stereographic projection from great-circle equilateral triangles on S.

Comparison

Let us now compare our two versal families of degenerate triangles: the one parametrized by E and the one parametrized by $\Lambda = \mathbb{C} \cup \{\infty\} \cong \overline{N}^{\pm}$. (We use the notation Λ for the Riemann sphere, to emphasize its role as parameter space for the Legendre family of triangles.)

Let us now restrict ourselves to unpinched triangles because for these our two notions of degenerate triangles agree. Thus no vertices are allowed to coincide, no vertices are allowed to "escape to infinity." Therefore, our restriction leads to

$$\Lambda^0 = \mathbb{C} - \{0, 1\} \cong \overline{N}^{\pm} - \{(0, 1, 1), (1, 0, 1), (1, 1, 0)\}$$

and

$$E^0 = E - \{\text{great-circle triangles in } S \text{ with a vertex at the north pole}\}.$$

Now the family $\mathscr{E}|_{E^0}$ has a generalized moduli map to Λ^0, and the family $\overline{\mathscr{N}}^{\pm}|_{\Lambda^0}$ has a generalized moduli map to E^0. The key is that these generalized moduli maps are *"the same."* In fact, consider the space $Q^0 = \underline{\mathrm{Isom}}(\overline{\mathscr{N}}^{\pm}|_{\Lambda^0}, \mathscr{E}|_{E^0})$ of isomorphisms between the two families. It fits into the following diagram:

$$
\begin{array}{ccc}
Q^0 & \xrightarrow[]{\substack{S^1\text{-equivariant}\\ S_3\text{-bundle}}} & E^0 \\[2ex]
\Big\downarrow{\scriptstyle\substack{S_3\text{-equivariant}\\ S^1\text{-bundle}}} & & \\[2ex]
\Lambda^0 & &
\end{array}
\qquad (1.26)
$$

There are two group actions on Q^0, one by S_3 and one by S^1, and these actions commute with each other. This makes $Q^0 \to \Lambda^0$ an S^1-bundle and $Q^0 \to E^0$ an S_3-bundle. The map $Q^0 \to E^0$ is S^1-equivariant; the map $Q^0 \to \Lambda^0$ is S_3-equivariant.

Thus, diagram (1.26) displays both generalized moduli maps at the same time. We see that different versal families for the same moduli problem will always have a common generalized moduli map intertwining the two parameter spaces in this way.

A key fact is that the converse is true: if you have two a-priori different moduli problems, and a versal family for each, and if you can intertwine the two symmetry groupoids in this way, the two moduli problems are actually equivalent. We will discuss this next.

The comparison theorem

Suppose Q is a space with commuting actions by two topological groups, G and H. Suppose further that the quotient maps $\pi : Q \to X = Q/G$ and $\rho : Q \to Y = Q/H$ are principal bundles. Then the quotient maps are equivariant:

$$
\begin{array}{ccc}
Q & \xrightarrow[\rho]{\substack{G\text{-equivariant}\\ H\text{-bundle}}} & Y \\[2ex]
\Big\downarrow{\scriptstyle\substack{H\text{-equivariant}\\ G\text{-bundle}}}{\pi} & & \\[2ex]
X & &
\end{array}
\qquad (1.27)
$$

Theorem 1.51. *In this situation, a G-bundle T'/T and a G-equivariant map $f : T' \to Y$ are essentially the same thing as a $G \times H$-bundle $\widetilde{T}' \to T$ and a $G \times H$-equivariant map $F : \widetilde{T}' \to Q$.*

Proof. Given T'/T and $f : T' \to Y$, we let $\widetilde{T}' = Q \times_Y T'$ be the fibered product and $F : \widetilde{T}' \to Q$ the first projection:

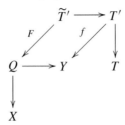

Then $\widetilde{T}' \to T$ is a $G \times H$-bundle because the G- and H-actions on $\widetilde{T}'$ commute and $\widetilde{T}' \to T'$ is an H-bundle.

Conversely, given $\widetilde{T}'/T$ and $F : \widetilde{T}' \to Q$, define $T' = \widetilde{T}'/H$ to be the quotient space and $f : T' \to Y$ the map induced by F on quotient spaces, noting that $Y = Q/H$.

One can check that these two processes are essentially inverses of one another. □

Corollary 1.52. *In this situation, a G-bundle T'/T and a G-equivariant map $f : T' \to Y$ are essentially the same thing as an H-bundle $\widetilde{T} \to T$ and an H-equivariant map $g : \widetilde{T} \to X$.*

Proof. This follows from the theorem by the symmetry of the situation. □

This corollary allows us to prove that moduli problems are equivalent by comparing the symmetry groupoids of versal families.

We apply this principle to the two notions of family of degenerate triangles.

Let us fix some more data. Let the sphere S have diameter 1, so that under the stereographic projection the equator of S corresponds to the unit circle in $\mathbb{C}$. Fix e_0 to be the great-circle equilateral triangle on S, which lies on the real great circle and has a vertex at the north pole ∞. The stereographic projection of e_0 is the pinched triangle with vertices at $-\frac{1}{2}, \frac{1}{2}$, and ∞.

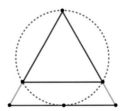

We use the stereographic projection to identify Λ with S.

Now construct the diagram

$$
\begin{array}{ccc}
SO_3 & \xrightarrow{\;\rho\;} & E \\
{\scriptstyle\pi}\Big\downarrow & & \\
\Lambda & &
\end{array}
\tag{1.28}
$$

The map ρ is defined by $\rho(R) = Re_0$, for every rotation $R \in SO_3$. The map π is defined by $\pi(R) = AR^{-1}\infty$, where $A : S \to S$ is the similarity transformation of S, defined by

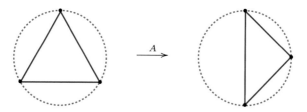

In the complex plane, A corresponds to the translation $z \mapsto z + \frac{1}{2}$, for $z \in \mathbb{C} \cup \{\infty\}$. Then we claim that

(i) diagram (1.28) is an $(S_3 \times \Lambda, S^1 \times E)$-bibundle,
(ii) restricting to Λ^0 and E^0, we get diagram (1.26).

The first claim says that the two moduli problems are equivalent, the second that this equivalence of moduli problems is compatible with the obvious one for non-pinched triangles. We conclude that E supports a versal family of degenerate triangles in the coincident sense, and that its symmetry groupoid is equal to the transformation groupoid $S^1 \times E$.

To see the second claim, note that, by Exercise 1.45, the generalized moduli map $\rho^{-1}(E_0) \to \Lambda$ of $\mathscr{E}|_{E_0}$ is given by $R \mapsto cr(Re_0, \infty) = cr(e_0, R^{-1}\infty) = cr(Ae_0, AR^{-1}\infty) = AR^{-1}\infty$.

Exercise 1.53. Prove that (1.28) is isomorphic to

$$
\begin{array}{ccc}
SO_3 & \xrightarrow{\hspace{1.5cm}} & S_3\backslash SO_3 \\
\Big\downarrow & & \\
SO_3/S_1 & &
\end{array}
$$

The stack of triangles is a stacky version of the double quotient $S_3\backslash SO_3/S^1$, or, equivalently, $S_3\backslash PSL_2\mathbb{C}/(\mathbb{C}^+ \rtimes \mathbb{C}^\times)$. $\square$

1.1.12 Weierstrass compactification

We now discuss a different way to compactify the stack of oriented triangles. The notion of family containing a pinched triangle will be different.

This new point of view comes about by viewing an oriented degenerate triangle as the zero locus (in $\mathbb{C}$) of a degree 3 polynomial with complex coefficients. If $a_0 z^3 + a_1 z^2 + a_2 z + a_3$, with $a_0, a_1, a_2, a_3 \in \mathbb{C}$ and $a_0 \neq 0$, is a degree 3 polynomial, factor

$$a_0 z^3 + a_1 z^2 + a_2 z + a_3 = a_0(z - e_1)(z - e_2)(z - e_3),$$

and the corresponding triangle has vertices at $e_1, e_2, e_3 \in \mathbb{C}$. We ask of the polynomial that not all three roots coincide.

Two polynomials give rise to the same triangle if they differ by an overall multiplication by an element of $\mathbb{C}^*$; they give rise to similar triangles if one can be transformed into the other by substituting z by $\alpha z + \beta$, for $\alpha \in \mathbb{C}^*$, $\beta \in \mathbb{C}$.

To simplify, we can put degree 3 polynomials into *Weierstrass normal form*,

$$4z^3 - g_2 z - g_3, \tag{1.29}$$

and consider only polynomials of this form. Factoring this polynomial,

$$4z^3 - g_2 z - g_3 = 4(z - e_1)(z - e_2)(z - e_3),$$

we see that $e_1 + e_2 + e_3 = 0$, and so the centroid (or center of mass) of the triangle is at the origin.

The family of all degree 3 polynomials in Weierstrass normal form is parametrized by $W = \mathbb{C}^2 - \{(0, 0)\}$, where the coordinates in W are named g_2 and g_3. The polynomial

$$4z^3 - g_2 z - g_3 \in \mathbb{C}[z, g_2, g_3]$$

can be thought of as a family of degree 3 polynomials parametrized by W.

Exercise 1.54. We claim that this is a versal family of polynomials. To prove this, we have to agree on what a family of polynomials is locally and what an isomorphism of local families of polynomials is.

Let us agree that a family of degree 3 polynomials without triple root is locally given by four continuous functions $a_0, a_1, a_2, a_3 : T \to \mathbb{C}$, where a_0 does not vanish anywhere in T, and the polynomial $a_0(t)z^3 + a_1(t)z^2 + a_2(t)z + a_3(t)$ does not have a triple root, for any value $t \in T$.

Prove that, locally, you can make continuous coordinate changes and continuous rescalings of the coefficients, to put the polynomial $a_0 z^3 + a_1 z^2 + a_2 z + a_3$ into *Weierstrass normal form* $4z^3 - g_2 z - g_3$. This takes care of Definition 1.33(i).

Now suppose we are given two continuous families g, f of polynomials in Weierstrass normal form, both parametrized by the space T. Let us agree that an isomorphism $g \to f$ is given by a continuous map $(a, b) : T \to \mathbb{C}^* \times \mathbb{C}$, such that $4z^3 - g_2z - g_3$ and $4(az + b)^3 - f_2(az + b) - f_3$ differ by an overall rescaling of the coefficients. Show that this implies that $b = 0$ and that $(f_2, f_3) = (a^2 g_2, a^3 g_3)$. Conclude that the symmetry groupoid of the Weierstrass family, with its canonical topology, is isomorphic to the transformation groupoid $W \times \mathbb{C}^*$, where $\mathbb{C}^*$ acts on W with weights 2 and 3. This takes care of Definition 1.33(ii).
□

Corollary 1.55. *From Exercise 1.54, as well as Exercise 1.50, it follows that a continuous family of degree 3 polynomials, up to linear coordinate changes, without triple root, is a pair (P, g), where P is a principal $\mathbb{C}^*$-bundle and $g : P \to W$ is a $\mathbb{C}^*$-equivariant map. Equivalently, it is given by a complex line bundle $\mathscr{L}/T$ with two sections, $g_2 \in \mathscr{L}^{\otimes 2}$ and $g_3 \in \mathscr{L}^{\otimes 3}$, that do not vanish simultaneously.*

The question now arises whether or not this point of view using normalized degree 3 polynomials up to linear substitutions is equivalent to the previous point of view on triangles as a set of three points together with a map to $\mathbb{C}$; see Definition 1.40 and Exercise 1.44.

In the polynomial picture, the only two symmetric oriented triangles are the equilateral one, given by $4z^3 - g_3$, and the bisected line segment, given by $4z^3 - g_2z$. They have 3 and 2 symmetries, respectively. All other triangles are completely asymmetric in this picture because $\mathbb{C}^*$ has non-trivial stabilizers on W only if one of the coordinates vanishes. In particular, the pinched triangle, given by any polynomial with $g_2^3 = 27g_3^2$, is completely asymmetric. (Recall that previously the pinched triangle had a non-trivial symmetry given by swapping the two points that map to identical points in $\mathbb{C}$.)

Thus, the stack $\mathfrak{W}$ of degree 3 polynomials cannot be isomorphic to the stack $\mathfrak{L}$ of oriented degenerate triangles. Let us see if there is at least a morphism in either direction.

To define a morphism $\mathfrak{W} \to \mathfrak{L}$ would mean to turn any family of degree 3 polynomials into a family of oriented triangles. In particular, we would have to convert the Weierstrass family of polynomials parametrized by W into a family of oriented degenerate triangles in the sense of Definition 1.40 and Exercise 1.44. This almost works.

The subspace $W' \subset \mathbb{C} \times W$, defined by

$$W' = \{(z, g_2, g_3) \in \mathbb{C} \times W \mid 4z^3 - g_2z - g_3 = 0\},$$

has a projection map $\pi : W' \to W$ and also a map $W' \to \mathbb{C}$. In general, over every point $(g_2, g_3) \in W$, there are three points e_1, e_2, e_3 in W' lying over it, and the images of these points in $\mathbb{C}$ form the vertices of an oriented triangle in $\mathbb{C}$. If $\pi : W' \to W$ were a topological covering map, this would be a family of oriented triangles (because in that case we could, locally, label the three roots of the polynomial by e_1, e_2, e_3). But over the discriminant locus of W, where

$$\Delta = g_2^3 - 27g_3^2$$

vanishes, the covering $W' \to W$ is ramified, and this is impossible.

Exercise 1.56. Prove that, in fact, W does not support any family of degenerate oriented triangles in the sense of Exercise 1.44, which restricts to the Weierstrass family over $W \setminus \{\Delta = 0\}$. □

So let us try to define a morphism $\mathfrak{L} \to \mathfrak{W}$. This would amount to converting any family of triangles into a family of polynomials. This is, in fact, possible: a family of oriented triangles is locally given by three continuous functions $A, B, C : T \to \mathbb{C}$, and to these we can associate the continuous family of polynomials $(z - A)(z - B)(z - C)$, whose (continuous!) coefficient functions are given by $a_0 = 1, a_1 = -(A+B+C), a_2 = (AB+AC+BC), a_3 = -ABC$. We also have to check compatibility with isomorphisms of families. Indeed, if the family of triangles given by A, B, C is isomorphic to the one given by $A', B'C'$, via $r : T \to \mathbb{C}$ and $s : T \to \mathbb{C}^*$, as in Definition 1.40, then

$$s^3(z - A)(z - B)(z - C) = (sz + r - A')(sz + r - B')(sz + r - C'),$$

and so, indeed, the corresponding families of polynomials are isomorphic.

We have defined a morphism of stacks $\mathfrak{L} \to \mathfrak{W}$. We can also form a more global geometric picture of this morphism by considering the space

$$Q = \mathbb{C}^2 \setminus \{(0, 0)\},$$

whose coordinates we write as (μ, λ), together with two commuting actions by the groups S_3 and $\mathbb{C}^*$. The group S_3 acts by the six substitutions

$$(\mu, \lambda) \longmapsto$$
$$(\mu, \lambda),\ (\lambda, \mu),\ (-\mu, \lambda - \mu),\ (\lambda - \mu, -\mu),\ (-\lambda, \mu - \lambda),\ (\mu - \lambda, -\lambda),$$
$$(1.30)$$

and the group $\mathbb{C}^*$ acts by rescaling: $(\mu, \lambda) \cdot \alpha = (\mu\alpha, \lambda\alpha)$, for $\alpha \in \mathbb{C}^*$. We have a diagram of continuous maps:

$$
\begin{array}{ccc}
Q & \xrightarrow{\substack{g_2(\mu,\lambda),\, g_3(\mu,\lambda) \\ \mathbb{C}^*\text{-equivariant} \\ S_3\text{-invariant map}}} & W \\
\scriptstyle{S_3\text{-equivariant}} \downarrow \scriptstyle{\mathbb{C}^*\text{-bundle}} & & \\
\Lambda & &
\end{array}
\tag{1.31}
$$

Here the vertical map $Q \to \Lambda$ is the map $(\mu, \lambda) \longmapsto \lambda/\mu$, and the horizontal map $Q \to W$ is given by the formulas

$$
g_2 = 2(\lambda^2 + \mu^2) - \frac{2}{3}(\lambda + \mu)^2, \qquad g_3 = \frac{4}{9}(\lambda^3 + \mu^3) - \frac{4}{27}(\lambda + \mu)^3. \tag{1.32}
$$

The vertical map $Q \to \Lambda$ is a principal homogeneous $\mathbb{C}^*$-bundle. (It is the tautological $\mathbb{C}^*$-bundle of the Riemann sphere considered as the complex projective line.) Moreover, the map $Q \to \Lambda$ commutes with the S_3-actions on Q and Λ, which one sees by comparing (1.30) with (1.25).

The horizontal map is invariant under the S_3-action on Q: the formulas (1.32) are invariant under the substitutions (1.30). It is also equivariant with respect to the $\mathbb{C}^*$-actions on Q and W, because, in (1.32), g_2 is quadratic in μ and λ, whereas g_3 is cubic in μ and λ.

Exercise 1.57. Suppose that G and H are topological groups and that Q is a space with commuting actions by G and H. Suppose that $Q \to X$ is an H-equivariant G-bundle and that $Q \to Y$ is a G-equivariant map which is H-invariant. Then there is a natural construction which associates to every pair (P, f), where P is an H-bundle and $f : P \to X$ is an H-equivariant map, a pair (P', f'), where P' is a G-bundle and $f' : P' \to Y$ is a G-equivariant map. $\qquad\square$

We see that we can associate to every global family of triangles, in the sense of Exercise 1.44, a global family of polynomials in the sense of Corollary 1.55. So we get another construction of a morphism of stacks $\mathfrak{L} \to \mathfrak{W}$. (Of course, it is the same morphism, because on substituting (1.32) into (1.29) and dividing by μ we get $z(z - 1)(z - \lambda)$, up to normalization.)

In fact, every morphism of stacks comes about in this way.

Theorem 1.58. *Suppose that X parametrizes a versal family $\mathscr{F}$ for the stack $\mathfrak{X}$, with symmetry groupoid $X \times H$, and that Y parametrizes a versal family $\mathscr{G}$ for the stack $\mathfrak{Y}$ with symmetry groupoid $G \times Y$. Any diagram*

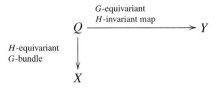

gives rise to a morphism of stacks $\mathfrak{X} \to \mathfrak{Y}$. *(It is an isomorphism if and only if the H-invariant map $Q \to Y$ is a principal H-bundle.) Conversely, every morphism of stacks* $\mathfrak{X} \to \mathfrak{Y}$ *comes from such a diagram.*

Proof. That such a space Q gives rise to a morphism of stacks was proved in Exercise 1.57. For the converse, we would need the formal definition of stacks, see Section 1.2, so let us just remark for now that, given the morphism of stacks $F : \mathfrak{X} \to \mathfrak{Y}$, we can define Q to be the space of triples (x, ϕ, y), where $x \in X$, $y \in Y$, and $\phi : F(\mathscr{F}|_x) \to \mathscr{G}|_y$ is an isomorphism in $\mathfrak{Y}$. The space Q can be endowed with a canonical topology. The H-action on Q is given by $(x, \phi, y) \cdot h = \left(xh, F(h^{-1} : \mathscr{F}|_{xh} \to \mathscr{F}|_x) * \phi, y\right)$; the G-action is given by $(x, \phi, y) \cdot g = \left(x, \phi * (g : \mathscr{G}|_y \to \mathscr{G}|_{yg}), yg\right)$. $\square$

This theorem only applies to the case where there exist versal families whose symmetry groupoids are transformation groupoids. For the general case, see Exercise 1.95.

We can informally write our morphism $\mathfrak{L} \to \mathfrak{W}$ as

$$g_2 = 2(\lambda^2 + 1) - \frac{2}{3}(\lambda + 1)^2, \qquad g_3 = \frac{4}{9}(\lambda^3 + 1) - \frac{4}{27}(\lambda + 1)^3,$$

but we should keep in mind that it is really defined by the diagram (1.31).

The coarse moduli spaces of $\mathfrak{L}$ and $\mathfrak{W}$ are isomorphic. This common moduli space is another copy of the Riemann sphere, denoted J, and it is customary to write the coordinate as j, and normalize j in such a way that

$$j = 1728 \frac{g_2^3}{g_2^3 - 27g_3^2} = 256 \frac{(\lambda^2 - \lambda + 1)^2}{\lambda^2(\lambda - 1)^2}.$$

The point $j = 0$ gives the equilateral triangle, the point $j = 1728$ gives the bisected line segment, and the point $j = \infty$ corresponds to the pinched triangle. The real axis in the j-plane contains both the isosceles triangles (for $j < 1728$) and the properly degenerate triangles (for $j \geq 1728$).

We have seen that both $\mathfrak{L}$, the stack of three points on the Riemann sphere, up to affine linear transformations, and $\mathfrak{W}$, the stack of degree 3 polynomials up to affine linear substitutions, are compactifications of the stack of oriented triangles. The morphism $\mathfrak{L} \to \mathfrak{W}$ is an isomorphism away from the point corresponding to the pinched triangle, or $j = \infty$.

A pictorial representation of the commutative diagram

follows:

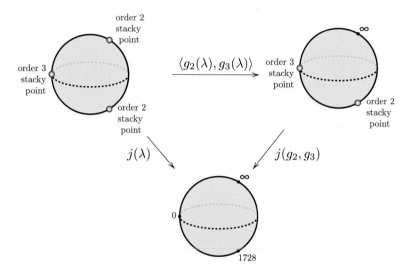

If we remove from $\mathfrak{L}$ or $\mathfrak{W}$ the line segment corresponding in the j-sphere to the real segment $[1728, \infty]$, we obtain the moduli stack $\widetilde{\mathfrak{M}}$ with a single order 3 stacky point in the middle, which we obtained from the length-of-sides point of view, see (1.16).

The sphere J can also be viewed as having been obtained by sewing together the edge of the cone $\widetilde{M}$ of (1.12). It is the topological quotient of the bipyramid $\overline{N}^{\pm}$ of (1.21) or the sphere (1.24) by S_3.

The j-plane

We have seen that the Riemann sphere is a coarse moduli space for oriented triangles. There does not exist a modular family parametrized by the j-sphere. That is why there are discontinuities in the triangles corresponding to various j-values of Figure 1.4.

To see the behavior near $j = 0$, we pass to a neighborhood of $j = 0$ by removing $j = 1728$. This corresponds to setting $g_3 = 1$ and going up to the Riemann surface defined by solving the equation

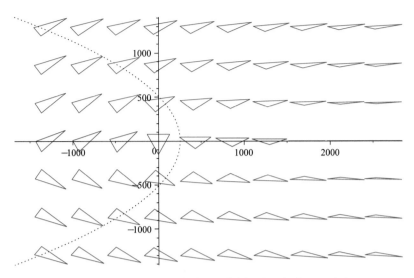

Figure 1.4. The j-plane. The locus of right triangles is a parabola.

Figure 1.5. The g_2-plane.

$$j = 1728\frac{g_2^3}{g_2^3 - 27}$$

for g_2, which is of degree 3 over the j-plane. We obtain Figure 1.5. The pinched triangle $j = \infty$ appears three times in this figure, namely at $g_2 = 3e^{2\pi in/3}$, $n = 0, 1, 2$. No oriented triangle appears more than once near these points. This corresponds to the fact that the pinched triangle is asymmetric in this picture.

To see the behavior near $j = 1728$, we pass to a neighborhood of $j = 1728$ by setting $g_2 = 1$. This means going up to the degree 2 Riemann surface defined by solving

$$j = 1728 \frac{1}{1 - 27g_3^2}$$

for g_3. We get Figure 1.6. This time there are two pinched triangles, at $g_3 = \pm\frac{1}{9}\sqrt{3}$. At the center is the bisected line segment which has order 2 symmetry group; it appears once in the picture. All other triangles appear twice in the picture; they are asymmetric.

For further discussion of the g_2- and g_3-planes, see Example 1.109.

Note the difference between the neighborhoods of the pinched triangle in these pictures and in the $\mathfrak{L}$-picture, (1.22). In the latter, the three vertices can be consistently labeled near $\lambda = \infty$, but in the current ones this is impossible: small loops around the pinched triangles give rise to Moebius strips and Klein bottles, as in (1.5). So, from the polynomial point of view, there is no local labeling of vertices.

It is a matter of taste which of the two completions $\mathfrak{L}$ or $\mathfrak{W}$ of $\widetilde{\mathfrak{M}}$ one considers to be the "right one." It also depends on applications which one of the two could be more useful.

Figure 1.6. The g_3-plane.

Exercise 1.59. We can enlarge the stack $\mathfrak{W}$ to include the triangle whose three vertices coincide. In Corollary 1.55, replace W by $\mathbb{C}^2$, or equivalently drop the requirement that the sections g_2 and g_3 cannot vanish simultaneously. The Weierstrass family of polynomials extends to $\mathbb{C}^2$, and is versal. It has symmetry groupoid given by the action of $\mathbb{C}^*$ on $\mathbb{C}^2$, rather than $W = \mathbb{C}^2 \setminus \{(0,0)\}$.

The coarse moduli space of this triangle has one more point than the j-sphere. Every point in the j-sphere is in the closure of this additional point. Thus, the coarse moduli space is no longer Hausdorff. This is the main reason for excluding the triangle reduced to a point.   □

1.2 Formalism

Let us now make the theory we have developed completely rigorous. This will require some formalism.

1.2.1 Objects in continuous families: categories fibered in groupoids

We will first make precise what we mean by mathematical objects that can vary in continuous families.

We start with the category $\mathscr{T}$ of topological spaces. This category consists of all topological spaces and all continuous maps. The class of topological spaces forms the class of **objects** of $\mathscr{T}$, and, for every two topological spaces S, T, the set of continuous maps from S to T forms the set of **morphisms** from the object S to the object T. Every object has an identity morphism (the identity map, which is continuous), and composition of morphisms (i.e. composition of continuous maps) is associative.

Definition 1.60. A **groupoid fibration** (or a *category fibered in groupoids*) over $\mathscr{T}$ is another category $\mathfrak{X}$, together with a functor $\mathfrak{X} \to \mathscr{T}$, such that two axioms, specified below, are satisfied. If the functor $\mathfrak{X} \to \mathscr{T}$ maps the object x of $\mathfrak{X}$ to the topological space T, we say that x *lies over* T, or that x is an $\mathfrak{X}$-*family parametrized* by T, and we write x/T. If the morphism $\eta : x \to y$ in $\mathfrak{X}$ maps to the continuous map $f : T \to S$, we say that η *lies over* f, or *covers* f. The two groupoid fibration axioms are

(i) for every continuous map $T' \to T$, and every $\mathfrak{X}$-family x/T, there exists an $\mathfrak{X}$-family x'/T' and an $\mathfrak{X}$-morphism $x' \to x$ covering $T' \to T$,

$$
\begin{array}{ccc}
x' & \longrightarrow & x \\
\downarrow & & \downarrow \\
T' & \longrightarrow & T
\end{array}
\tag{1.33}
$$

(ii) the object x'/T' together with the morphism $x' \to x$ is unique up to a unique isomorphism,

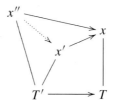

which means that if $x'' \to x$ is another $\mathfrak{X}$-morphism covering $T' \to T$, there exists a unique $\mathfrak{X}$-morphism $x' \to x''$ covering the identity of T', and making the diagram

in $\mathfrak{X}$ commute. □

If we have a diagram (1.33), the family x'/T' is said to be the *pullback*, or *restriction*, of the family x/T, via the continuous map $f : T' \to T$. Using the definite article is justified by the fact that x' is, up to isomorphism, completely determined by x/T and $T' \to T$. We use the notation $x' = f^*x$, or $x' = x|_{T'}$. Sometimes the word restriction is reserved for the case when $T' \to T$ is the inclusion map of a subspace.

The notion of groupoid fibration over $\mathscr{T}$ captures two notions at once: isomorphisms of families and pullbacks of families. The first axiom says that restriction/pullback always exists, and the second says that restriction/pullback is essentially unique. Note that the axioms also imply that pullback is associative: $f^*g^*x = (gf)^*x$. The equality sign stands for *canonically isomorphic*.

For isomorphisms of families, see Exercise 1.61:

Exercise 1.61. Let $\mathfrak{X} \to \mathscr{T}$ be a groupoid fibration. Let T be a topological space. The *fiber* of $\mathfrak{X}$ over T, notation $\mathfrak{X}(T)$, consists of all objects of $\mathfrak{X}$ lying over T and all morphisms of $\mathfrak{X}$ lying over the identity map of T. Prove that $\mathfrak{X}(T)$ is a groupoid, i.e. a category in which all morphisms are invertible. □

Suppose x/T is an $\mathfrak{X}$-family parametrized by T. Let $t \in T$ be a point of t. If we think of t as a continuous map $t : * \to T$, from the one point space $*$ to T, we see that we have a pullback object $x_t = t^*x$ in the category $\mathfrak{X}(*)$. As we vary $t \in T$, the various x_t form the *family members* of x.

Exercise 1.62. For $\mathfrak{M}$, the stack of triangles formalized in Exercise 1.11, the corresponding category fibered in groupoids has objects (T, T', a), where T is a topological space, $T' \to T$ is a degree 3 covering, and $a : T' \to \mathbb{R}_{>0}$ is continuous (such that the triangle inequality is satisfied). A morphism in $\mathfrak{M}$ from (S, S', b) to (T, T', a) is a pair (f, ϕ), where f and ϕ are continuous maps making the triangle in (1.34) commute,

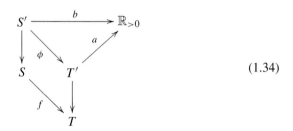

$$(1.34)$$

and the parallelogram in (1.34) a pullback diagram. Composition in the category $\mathfrak{M}$ is defined in a straightforward manner, and the functor $\mathfrak{M} \to \mathscr{T}$ is defined by projecting onto the first component: $(T, T', a) \longmapsto T$ and $(f, \phi) \longmapsto f$.

The fact that $\mathfrak{M}$ is a groupoid fibration follows from the requirement that the morphisms in $\mathfrak{M}$ define pullback diagrams.   □

Exercise 1.63. It is sometimes convenient to choose, for every x/T and for every $T' \to T$, a pullback. (This could be done, for example, by specifying a particular construction of the pullback family, but in general requires the axiom of choice for classes.) The chosen pullbacks give rise to a pullback functor $f^* : \mathfrak{X}(T) \to \mathfrak{X}(T')$, for every $f : T' \to T$. They also give rise, for every composition of continuous maps $T'' \xrightarrow{f} T' \xrightarrow{g} T$, to a natural transformation $\theta_{fg} : (gf)^* \Rightarrow f^* \circ g^*$. The θ have to satisfy an obvious compatibility condition, with respect to composition of continuous maps. So a groupoid fibration with chosen pullbacks gives rise to a *lax functor* from $\mathscr{T}$ to the 2-category of groupoids.   □

Exercise 1.64. Every topological space X gives rise to a tautological groupoid fibration $\underline{X}$, in such a way that X is the fine moduli space of $\underline{X}$. So $\underline{X}$-families parametrized by the topological space T are continuous maps $T \to X$, and morphisms are commutative triangles

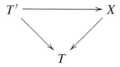

The structure functor $\underline{X} \to \mathscr{T}$ maps $T \to X$ to T. ☐

Definition 1.65. Let $\mathfrak{X}$ and $\mathfrak{Y}$ be groupoid fibrations over $\mathscr{T}$. A **morphism** of groupoid fibrations $F : \mathfrak{X} \to \mathfrak{Y}$ is a functor F, compatible with the structure functors to $\mathscr{T}$. This means that

(i) for every object x of $\mathfrak{X}$, lying over the topological space T, the object $F(x)$ of $\mathfrak{Y}$ also lies over T,
(ii) for every morphism $\eta : x' \to x$ of $\mathfrak{X}$ lying over $f : T' \to T$, the morphism $F(\eta) : F(x') \to F(x)$ also lies over f.

Therefore a morphism of groupoid fibrations turns $\mathfrak{X}$-families into $\mathfrak{Y}$-families, in a way compatible with pullback of families: $F(f^*x) = f^*F(x)$. (Again, the equality stands for canonically isomorphic).

Definition 1.66. Let $\mathfrak{X}$ and $\mathfrak{Y}$ be groupoid fibrations over $\mathscr{T}$, and $F, G : \mathfrak{X} \to \mathfrak{Y}$ morphisms. Then a **2-isomorphism** from F to G

$$\mathfrak{X} \overset{F}{\underset{G}{\Longrightarrow\!\!\!\theta}} \mathfrak{Y}$$

is a natural transformation $\theta : F \to G$ such that, for every object $x/T \in \mathfrak{X}$, the morphism $\theta(x) : F(x) \to G(x)$ in $\mathfrak{Y}$ lies over the identity of T.

The hierarchy (groupoid fibrations, morphisms, 2-isomorphisms) forms a 2-category, formally identical to the hierarchy (categories, functors, natural transformations), with the added benefit that all 2-isomorphisms are invertible.

Two groupoid fibrations $\mathfrak{X}$ and $\mathfrak{Y}$ are called **isomorphic**, or **equivalent**, if there exist morphisms $F : \mathfrak{X} \to \mathfrak{Y}$ and $G : \mathfrak{Y} \to \mathfrak{X}$, and 2-isomorphisms $\theta : G \circ F \Rightarrow \mathrm{id}_{\mathfrak{X}}$ and $\eta : F \circ G \Rightarrow \mathrm{id}_{\mathfrak{Y}}$. In this case, both F and G are called **isomorphisms** or *equivalences* of groupoid fibrations.

Exercise 1.67. Prove that a morphism of groupoid fibrations $F : \mathfrak{X} \to \mathfrak{Y}$ is an isomorphism if it is an equivalence of categories. This is the case if F is *fully faithful* and *essentially surjective*. ☐

Exercise 1.68. Let $\mathfrak{X}$ be a groupoid fibration and let T be a topological space. Show that a morphism $\underline{T} \to \mathfrak{X}$ is the same thing as an $\mathfrak{X}$-family x, parametrized by T, together with a chosen pullback family f^*x, for every continuous map $f : T' \to T$. So, if $\mathfrak{X}$ is endowed with chosen pullbacks, as in Exercise 1.63, then a morphism $\underline{T} \to \mathfrak{X}$ is the same thing as an $\mathfrak{X}$-family over T. Moreover, a 2-isomorphism

$$\underline{T} \underset{y}{\overset{x}{\rightrightarrows}} \Downarrow\theta \; \mathfrak{X}$$

is the same thing as an isomorphism of $\mathfrak{X}$-families $\theta : x \to y$.

One should think of the morphism $\underline{T} \to \mathfrak{X}$ as the moduli map corresponding to the family x/T. Viewing an $\mathfrak{X}$-family over T as a morphism $\underline{T} \to \mathfrak{X}$ is a very powerful way of thinking, but all arguments can always be formulated purely in the language of groupoid fibrations and $\mathfrak{X}$-families; no result depends on choices of pullbacks existing. □

Fibered products of groupoid fibrations

Let $F : \mathfrak{X} \to \mathfrak{Z}$ and $G : \mathfrak{Y} \to \mathfrak{Z}$ be morphisms of groupoid fibrations over $\mathcal{T}$. The **fibered product** of $\mathfrak{X}$ and $\mathfrak{Y}$ over $\mathfrak{Z}$ is the groupoid fibration $\mathfrak{W}$ defined as follows: $\mathfrak{W}$-families parametrized by T are triples (x, ϕ, y), where x/T is an $\mathfrak{X}$-family, y/T is a $\mathfrak{Y}$-family, and $\phi : F(x) \to G(y)$ is an isomorphism of $\mathfrak{Z}$-families. A morphism from (x', ϕ', y') over T' to (x, ϕ, y) over T, covering the continuous map $f : T' \to T$, is a pair (α, β), where $\alpha : x' \to x$ is a morphism in $\mathfrak{X}$ covering f and $\beta : y' \to y$ is a morphism in $\mathfrak{Y}$ covering f, such that

$$\begin{array}{ccc} F(x') & \xrightarrow{\;\phi'\;} & G(y') \\ {\scriptstyle F(\alpha)}\downarrow & & \downarrow{\scriptstyle G(\beta)} \\ F(x) & \xrightarrow{\;\phi\;} & G(y) \end{array}$$

commutes in $\mathfrak{Z}$.

There is a 2-commutative diagram

$$\begin{array}{ccc} \mathfrak{W} & \xrightarrow{\;\mathrm{pr}_\mathfrak{Y}\;} & \mathfrak{Y} \\ {\scriptstyle \mathrm{pr}_\mathfrak{X}}\downarrow & \overset{\phi}{\nearrow} & \downarrow{\scriptstyle G} \\ \mathfrak{X} & \xrightarrow{\;F\;} & \mathfrak{Z} \end{array}$$

This means that ϕ is a 2-isomorphism from $F \circ \mathrm{pr}_\mathfrak{X}$ to $G \circ \mathrm{pr}_\mathfrak{Y}$. It is defined by $(x, \phi, y) \mapsto \phi$.

Given an arbitrary 2-commutative diagram

$$
\begin{array}{ccc}
\mathfrak{U} & \xrightarrow{\;Q\;} & \mathfrak{Y} \\
{\scriptstyle P}\big\downarrow & {\scriptstyle \psi}\!\!\nearrow & \big\downarrow{\scriptstyle G} \\
\mathfrak{X} & \xrightarrow[\;F\;]{} & \mathfrak{Z}
\end{array}
\tag{1.35}
$$

there is an induced morphism $\mathfrak{U} \to \mathfrak{W}$ to the fibered product, given on objects by $u \longmapsto \big(P(u),\ \psi(u),\ Q(u) \big)$. The diagram (1.35) is called **2-cartesian** if $\mathfrak{U} \to \mathfrak{W}$ is an equivalence of groupoid fibrations.

Example 1.69. The main example is the symmetry groupoid of a family: let x/T be an $\mathfrak{X}$-family parametrized by T, and assume that its symmetry groupoid $\Gamma \rightrightarrows T$ satisfies the technical condition of Definition 1.33. Think of x as a morphism $\underline{T} \to \mathfrak{X}$, as in Exercise 1.68. Then there is a 2-cartesian diagram of groupoid fibrations

$$
\begin{array}{ccc}
\Gamma & \xrightarrow{\;t\;} & \underline{T} \\
{\scriptstyle s}\big\downarrow & {\scriptstyle \phi}\!\!\nearrow & \big\downarrow{\scriptstyle x} \\
\underline{T} & \xrightarrow[\;x\;]{} & \mathfrak{X}
\end{array}
\tag{1.36}
$$

where $\phi : s^*x \to t^*x$ is the tautological isomorphism.

1.2.2 Families characterized locally: prestacks

We have discussed several examples in which we characterized a moduli problem by specifying what continuous families looked like locally, and what isomorphisms between families looked like. These were examples of prestacks. Also, it is in the context of prestacks that symmetry groupoids behave well.

Remark 1.70. For experts, we should note that our definition of prestack is stronger than the usual one: we define a prestack to be a groupoid fibration with representable diagonal. This implies that all isomorphism functors are sheaves. In practice, the stronger condition is often verified, so our non-standard terminology seems justified.

We need some terminology. Let $\mathfrak{X}$ be a groupoid fibration over $\mathscr{T}$. Let x/T and y/S be $\mathfrak{X}$-families. The *space of isomorphisms* $\underline{\mathrm{Isom}}(x, y)$ is the set of all triples (t, ϕ, s), where $t \in T$, $s \in S$, and $\phi : x_t \to y_s$ is an isomorphism in $\mathfrak{X}(*)$. The topology on this space of isomorphisms is, by definition, the finest topology such that for every pair (U, ϕ), where U is a topological space, endowed with maps $U \to T$ and $U \to S$, and $\phi : x|_U \to y|_U$ is an isomorphism in the category $\mathfrak{X}(U)$, the induced map $U \to \underline{\mathrm{Isom}}(x, y)$, defined by $u \longmapsto \phi_u$, is continuous.

Definition 1.71. A **prestack** is a groupoid fibration $\mathfrak{X}$ over $\mathcal{T}$ such that, for any two objects x/T and y/S, the following conditions are satisfied:

(i) the canonical maps $\underline{\mathrm{Isom}}(x, y) \to T$ and $\underline{\mathrm{Isom}}(x, y) \to S$ are continuous,
(ii) for any continuous map $\alpha : U \to \underline{\mathrm{Isom}}(x, y)$, there exists a unique isomorphism of families $\phi : x|_U \to y|_U$ giving rise to α.

For the second condition, it suffices that the tautological isomorphism over $\underline{\mathrm{Isom}}(x, y)$ is continuous (i.e. occurs in the groupoid fibration $\mathfrak{X}$).

Exercise 1.72. If the prestack $\mathfrak{X}$ has chosen pullbacks, we can think of x/T and y/S as morphisms, as in Exercise 1.68. Then the diagram

$$\begin{array}{ccc} \underline{\mathrm{Isom}}(x, y) & \longrightarrow & S \\ \downarrow & \nearrow & \downarrow y \\ T & \xrightarrow{x} & \mathfrak{X} \end{array}$$

is 2-cartesian. □

Exercise 1.73. Conversely, suppose that $\mathfrak{X}$ is a prestack and that $F : \underline{U} \to \mathfrak{X}$ is a morphism, where U is a topological space. Then there exist a topological space R and a 2-cartesian diagram

$$\begin{array}{ccc} \underline{R} & \longrightarrow & \underline{U} \\ \downarrow & \nearrow & \downarrow F \\ \underline{U} & \xrightarrow{F} & \mathfrak{X} \end{array}$$

Moreover, $R \rightrightarrows U$ is a topological groupoid. (It is isomorphic to the symmetry groupoid of the $\mathfrak{X}$-family $F(\mathrm{id}_U)$.) □

Versal families
The following repeats Definition 1.33:

Definition 1.74. Let $\mathfrak{X}$ be a groupoid fibration. An $\mathfrak{X}$-family x/T is called **versal** if every family can be locally pulled back from x/T and if $\underline{\mathrm{Isom}}(x, x)$ satisfies the conditions of Definition 1.71.

So, if $\mathfrak{X}$ is a prestack, a family x/T is versal if every family can be locally pulled back from x/T. The following converse is more useful:

Lemma 1.75. *Suppose that a groupoid fibration admits a versal family. Then it is a prestack.*

Proof. Let x/T and y/S be $\mathfrak{X}$-families. We have to prove that there exist a topological space I and a 2-cartesian diagram

$$
\begin{array}{ccc}
\underline{I} & \longrightarrow & \underline{T} \times \underline{S} \\
\downarrow & & \downarrow{\scriptstyle x \times y} \\
\mathfrak{X} & \xrightarrow{\ \Delta\ } & \mathfrak{X} \times \mathfrak{X}
\end{array}
$$

Because we can glue topological spaces along open subspaces, it is enough to cover T and S with open subspaces $T = \bigcup U_i$ and $S = \bigcup V_j$, and prove that there exist a topological space J_{ij} and a 2-cartesian diagram

$$
\begin{array}{ccc}
\underline{J}_{ij} & \longrightarrow & \underline{U}_i \times \underline{V}_j \\
\downarrow & & \downarrow{\scriptstyle x|_{U_i} \times y|_{V_j}} \\
\mathfrak{X} & \xrightarrow{\ \Delta\ } & \mathfrak{X} \times \mathfrak{X}
\end{array}
$$

for all i, j. Now we know that there is a 2-cartesian diagram

$$
\begin{array}{ccc}
\underline{\Gamma}_1 & \longrightarrow & \underline{\Gamma}_0 \times \underline{\Gamma}_0 \\
\downarrow & & \downarrow \\
\mathfrak{X} & \xrightarrow{\ \Delta\ } & \mathfrak{X} \times \mathfrak{X}
\end{array}
$$

where $\Gamma_1 \rightrightarrows \Gamma_0$ is the symmetry groupoid of the given versal family. By the first property of versal family, we can cover $T = \bigcup U_i$ and $S = \bigcup_j V_j$, and find 2-commutative diagrams

$$
\begin{array}{ccc}
U_i & \xrightarrow{\ f_i\ } & \Gamma_0 \\
\downarrow & \nearrow & \downarrow \\
T & \xrightarrow{\ x\ } & \mathfrak{X}
\end{array}
\qquad
\begin{array}{ccc}
V_j & \xrightarrow{\ g_j\ } & \Gamma_0 \\
\downarrow & \nearrow & \downarrow \\
S & \xrightarrow{\ y\ } & \mathfrak{X}
\end{array}
$$

Then we define J_{ij} to be the fibered product

$$
\begin{array}{ccc}
J_{ij} & \longrightarrow & U_i \times V_j \\
\downarrow & & \downarrow{\scriptstyle f_i \times g_j} \\
\Gamma_1 & \longrightarrow & \Gamma_0 \times \Gamma_0
\end{array}
$$

and glue the J_{ij} to obtain I.   □

Exercise 1.76. For a topological space X, a versal family for the groupoid fibration $\underline{X}$ is the same thing as a continuous map $f : T \to X$ which admits local sections, i.e. for every $x \in X$, there exists an open neighborhood $x \in U \subset X$ and a continuous map $s : U \to T$, such that $f \circ s$ is equal to the inclusion map $U \to X$.

The symmetry groupoid of $T \to X$ is the fibered product groupoid $T \times_X T \rightrightarrows T$. Such groupoids are called **banal groupoids**. Banal groupoids are equivalence relations.   □

1.2.3 Families which can be glued: stacks

Definition 1.77. A prestack is called a **stack** if it satisfies the gluing axiom of Exercise 1.12.

Example 1.78. In Section 1.1.10, (see also Definition 1.40) we defined a groupoid fibration which we shall call $\mathfrak{L}^{\mathrm{pre}}$. Objects of $\mathfrak{L}^{\mathrm{pre}}$ are quadruples (T, A_1, A_2, A_3), where T is the parameter space and $A_i : T \to \mathbb{C}$ are continuous functions (no more than two of which are ever allowed to coincide). A morphism from (T', A'_1, A'_2, A'_3) to (T, A_1, A_2, A_3) is a quadruple (f, σ, R, S), where $f : T' \to T$ is a continuous map between the parameter spaces, $\sigma \in S_3$ is a permutation of $\{1, 2, 3\}$, and $R : T' \to \mathbb{C}$ and $S : T' \to \mathbb{C}^*$ are continuous maps, such that $A'_{\sigma(i)} = S \cdot (A_i \circ f) + R$, for $i = 1, 2, 3$.

This groupoid fibration is a prestack, but not a stack. It is a prestack because it admits a versal family, as we saw in Section 1.1.10. But it is not a stack: it is possible to specify gluing data in $\mathfrak{L}^{\mathrm{pre}}$ which give rise to families with a twist, even though all $\mathfrak{L}^{\mathrm{pre}}$-families are untwisted.   □

1.2.4 Topological stacks

Definition 1.79. A stack $\mathfrak{X}$ which admits a versal family is called a **topological stack**.

If $\Gamma_1 \rightrightarrows \Gamma_0$ is the symmetry groupoid of a versal family for $\mathfrak{X}$, we say that $\Gamma_1 \rightrightarrows \Gamma_0$ is a **presentation** of $\mathfrak{X}$.

All our examples $\mathfrak{M}, \widetilde{\mathfrak{M}}, \overline{\mathfrak{M}}, \mathfrak{L}, \mathfrak{W}$, etc., are topological stacks.

In practice, the current definition is not strong enough: to be able to "do topology" on a topological stack, we have to put conditions on the spaces Γ_0 and Γ_1, or on the maps $s, t : \Gamma_1 \to \Gamma_0$ of the symmetry groupoid of a versal family. For example, to do homotopy theory, we need that s and t are *topological submersions*; see Section 1.2.7.

The nicest topological stacks, which are closest to topological spaces, are those of Deligne–Mumford type, see Definition 1.104. In this case, s and t are required to be local homeomorphisms. In particular, all symmetry groups are discrete. If they are *separated* (see Exercise 1.99), these stacks admit the structure of an *orbispace*; see Theorem 1.108.

Properties of $\Gamma_1 \to \Gamma_0 \times \Gamma_0$ are *separation properties* of $\mathfrak{X}$. See, for example, Exercise 1.99 or Proposition 1.110.

Exercise 1.80. Note that Definition 1.18 applies to any groupoid fibration. Prove that if $\Gamma_1 \rightrightarrows \Gamma_0$ is a presentation of a topological stack $\mathfrak{X}$, then the image of Γ_1 in $\Gamma_0 \times \Gamma_0$ defines an equivalence relation on Γ_0, and the topological quotient of Γ_0 by this equivalence relation is a coarse moduli space for $\mathfrak{X}$. □

Topological groupoids

The main general example of a topological stack is the stack of Γ-torsors, for a topological groupoid Γ.

Definition 1.81. A **topological groupoid** is a groupoid $\Gamma_1 \rightrightarrows \Gamma_0$, as in Definition 1.10, where Γ_1 and Γ_0 are also topological spaces, and all structure maps s, t, e, μ, ϕ are continuous. Here, $e : \Gamma_0 \to \Gamma_1$ is the identity map, $\mu : \Gamma_2 \to \Gamma_1$ is the composition map, and $\phi : \Gamma_1 \to \Gamma_1$ is the inverse map. Often we abbreviate the notation to $\Gamma_\bullet$ or simply Γ.

A **continuous morphism** of topological groupoids is a functor $\phi : \Gamma \to \Gamma'$, such that the two maps $\phi_0 : \Gamma_0 \to \Gamma'_0$ and $\phi_1 : \Gamma_1 \to \Gamma'_1$ are continuous.

Exercise 1.82. The symmetry groupoid of a family in a prestack is a topological groupoid. □

Exercise 1.83. Suppose Γ is the symmetry groupoid of the family x/T in a prestack, and that $f : T' \to T$ is a continuous map. Form the fibered product of topological spaces

$$\begin{array}{ccc} \Gamma' & \longrightarrow & T' \times T' \\ \downarrow & & \downarrow {\scriptstyle f \times f} \\ \Gamma & \xrightarrow{\;s \times t\;} & T \times T \end{array}$$

Prove that Γ' is a topological groupoid, and that it is isomorphic to the symmetry groupoid of the pullback family f^*x over T'. □

This exercise leads to the following definition.

Definition 1.84. Let $\Gamma_1 \rightrightarrows \Gamma_0$ be a topological groupoid and let $\Gamma_0' \to \Gamma_0$ be a continuous map. The fibered product

$$\begin{array}{ccc} \Gamma_1' & \longrightarrow & \Gamma_0' \times \Gamma_0' \\ \downarrow & & \downarrow \\ \Gamma_1 & \longrightarrow & \Gamma_0 \times \Gamma_0 \end{array}$$

defines another topological groupoid $\Gamma_1' \rightrightarrows \Gamma_0'$, called the **restriction** of the groupoid $\Gamma_1 \rightrightarrows \Gamma_0$ via the map $\Gamma_0' \to \Gamma_0$. It comes with a continuous morphism of groupoids $\Gamma' \to \Gamma$ which is *fully faithful*, in categorical terms.

Exercise 1.85. If G is a topological group acting continuously on the topological space X, the transformation groupoid $X \times G \rightrightarrows X$ is a topological groupoid. Note the special cases $G = \{e\}$ or $X = \{*\}$. □

Generalized moduli maps: groupoid torsors

Definition 1.86. Let $\Gamma_\bullet$ be a topological groupoid. A $\Gamma_\bullet$-torsor over the topological space T is a pair (P_0, ϕ), where P_0 is a topological space endowed with a continuous map $\pi : P_0 \to T$, and $\phi : P_\bullet \to \Gamma_\bullet$ is a continuous morphism of topological groupoids. Here, $P_\bullet$ is the banal groupoid associated to $P_0 \to T$ (Exercise 1.76). Moreover, it is required that

(i) the diagram

$$\begin{array}{ccc} P_1 & \longrightarrow & \Gamma_1 \\ \downarrow & & \downarrow \\ P_0 & \longrightarrow & \Gamma_0 \end{array} \qquad (1.37)$$

is a pullback diagram of topological spaces,

(ii) the map $P_0 \to T$ admits local sections, as in Exercise 1.76.

A **morphism** of $\Gamma_\bullet$-torsors from (P_0', ϕ') over T' to (P, ϕ) over T consists of a pullback diagram of topological spaces

$$\begin{array}{ccc} P_0' & \longrightarrow & P_0 \\ \downarrow & & \downarrow \\ T' & \longrightarrow & T \end{array}$$

such that the induced diagram

is a commutative diagram of topological groupoids.

Exercise 1.87. The Γ-torsors form a stack. It is called the **stack associated to the topological groupoid Γ.** □

Exercise 1.88. Show that if the topological groupoid Γ is a transformation groupoid $X \times G \rightrightarrows X$, then a Γ-torsor over T is the same thing as a principal homogeneous G-bundle $P \to T$, together with a G-equivariant map $P \to X$. Note the special cases $G = \{e\}$ and $X = \{*\}$. □

Exercise 1.89. Let Γ be a topological groupoid. Prove that the stack of Γ-torsors $\mathfrak{X}$ is a topological stack by proving that there is a tautological Γ-torsor over the topological space Γ_0. The symmetry groupoid of the tautological Γ-torsor is the groupoid Γ itself. Thus Γ itself is a presentation of $\mathfrak{X}$. There is a 2-cartesian diagram

$$\begin{array}{ccc} \underline{\Gamma_1} & \longrightarrow & \underline{\Gamma_0} \\ \downarrow & \nearrow & \downarrow \\ \underline{\Gamma_0} & \longrightarrow & \mathfrak{X} \end{array} \qquad (1.38)$$

□

The following theorem generalizes Exercise 1.34 and Exercise 1.50.

Theorem 1.90. *If $\mathfrak{X}$ is a topological stack, and $\Gamma_1 \rightrightarrows \Gamma_0$ is the symmetry groupoid of a versal family x/X_0, then $\mathfrak{X}$ is isomorphic to the stack of $\Gamma_\bullet$-torsors.*

Proof. As we have studied the proof in detail in special cases, we will only say that, to define the morphism from $\mathfrak{X}$ to $\Gamma_\bullet$-torsors, we associate to an

$\mathfrak{X}$-family y/T the $\Gamma_\bullet$-torsor $\underline{\mathrm{Isom}}(x, y)$ (which is the generalized moduli map of y/T). □

In our examples, we often specified a moduli problem by giving a prestack, then constructing a versal family for the prestack, and finally replacing the prestack by the stack of torsors for the symmetry groupoid of the prestack. This process is known as **stackification**. We followed it, for example, when passing from Definition 1.40 to Proposition 1.43, or when going from Exercise 1.54 to Corollary 1.55.

Example 1.91. If $V/*$ is a versal family with only one family member, so that the symmetry groupoid of $V/*$ is just a topological group G, then families are the same thing as *twisted forms* of V, i.e. locally constant families all of whose family members are isomorphic to V. The stack of twisted forms of V is equivalent to the stack of G-torsors. Often, we say simply *forms* instead of twisted forms.

Change of versal family: Morita equivalence of groupoids

Suppose x/Γ_0 is a versal $\mathfrak{X}$-family with symmetry groupoid $\Gamma_\bullet$, and let y/S be an arbitrary $\mathfrak{X}$-family. There is a 2-cartesian diagram (we have stopped underlining)

$$
\begin{array}{ccc}
P_1 & \xrightarrow{\phi_1} & \Gamma_1 \\
\downdownarrows & & \downdownarrows \\
P_0 & \xrightarrow{\phi_0} & \Gamma_0 \\
\downarrow & \nearrow & \downarrow x \\
S & \xrightarrow{y} & \mathfrak{X}
\end{array}
\tag{1.39}
$$

in the shape of a cube (all six sides of the cube are cartesian). The right-hand edge of the diagram abbreviates (1.36), and $(P_\bullet, \phi_\bullet)$ is the generalized moduli map of y. Because this diagram is 2-cartesian, we have $P_0 = \underline{\mathrm{Isom}}(y, x)$.

Suppose x/Γ_0 is a versal family with symmetry groupoid $\Gamma_\bullet$, and y/Γ'_0 is a second versal family with symmetry groupoid $\Gamma'_\bullet$. Then we can form a larger 2-cartesian diagram

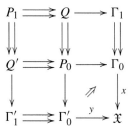

We see that P_0 is at the same time a $\Gamma_\bullet$-torsor over Γ'_0 and a $\Gamma'_\bullet$-torsor over Γ_0. We say that P_0 is a $\Gamma_\bullet$-$\Gamma'_\bullet$-bitorsor.

Exercise 1.92. Conversely, if there exists a $\Gamma_\bullet$-$\Gamma'_\bullet$-bitorsor, the stack of $\Gamma_\bullet$-torsors and the stack of $\Gamma'_\bullet$-torsors are isomorphic. This is the general case of Corollary 1.52. ☐

Definition 1.93. Two topological groupoids Γ and Γ' are called **Morita equivalent** if there exists a Γ-Γ'-bitorsor.

Thus we can say that stacks "are" *groupoids up to Morita equivalence*.

Exercise 1.94. Prove that two topological groupoids Γ and Γ' are Morita equivalent if and only if there exists a third topological groupoid Γ'' and two morphisms $\Gamma'' \to \Gamma$ and $\Gamma'' \to \Gamma'$ which are *topological equivalences*. Here, a morphism $\Gamma'' \to \Gamma$ of topological groupoids is a **topological equivalence** if

(i) (topological full faithfulness) the diagram

$$
\begin{array}{ccc}
\Gamma''_1 & \longrightarrow & \Gamma''_0 \times \Gamma''_0 \\
\downarrow & & \downarrow \\
\Gamma_1 & \longrightarrow & \Gamma_0 \times \Gamma_0
\end{array}
$$

is a pullback diagram of topological spaces,

(ii) (topological essential surjectivity) the morphism

$$
\Gamma''_0 \times_{\Gamma_0,s} \Gamma_1 \xrightarrow{\ t\ } \Gamma_0
$$

admits local sections (see Exercise 1.76).

In fact, the 2-category of topological stacks is a localization of the category of topological groupoids at the topological equivalences. ☐

More generally, let $\mathfrak{X} \to \mathfrak{Y}$ be a morphism of topological stacks, and let $X_\bullet$ be a groupoid presentation of $\mathfrak{X}$ and $Y_\bullet$ one of $\mathfrak{Y}$. Form the larger 2-cartesian diagram:

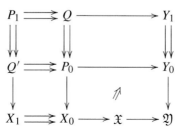

Then P_0 is a $Y_\bullet$-torsor over X_0, and an $X_\bullet$-equivariant map to Y_0.

Exercise 1.95. State and prove the general case of Theorem 1.58.   □

Quotient stacks

Suppose the topological group G acts on the topological space X. The associated stack of pairs (P, ϕ), where P is a G-bundle and $\phi : P \to X$ is an equivariant continuous map, is usually denoted by $[X/G]$ and called the **quotient stack** of X by G. There is a 2-cartesian diagram of groupoid fibrations (1.38)

We have seen that if a topological stack $\mathfrak{X}$ admits a versal family whose symmetry groupoid is the transformation groupoid $X \times G \rightrightarrows X$, then $\mathfrak{X}$ is isomorphic to the quotient stack $[X/G]$.

For every (P, ϕ) as given, parametrized by T, there is a 2-cartesian diagram (1.39)

If $X = *$ is the one-point space, the quotient stack $[*/G]$ is denoted by BG, and is called the *classifying stack* of G. There is a 2-cartesian diagram

and for every principal G-bundle G/T a 2-cartesian diagram

$$\begin{array}{ccc} \underline{P} & \longrightarrow & * \\ \downarrow & & \downarrow \\ \underline{T} & \longrightarrow & BG \end{array}$$

Therefore $\underline{*} \to BG$ is known as the *universal principal G-bundle*.

Exercise 1.96. Let G be a topological group acting on the topological space X, and let $P \to T$ be a G-bundle. Then there is a 2-cartesian diagram

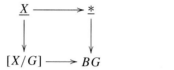

Therefore, $[X/G] \to BG$ is called the *universal fiber bundle with fiber X*. For example, there is always a 2-cartesian diagram

$$\begin{array}{ccc} X & \longrightarrow & * \\ \downarrow & & \downarrow \\ [X/G] & \longrightarrow & BG \end{array}$$

$\square$

Exercise 1.97. The quotient space X/G admits a morphism $[X/G] \to X/G$, which turns X/G into the coarse moduli space of $[X/G]$.

$\square$

Exercise 1.98. Suppose that G acts trivially on X. Then $[X/G] = X \times BG$.

$\square$

Separated topological stacks

Many properties of topological stacks can be defined in terms of presenting groupoids, if these properties are invariant under Morita equivalence. The following exercise treats an example.

Exercise 1.99. We call a topological groupoid Γ **separated** if the map $\Gamma_1 \to \Gamma_0 \times \Gamma_0$ is *universally closed*, i.e. *proper* in the sense of Bourbaki [5]. Prove that if Γ' is Morita equivalent to Γ, then Γ' is separated if and only if Γ is. Therefore, we call a topological stack **separated** if any groupoid presentation of it is separated. Being separated is the analog of the Hausdorff property for stacks. Prove that separated topological stacks have Hausdorff coarse moduli spaces.

When working with separated topological stacks, additional assumptions (such as the parameter space of a versal family being Hausdorff or at least locally Hausdorff) may be necessary. See, for example, Theorem 1.108. $\square$

1.2.5 Deligne–Mumford topological stacks

We now introduce the important idea that the parameter space of a versal family should be thought of as a local model for a topological stack. For this to hold true, the versal family has to have additional properties. We introduce the most basic of these in this section. It comes about in analogy to gluing data for manifolds.

Suppose X is a topological manifold, with an atlas $\{U_i\}_{i \in I}$ of local charts $U_i \to X$. The atlas gives rise to a versal family for the groupoid fibration $\underline{X}$ (see Exercise 1.76). The parameter space

$$\Gamma_0 = \coprod_{i \in I} U_i$$

is the disjoint union of the charts in the atlas, and the versal family is the induced continuous map $\Gamma_0 \to X$. The symmetry groupoid $\Gamma_1 \rightrightarrows \Gamma_0$ has morphism space

$$\Gamma_1 = \coprod_{(i,j) \in I \times I} U_i \cap U_j .$$

We write $U_{ij} = U_i \cap U_j$. This symmetry groupoid is an equivalence relation.

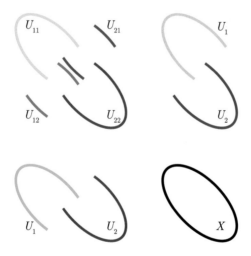

Exercise 1.100. The groupoid $\Gamma_1 \rightrightarrows \Gamma_0$ is the restriction (Definition 1.84) of the trivial groupoid $X \rightrightarrows X$ via the map $\coprod U_i \to X$. The morphism from $\Gamma_1 \rightrightarrows \Gamma_0$ to $X \rightrightarrows X$ is a topological equivalence of topological groupoids (Exercise 1.94). This expresses the fact that $\{U_i\}$ is an atlas of the manifold X in groupoid language. The groupoid $\coprod U_{ij} \rightrightarrows \coprod U_i$ encodes the way that X is obtained by gluing the U_i. Morita equivalence encodes the way different atlases for the same manifold relate to one another. □

By analogy, a general topological groupoid Γ presenting a topological stack $\mathfrak{X}$ should be thought of as an atlas for $\mathfrak{X}$; in fact, **atlas** is a commonly used synonym for "presentation."

There are many topological equivalence relations giving rise to $\underline{X}$ as an associated topological stack: the banal groupoid associated to any continuous map $Y \to X$ admitting local sections will do. For example, we could take a point $P : * \to X$ and pass to the equivalence relation $\Gamma'_1 \rightrightarrows \Gamma'_0$, with $\Gamma'_0 = \Gamma_0 \amalg *$. But, unless X is a manifold of dimension 0, this equivalence relation Γ' no longer reflects the local structure of X.

The morphism $\Gamma_0 \to X$ is a local homeomorphism (every point of Γ_0 has an open neighborhood which maps homeomorphically to an open neighborhood of the image point in X). Because $\Gamma_0 \to X$ has local sections, this is equivalent to source and/or target maps $\Gamma_1 \to \Gamma_0$ being local homeomorphisms.

As another example, consider a discrete group G acting on a topological space Y in such a way that every point of Y has an open neighborhood U such that all Ug, $g \in G$, are disjoint. The quotient map $Y \to X$ is a local homeomorphism, and $Y \to X$ is a versal family for $\underline{X}$.

The property that $\Gamma_0 \to X$ is a local homeomorphism also makes sense for the morphism $\Gamma_0 \to \mathfrak{X}$ of a groupoid presentation for a topological stack, and gives rise to the notion of étale versal family.

Definition 1.101. A family x/T is **étale at the point** $t \in T$ if, for every family y/S, point $s \in S$, and isomorphism $\phi : y_s \to x_t$,

(i) there exists an open neighborhood U of s in S, a continuous map $f : U \to T$, and an isomorphism of continuous families $\Phi : y|_U \to f^*x$, such that $\Phi_s = \phi$.

(ii) Given (U, f, Φ) and (U', f', Φ') as in (i), there exists a third open neighborhood $V \subset U \cap U'$ of s, such that $f|_V = f'|_V$ and $\Phi|_V = \Phi'|_V$.

The family x/T is **étale**, if it is étale at every point of T.

A topological groupoid is called *étale* if source and target maps are local homeomorphisms.

Exercise 1.102. Every étale family has an étale symmetry groupoid. Every versal family with étale symmetry groupoid is étale. (So, most of the versal families we constructed are étale. Exceptions are the versal family of oriented triangles parametrized by the space of great-circle equilateral triangles, and the versal family of degree 3 polynomials parametrized by W.) Having an étale symmetry groupoid is by itself not sufficient for being an étale family. □

Exercise 1.103. If an $\mathfrak{X}$-family x/T is étale, and every object of $\mathfrak{X}(*)$ is isomorphic to x_t, for some point $t \in T$, then x/T is versal.   □

Definition 1.104. If the topological stack $\mathfrak{X}$ admits an étale versal family, it is called a **Deligne–Mumford topological stack**.

Thus, Deligne–Mumford topological stacks "look like" topological spaces, locally. If Γ is an étale groupoid presentation for $\mathfrak{X}$, then $\mathfrak{X}$ looks locally like Γ_0 (and also Γ_1).

Exercise 1.105. Let us make this statement precise.

Surjective local homeomorphisms are *local in the base*. This means that if $X \to Y$ is a continuous map of topological spaces, and $Y' \to Y$ is a continuos map admitting local sections, then $X \to Y$ is a surjective local homeomorphism if and only if the base change $X' \to Y'$ defined by the pullback diagram

is a surjective local homeomorphism.

The morphism $\Gamma_0 \to \mathfrak{X}$ given by an étale versal family for $\mathfrak{X}$, is considered to admit local sections, because for every morphism $T \to \mathfrak{X}$, the base change $T \times_{\mathfrak{X}} \Gamma_0 \to T$ admits local sections.

Therefore, the morphism $\Gamma_0 \to \mathfrak{X}$ is considered to be a surjective local homeomorphism:

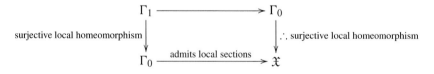

                                                              □

Example 1.106. The topological stacks $\mathfrak{M}$, $\widetilde{\mathfrak{M}}$, $\overline{\mathfrak{M}}$, and $\mathfrak{L}$ are topological Deligne–Mumford stacks. We will see below (Example 1.109) that $\mathfrak{W}$ is of Deligne–Mumford type, too.

Example 1.107. If $X_1 \subset X_0 \times X_0$ is an étale equivalence relation, where X_1 has the subspace topology of the product topology, then the associated

Deligne–Mumford topological stack is equal to $\underline{X}$, where X is the quotient of X_0 by X_1 with the quotient topology.

For example, consider the equivalence relation on $\mathbb{R}$, defined by the action of $\mathbb{Q}$ by translation. If we endow $\mathbb{Q}$ with the discrete topology, the equivalence relation is étale and we obtain a Deligne–Mumford topological quotient stack $[\mathbb{R}/\mathbb{Q}]$. If we endow $\mathbb{Q}$ with the subspace topology, we obtain a topological stack not of Deligne–Mumford type $[\mathbb{R}/\mathbb{Q}]'$. There are morphisms

$$[\mathbb{R}/\mathbb{Q}] \longrightarrow [\mathbb{R}/\mathbb{Q}]' \longrightarrow \underline{\mathbb{R}/\mathbb{Q}},$$

neither of which are an isomorphism.

In particular, $[\mathbb{R}/\mathbb{Q}]$ and $[\mathbb{R}/\mathbb{Q}]'$ are examples of moduli problems without symmetries, which still do not admit fine moduli spaces.

Structure theorem

Let us call a topological Deligne–Mumford stack *separated* if it is separated according to Exercise 1.99, and admits an étale versal family with Hausdorff parameter space.

Theorem 1.108. *Every separated Deligne–Mumford topological stack is locally a quotient stack by a finite group.*

Proof. Let Γ be an étale groupoid presenting the stack $\mathfrak{X}$. We may assume that Γ_0 is Hausdorff, and that $s \times t : \Gamma_1 \to \Gamma_0 \times \Gamma_0$ is proper. Then Γ_1 is Hausdorff as well. Let $P_0 \in \Gamma_0$ be a point, and let G be its automorphism group. Then G is a compact subspace of the discrete space $s^{-1}(P_0) \subset \Gamma_1$, and is therefore finite.

We start by choosing disjoint open neighborhoods of the points of $G \subset \Gamma_1$ that, via s, map homeomorphically to an open neighborhood U_0 of P_0 in Γ_0. (This is possible because s is a local homeomorphism, and G is finite.) This identifies $U_0 \times G$ with an open neighborhood of G in Γ_1. Hence, we have a commutative diagram

$$
\begin{array}{ccccc}
G & \longrightarrow & U_0 \times G & \longrightarrow & \Gamma_1 \\
\downarrow & & {\scriptstyle p_1}\downarrow & & \downarrow{\scriptstyle s} \\
P_0 & \longrightarrow & U_0 & \longrightarrow & \Gamma_0
\end{array}
$$

Now, using the closedness of $s \times t : \Gamma_1 \to \Gamma_0 \times \Gamma_0$, we choose an open neighborhood V_0 of P_0 in U_0 such that $V_1 = (s \times t)^{-1}(V_0 \times V_0) \subset U_0 \times G \subset \Gamma_1$. Then $V_1 \rightrightarrows V_0$ is a subgroupoid of $\Gamma_1 \rightrightarrows \Gamma_0$, and the arrows in V_1 are pairs (u, g), with $u \in V_0 \subset U_0$ and $g \in G$.

Consider the diagram

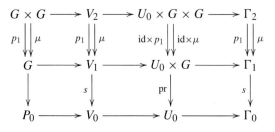

There are four vertical cartesian squares in this diagram, and the upper horizontal arrows are uniquely determined by the lower commutative diagrams. There is, a priori, no reason for

$$
\begin{array}{ccc}
V_2 & \longrightarrow & U_0 \times G \times G \\
\downarrow{\scriptstyle p_2} & & \downarrow{\scriptstyle t \times \mathrm{id}} \\
V_1 \longrightarrow U_0 \times G & \longrightarrow & \Gamma_0 \times G
\end{array}
\qquad (1.40)
$$

to commute, although, after projecting onto Γ_0, the two induced maps $V_2 \to \Gamma_0$ are equal to the projection onto the "middle" object. Therefore, the locus in V_2, where the two maps to $\Gamma_0 \times G$ are equal, is an open neighborhood $V_2' \subset V_2$ of $G \times G$. Now, using the closedness of $\Gamma_2 \to \Gamma_0 \times \Gamma_0 \times \Gamma_0$, we can find an open neighborhood V_0' of P in V_0 such that the preimage of $V_0 \times V_0 \times V_0$ is contained in V_2'. Restricting the groupoid Γ further to V_0', we get a subgroupoid $V' \subset V \subset \Gamma$, and on replacing V with V' we may assume that, in fact, diagram (1.40) does commute. This means that, for all $(u, g, h) \in V_2$, we have $p_1(u, g, h) = (u, g)$, $\mu(u, g, h) = (u, gh)$, and $p_2(u, g, h) = (ug, h)$, where we have written ug for $t(u, g)$. In other words, we have

$$
(u, g) * (ug, h) = (u, gh), \qquad \text{for all } (u, g, h) \in V_2.
$$

For $u \in V_0$ and $g \in G$, define $ug = t(u, g)$. Then let

$$
W_0 = \{u \in V_0 \mid \forall g \in G : ug \in V_0\}.
$$

Then W_0 is an open neighborhood of P_0 in V_0, and for $u \in W_0$ we have $(ug)h = u(gh)$. Restricting our groupoid to $W_0 \subset V_0$, we see that W is the transformation groupoid of the G-action on W_0 previously defined. We have a morphism of groupoids

$$
\begin{array}{ccc}
W_0 \times G & \longrightarrow & \Gamma_1 \\
\downdownarrows & & \downdownarrows \\
W_0 & \longrightarrow & \Gamma_1
\end{array}
$$

which induces an open immersion of topological stacks $[W_0/G] \to \mathfrak{X}$. $\qquad \square$

Example 1.109. The Weierstrass stack $\mathfrak{W}$ is a separated Deligne–Mumford topological stack. An étale versal family is parametrized by the disjoint union of two copies of $\mathbb{C}$. This family is the union of the two families $4z^3 - g_2 z - 1$ and $4z^3 - z - g_3$. A picture of this family is the union of (1.5) and (1.6). The symmetry groupoid of this étale family is the restriction of the transformation groupoid $W \times \mathbb{C}^*$ via the map $\mathbb{C} \amalg \mathbb{C} \to W$, which is $g_2 \mapsto (g_2, 1)$ on one copy of $\mathbb{C}$ and $g_3 \mapsto (1, g_3)$ on the other copy of $\mathbb{C}$. It is not a transformation groupoid.

Thus, on removing the bisected line segment we get $\mathfrak{W} \setminus B\mathbb{Z}_2 \cong [\mathbb{C}/\mathbb{Z}_3]$, and on removing the equilateral triangle we get $\mathfrak{W} \setminus B\mathbb{Z}_3 \cong [\mathbb{C}/\mathbb{Z}_2]$. We have $\mathfrak{W} = [\mathbb{C}/\mathbb{Z}_3] \cup [\mathbb{C}/\mathbb{Z}_2]$, a union of two open substacks.

On the other hand, $\mathfrak{W}$ is not globally a finite group quotient, because it is simply connected; see Example 1.123. $\qquad\qquad\square$

Orbifolds

Working in the category of differentiable manifolds and differentiable maps gives rise to the notion of *differentiable stack*. Care needs to be taken, because not all fibered products exist in this category (although pullbacks via differentiable submersions exist, which is sufficient).

In this context, a stack over the category of differentiable manifolds is said to be *differentiable* if it admits a versal family whose symmetry groupoid is a *Lie groupoid*. A Lie groupoid is a topological groupoid $\Gamma_1 \rightrightarrows \Gamma_0$, where both Γ_1 and Γ_0 are endowed with the structure of differentiable manifold, s and t are differentiable submersions, and all structure maps are differentiable.

Different presentations of a differentiable stack give rise to Lie groupoids which are differentiably Morita equivalent (this means that the structure maps of a bitorsor have to be differentiable submersions). Lie groupoids form a classical subject in differential geometry; see, for example, [17].

By abuse of terminology, we call a differentiable stack an **orbifold** if it admits a presentation by an étale Lie groupoid. (A Lie groupoid is *étale* if its source and target maps are local diffeomorphisms.) An analog of Theorem 1.108 (with the same proof) shows that every orbifold is locally the quotient of a finite group acting by diffeomorphisms on an open subset of $\mathbb{R}^n$. All the examples of topological Deligne–Mumford stacks we have encountered are naturally orbifolds. There is a vast literature on orbifolds; see, for example, [7] and the unpublished notes by William Thurston.[2]

If $U \subset \mathbb{R}^n$ is open, and endowed with an action by a finite group G, and if $[U/G]$ is an open substack of an orbifold $\mathfrak{X}$, then it is called a local *orbifold*

---

[2] William Thurston, The geometry and topology of three-manifolds, Princeton University lecture notes (unpublished, 1980). Available from http://library.msri.org/books/gt3m.

chart of $\mathfrak{X}$. In the literature, the term orbifold is usually reserved for those $\mathfrak{X}$ that admit orbifold charts $[U/G]$, where G acts effectively on U.

The language of group actions is not well suited for the global description of orbifolds. The way different orbifold charts are glued together is described by a groupoid presentation as in Example 1.109. Moreover, Morita equivalence of groupoids encodes what happens when two different orbifold atlases describe the same orbifold.

The following result shows that the orbifold property for differentiable stacks is a property of the diagonal, hence a separation property. We say that a Lie groupoid $X_1 \rightrightarrows X_0$ has *immersive diagonal* if $s \times t : X_1 \to X_0 \times X_0$ is injective on tangent spaces. This property is invariant under differentiable Morita equivalence, and hence gives rise to a separation property of differentiable stacks.

Proposition 1.110. *Every differentiable stack with immersive diagonal is an orbifold.*

Proof. We have to show that every Lie groupoid with immersive diagonal is Morita equivalent to an étale Lie groupoid. The fact that $X_\bullet$ has immersive diagonal allows us to construct a foliation $T_{X_0/\mathfrak{X}} \hookrightarrow T_{X_0}$, by taking $T_{X_0/\mathfrak{X}}$ to be equal to the normal bundle N_{X_0/X_1} of the identity section $X_0 \to X_1$, and embedding it into T_{X_0} via the difference of the two maps Ds, $Dt : T_{X_1}|_{X_0} \to T_{X_0}$. Then we take $U_0 \to X_0$ to be transverse to the foliation $T_{X_0/\mathfrak{X}}$ and containing each isomorphism class in $\mathfrak{X}(*)$ at least once. Restricting the groupoid X_1 via $U_0 \to X_0$ gives the Morita equivalent groupoid $U_1 \rightrightarrows U_0$, which is an étale groupoid presenting $\mathfrak{X}$.

There is a theory of foliations using étale groupoids; see [18].   □

Corollary 1.111. *Every separated differentiable stack with immersive diagonal is locally a quotient of $\mathbb{R}^n$ by a finite group.*

This result explains that finite group actions are, in fact, quite typical for stacks, and justifies the heavy reliance on them in our examples.

1.2.6 Lattices up to homothety

We very briefly cover the classical moduli problems of lattices and elliptic curves, and see how they are related to the moduli problem of oriented triangles. For a more detailed account, see [14].

A *lattice* is a subgroup of $\mathbb{C}^+$, which is a free abelian group of rank 2 and which generates $\mathbb{C}$ as $\mathbb{R}$-vector space. Two lattices $\Lambda_1 \subset \mathbb{C}$ and $\Lambda_2 \subset \mathbb{C}$ are *homothetic* if there exits a non-zero complex number ϕ such that $\Lambda_2 = \phi \cdot \Lambda_1$.

A *local* continuous family of lattices, parametrized by the topological space T, is given by two continuous functions $\omega_1, \omega_2 : T \to \mathbb{C}^*$, which are not real multiples of one another anywhere in T. The corresponding family of lattices is $\mathbb{Z}\omega_1 + \mathbb{Z}\omega_2 \subset T \times \mathbb{C}$. A *homothety* between two local families $\phi : (\omega_1, \omega_2) \to (\tau_1, \tau_2)$ is a continuous map $\phi : T \to \mathbb{C}^*$ such that $\mathbb{Z}\tau_1 + \mathbb{Z}\tau_2 = \phi \cdot (\mathbb{Z}\omega_1 + \mathbb{Z}\omega_2)$. This defines the prestack of lattices up to homothety.

Obviously, $S = \{(\omega_1, \omega_2) \in (\mathbb{C}^*)^2 \mid \mathbb{R}\omega_1 \neq \mathbb{R}\omega_2\}$ parametrizes a local continuous family of lattices in a tautological fashion: the two functions ω_1, ω_2 are simply the coordinate projections. Just as obviously, every local family of lattices is pulled back from this tautological one. Thus the family parametrized by S is versal. The symmetry groupoid of this family is the transformation groupoid of $\mathbb{C}^* \times GL_2(\mathbb{Z})$ acting on S.

By Theorem 1.90, i.e. stackifying our prestack, a (global) continuous family of lattices parametrized by T is a complex line bundle L/T together with a rank 2 local system $\Lambda \subset L$. We will call the stack of lattices up to homothety $\mathfrak{E}$.

To every compact Riemann surface E of genus 1, we associate the lattice

$$H_1(E, \mathbb{Z}) \longrightarrow \Gamma(E, \Omega_E)^*$$

$$\gamma \longmapsto \int_\gamma .$$

It is a lattice in the 1-dimensional complex vector space dual to $\Gamma(E, \Omega_E)$, the space of holomorphic 1-forms on E. It is known as the *period lattice*.

Conversely, to a lattice $\Lambda \in \mathbb{C}$ we associate the compact Riemann surface $\mathbb{C}/\Lambda$. These two processes define an equivalence of groupoids between elliptic curves (compact Riemann surfaces of genus 1 with a choice of base point serving as zero for the group law) and lattices up to homothety. We are therefore justified in declaring a *continuous family of elliptic curves* to be a continuous family of lattices. Thus we refer to $\mathfrak{E}$ also as the (topological) stack of elliptic curves.

Exercise 1.112. The upper half plane parametrizes a versal family of lattices with symmetry groupoid given by $SL_2(\mathbb{Z})$ acting by linear fractional transformations. The lattice at the point $\tau \in \mathbb{H}$ is $\mathbb{Z} + \tau\mathbb{Z}$. The τ-value of an elliptic curve is the quotient of its two periods. The corresponding elliptic curve is $\mathbb{C}/(\mathbb{Z} + \tau\mathbb{Z})$. $\quad\square$

Compactification

Let D denote the open disc in $\mathbb{C}$ of radius $e^{-2\pi}$ centered at the origin. Let $D^* \subset D$ be the pointed disc. Then D^* parametrizes a continuous family of lattices: over $q \in D^*$, the corresponding lattice is $\Lambda_q = \mathbb{Z} + \tau\mathbb{Z}$, where $\tau \in \mathbb{C}$ is any complex number such that $e^{2\pi i \tau} = q$. The corresponding family of elliptic curves can also be written as $\mathbb{C}/\Lambda_q = \mathbb{C}^*/q^{\mathbb{Z}}$.

Exercise 1.113. Prove that this is, indeed, a continuous family of lattices. Prove that, for different points in D^*, the corresponding lattices are not homothetic. Conclude that the symmetry groupoid of this family of lattices over D^* is the family of groups $D^* \times \mathbb{Z}_2$ over D^*, or, in other words, the transformation groupoid $D^* \times \mathbb{Z}_2$, where $\mathbb{Z}_2$ acts trivially on D^*. This uses the fact that we have restricted to $|q| < e^{-2\pi}$, and can be deduced from Exercise 1.112.   □

We therefore have a morphism of topological stacks $D^* \times B\mathbb{Z}_2 \to \mathfrak{E}$. This is, in fact, an open substack. We will compactify $\mathfrak{E}$ by gluing in a copy of $D \times B\mathbb{Z}_2$ along $D^* \times B\mathbb{Z}_2 \subset \mathfrak{E}$.

To make this rigorous, we construct a groupoid as follows: start with the symmetry groupoid $\Gamma_1 \rightrightarrows \Gamma_0$ of the family of lattices parametrized by the disjoint union $\mathbb{H} \sqcup D^*$. This has, as subgroupoid, the symmetry groupoid $D^* \times \mathbb{Z}_2 \rightrightarrows D^*$ of the family over D^*. To construct $\overline{\Gamma}_1 \rightrightarrows \overline{\Gamma}_0$ from $\Gamma_1 \rightrightarrows \Gamma_0$, take out $D^* \times \mathbb{Z}_2 \rightrightarrows D^*$, and replace it by $D \times \mathbb{Z}_2 \rightrightarrows D$, the groupoid given by the trivial action of $\mathbb{Z}_2$ on D.

Exercise 1.114. The object space $\overline{\Gamma}_0$ is the disjoint union $\mathbb{H} \sqcup D$, and the morphism space $\overline{\Gamma}_1$ is the disjoint union of $\mathbb{H} \times SL_2\mathbb{Z}$, $D \times \mathbb{Z}_2$, and four more components, which are all homeomorphic to D^* (or the part of $\mathbb{H}$ with imaginary part larger than 1).   □

Then we let $\overline{\mathfrak{E}}$ be the stack associated to $\overline{\Gamma}_1 \rightrightarrows \overline{\Gamma}_0$. This is the stack of degenerate elliptic curves.

Exercise 1.115. A family of degenerate elliptic curves over T is therefore given by

  (i) a cover of T by two open subsets, U and V,
 (ii) over U, a family of lattices, $\Lambda \subset L$,
(iii) over V, a continuous map $q : V \to D$ and a degree 2 covering space $V' \to V$,
 (iv) over $U \cap V$, an isomorphism of families of lattices $\Lambda|_{U \cap V} \cong V' \times_{\mathbb{Z}_2} \Lambda_q$,

with a natural notion of isomorphism.   □

Exercise 1.116. The disc D parametrizes a family of groups: the quotient of $D \times \mathbb{C}^*$ by the subgroup of all $(q, q^n) \in D \times \mathbb{C}^*$, for $q \in D^*$, $n \in \mathbb{Z}$. (The fiber of this family of groups over the origin is $\mathbb{C}^*$.) The groupoid $D \times \mathbb{Z}_2$ is a groupoid of symmetries of this family of groups. The stack $\overline{\mathfrak{E}}$ supports a family of groups, the *universal degenerate elliptic curve*, denoted $\overline{\mathfrak{F}}$. □

Exercise 1.117. There is a morphism of stacks $\overline{\mathfrak{E}} \to \mathfrak{W}$, defined by mapping a lattice $\Lambda \subset \mathbb{C}$ to the triangle $\wp(\frac{1}{2}\Lambda)$, where $\wp$ is the Weierstrass $\wp$-function corresponding to the lattice Λ. This morphism of stacks induces a homeomorphism on coarse moduli spaces. The fibers of this morphism are all isomorphic to $B\mathbb{Z}_2$. This means that, for every triangle δ, there is a 2-cartesian diagram

We say that $\overline{\mathfrak{E}}$ is a $\mathbb{Z}_2$-**gerbe** over $\mathfrak{W}$. □

Exercise 1.118. Every $\overline{\mathfrak{E}}$-family over T comes with a complex line bundle L/T. In the notation of Exercise 1.115, this line bundle is equal to L over U, and equal to $V' \times_{\mathbb{Z}_2} \mathbb{C}$ over V, where $\mathbb{Z}_2$ acts by multiplication by -1 on $\mathbb{C}$. These line bundles assemble to a line bundle $\mathscr{L}$ over $\overline{\mathfrak{E}}$. The bundle $\mathscr{L}^{\otimes -k}$ is called the *bundle of modular forms* of weight k. Global sections are called *continuous modular forms* of weight k. (The term "modular form" is usually reserved for holomorphic or algebraic modular forms; see Example 1.197.)

Prove that a modular form of weight k is the same thing as a continuous map $f : \mathbb{H} \to \mathbb{C}$ that satisfies the functional equation

$$f\left(\frac{a\tau + b}{c\tau + d}\right) = (c\tau + d)^k f(\tau),$$

for all $\left(\begin{smallmatrix} a & b \\ c & d \end{smallmatrix}\right) \in SL_2(\mathbb{Z})$, and which is continuous at $\mathrm{Re}(\tau) = \infty$.

Pulling back a modular form via $D \to \overline{\mathfrak{E}}$ gives rise to its *q-expansion*. □

1.2.7 Fundamental groups of topological stacks

As an example of the topology of topological stacks, we give a brief introduction to the fundamental group. For details, see [21].

Let $\mathfrak{X}$ be a topological stack that admits a versal family whose symmetry groupoid $X_1 \rightrightarrows X_0$ has the property that both source and target maps are topological submersions (locally in X_1, the map $X_1 \to X_0$ is homeomorphic to a product of the base times another topological space). This property will ensure that $\mathfrak{X}$ has the gluing property along closed subsets, Exercise 1.11.

Let ξ be an object of the groupoid $\mathfrak{X}(*)$, where $*$ is the one-point space. The fundamental group of $\mathfrak{X}$ with respect to the base point ξ is defined as follows. Denote the base point of S^1 by e.

A *loop* in $\mathfrak{X}$, based at ξ, is an $\mathfrak{X}$-family x/S^1, parametrized by the circle S^1, together with an isomorphism $\xi \to x_e$, where x_e is the family member at the base point $e \in S^1$. Equivalently, a loop is a diagram

$$
\begin{array}{ccc}
\xi & \xrightarrow{\ \alpha\ } & x \\
\downarrow & & \downarrow \\
* & \xrightarrow{\ e\ } & S^1
\end{array}
$$

Imagine a "film," or a "movie," of an $\mathfrak{X}$-loop as it changes over time. The film shows the loop varying continuously as time passes. Throughout the duration of the film, the family member over the base point $e \in S^1$ is always ξ. Such a film is called a *homotopy* between the loop depicted on the first frame of the movie and the loop depicted on the last frame. The first and last loops shown in the movie are then called *homotopic*.

Formally, a homotopy from the $\mathfrak{X}$-loop (x, α) to the $\mathfrak{X}$-loop (y, β) is a quadruple (h, η, ϕ, ψ). Here, h is an $\mathfrak{X}$-family parametrized by $I \times S^1$, where $I = [0, 1]$ is the unit interval in $\mathbb{R}$. Moreover, η is an isomorphism $\xi_I \to (\mathrm{id}_I, e)^* h$, where ξ_I is the constant family over I obtained by pulling back ξ via $I \to *$

$$
\begin{array}{ccc}
\xi_I & \xrightarrow{\ \eta\ } & h \\
\downarrow & & \downarrow \\
I & \xrightarrow{\ \mathrm{id}\times e\ } & I \times S^1
\end{array}
$$

and ϕ, ψ are isomorphisms $x \to (0 \times \mathrm{id}_{S^1})^* h$ and $y \to (1 \times \mathrm{id}_{S^1})^* h$

$$
\begin{array}{ccccc}
x & \xrightarrow{\ \phi\ } & h & \xleftarrow{\ \psi\ } & y \\
\downarrow & & \downarrow & & \downarrow \\
S^1 & \xrightarrow{\ 0\times\mathrm{id}\ } & I \times S^1 & \xleftarrow{\ 1\times\mathrm{id}\ } & S^1
\end{array}
$$

The two diagrams

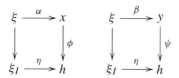

are required to commute.

The set of homotopy classes of $\mathfrak{X}$-loops based at ξ is denoted $\pi_1(\mathfrak{X}, \xi)$ and called the *fundamental group* of $\mathfrak{X}$, based at ξ. This is, in fact, a group: loops can be concatenated, by the gluing property, and homotopies can be constructed, which prove well-definedness, associativity, and existence of units and inverses.

The fundamental group of the stack of triangles

Let us compute the fundamental group of the stack $\mathfrak{M}$ of non-degenerate non-oriented triangles. Let us take the 3:4:5 right triangle as base point ξ. Let us label the edges of the base triangle 3, 4, and 5, according to their lengths.

Define a map

$$p : \pi_1(\mathfrak{M}, \xi)^{\mathrm{op}} \longrightarrow S_3 ,$$

where we think of S_3 as the group of permutations of the set $\{3, 4, 5\}$. For a given loop that starts and ends at the 3:4:5 triangle, we define the corresponding permutation of $\{3, 4, 5\}$ by following the labels around the loop in a counter-clockwise direction. For example, the loop (1.5) gives rise to the permutation (45) in cycle notation. Because concatenation of loops $x * y$ means that x is traversed before y, but composition of permutations $\pi \circ \sigma$ means that π is applied after σ, the map $p : \pi_1(\mathfrak{M}, \xi) \to S_3$ reverses the group operation, and is therefore a homomorphism of groups $\pi_1(\mathfrak{M}, \xi)^{\mathrm{op}} \to S_3$, where $\pi_1(\mathfrak{M}, \xi)^{\mathrm{op}}$ is the opposite group of $\pi_1(\mathfrak{M}, \xi)$.

We claim that p is an isomorphism of groups. To prove injectivity, assume that x/S^1 is a loop leading to the trivial permutation of $\{3, 4, 5\}$:

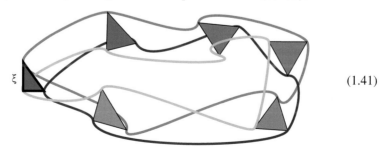

$$\text{(1.41)}$$

This means that the edges can be consistently labeled 3, 4, 5. To make a movie transforming (1.41) into the trivial family ξ_{S^1} (representing the identity element in $\pi_1(\mathfrak{X}, \xi)$), simply deform the triangles continuously until each side has length equal to its label:

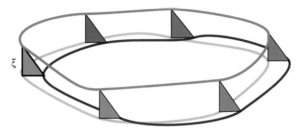

Then, at the end of the movie, all triangles in the family are 3:4:5 right triangles, and the family is isomorphic to the trivial family ξ_{S^1}, as there cannot be any non-trivial families of 3:4:5 triangles, the 3:4:5 right triangle being scalene.

To prove surjectivity of p, suppose σ is a given permutation of $\{3, 4, 5\}$. To construct a loop of triangles, based at ξ, giving rise to this permutation, take a family parametrized by an interval which deforms the 3:4:5 triangle in the middle to two equilateral triangles on either end

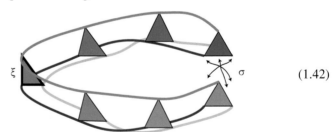

(1.42)

and then glue according to σ.

As any group is isomorphic to its opposite group, we see that the fundamental group of $\mathfrak{M}$ is S_3.

Exercise 1.119. Prove that the fundamental group of the stack of equilateral triangles is S_3. (More generally, the stack of forms of a single object with discrete symmetry group has as fundamental group the symmetry group of the object.)   □

Exercise 1.120. The stack $\widetilde{\mathfrak{M}}$ of oriented non-degenerate triangles has cyclic fundamental group with three elements.   □

More examples

The computation of the fundamental group of $\mathfrak{M}$ can be generalized to the following statement:

Theorem 1.121. *Suppose that $\mathfrak{X}$ admits a versal family whose symmetry groupoid is a transformation groupoid $X \times G$. Suppose that X is connected and simply connected, and that G is locally path connected. Then the fundamental group of $\mathfrak{X}$ is isomorphic to $\pi_0(G)$, the group of connected components of G.*

Proof. Given an $\mathfrak{X}$-loop, its generalized moduli map is a G-bundle $P \to S^1$, together with a G-equivariant continuous map $f : P \to X$. Divide P by G^0, the connected component of the identity, to obtain a $\pi_0(G) = G/G^0$-cover $\overline{P} \to S^1$. Then, going once around the loop inside $\overline{P}$ gives rise to an element of $\pi_0(G)$.

This process defines the homomorphism $\pi_1(\mathfrak{X})^{\mathrm{op}} \to \pi_0(G)$.

To prove injectivity, suppose that the element of $\pi_0(G)$ obtained from $\overline{P}$ is trivial. This means that the $\pi_0(G)$-cover $\overline{P}$ is trivial. Choosing a trivialization, the space P splits up into components indexed by $\pi_0(G)$. The component P^0 corresponding to the identity element is then a G^0-bundle over S^1.

Note that any H-bundle Q over S^1, for a topological group H, can be obtained by gluing the trivial bundle over an interval with an element h of H, similar to (1.42). If the group H is path connected, choosing a path connecting h to the identity element gives us a homotopy between Q and the trivial bundle.

Applying this to the above G^0-bundle P^0, we get a homotopy between P and the trivial G-bundle. So we may assume, without loss of generality, that the G-bundle P is trivial.

So then our map f is an equivariant map $f : S^1 \times G \to X$. Such a map is completely determined by a continuous map $S^1 \to X$, i.e. a loop in X. Contracting this loop in X gives rise to a second homotopy turning f into a trivial map $S^1 \times G \to X$, given by $(s, g) \mapsto x_0 g$, for a point $x_0 \in X$. Now our loop in $\mathfrak{X}$ is trivial.

We leave the surjectivity to the reader. $\square$

Example 1.122. The stack of degenerate triangles $\overline{\mathfrak{M}}$ has fundamental group $S_3 \times \mathbb{Z}_2$. The stack $\mathfrak{L}$ of oriented, degenerate triangles in the Legendre compactification has fundamental group S_3.

Example 1.123. The stack $\mathfrak{W}$ of oriented degenerate triangles in the Weierstrass compactification is simply connected. The stack of degenerate elliptic curves $\overline{\mathfrak{E}}$ is simply connected.

Example 1.124. The stack of non-degenerate lattices has fundamental group $SL_2(\mathbb{Z})$. The stack of non-pinched oriented triangles has fundamental group $PSL_2(\mathbb{Z})$.

1.3 Algebraic stacks

For algebraic stacks, the parameter spaces are not topological spaces, but rather algebraic varieties, or other algebro-geometric objects, such as schemes or algebraic spaces.

Let us work over a fixed base field k. The reader may assume that k is algebraically closed, or that $k = \mathbb{C}$. Let us take as the category of parameter spaces $\mathscr{S}$ the category of k-schemes with affine diagonal. (Group schemes are affine group schemes over k, and we will always tacitly assume that they are smooth.)

1.3.1 Groupoid fibrations

A groupoid fibration will now be a groupoid fibration $\mathfrak{X} \to \mathscr{S}$. The definition is the same as Definition 1.60, replacing $\mathscr{T}$ by $\mathscr{S}$, "topological space" by "k-scheme," and "continuous map" by "morphism of k-schemes." Morphisms are defined *mutatis mutandis* as in Definition 1.65.

Example 1.125. As an example, let k be a field of characteristic neither 2 nor 3, and consider $\mathfrak{E}$, the groupoid fibration of *elliptic curves*. An object of $\mathfrak{E}$ is a triple (T, E, P), where T is a k-scheme, E is a scheme endowed with a structure morphism $\pi : E \to T$, and $P : T \to E$ is a section of π, i.e. a morphism such that $\pi \circ P = \mathrm{id}_T$. Moreover, $\pi : E \to T$ is required to satisfy

(i) π is a smooth and proper morphism of finite presentation,
(ii) every geometric fiber of π is a curve of genus 1. This means that, for any algebraically closed field K and any morphism $t : \operatorname{Spec} K \to T$, the pullback E_t defined by the cartesian diagram

$$
\begin{array}{ccc}
E_t & \longrightarrow & E \\
\downarrow & & \downarrow \\
\operatorname{Spec} K & \xrightarrow{\ t\ } & T
\end{array}
$$

is a 1-dimensional irreducible (complete and non-singular by the first property) variety of genus 1, i.e. $\dim \Gamma(E_t, \Omega_{E_t}) = \dim H^1(E_t, \mathscr{O}_{E_t}) = 1$.

A morphism in $\mathfrak{E}$, from (T', E', P') to (T, E, P), is a cartesian diagram of k-schemes

$$
\begin{array}{ccc}
E' & \xrightarrow{\phi} & E \\
\downarrow & & \downarrow \\
T' & \xrightarrow{f} & T
\end{array}
\tag{1.43}
$$

such that $\phi \circ P' = P \circ f$.

The structure functor $\mathfrak{E} \to \mathscr{S}$ maps the object (T, E, P) to the k-scheme T, and the morphism (f, ϕ) to the morphism of k-schemes f.

Example 1.126. To define the groupoid fibration of *degenerate elliptic curves* (more precisely: with multiplicative reduction) $\overline{\mathfrak{E}}$, replace condition (i) in Example 1.125 by "π is a flat and proper morphism of finite presentation," and condition (ii) by "every geometric fiber of π is of one of two types: either a smooth curve of genus 1 as in Example 1.125, or an irreducible 1-dimensional scheme, non-singular except for a single node, whose arithmetic genus is 1, i.e. dim $H^1(E_t, \mathscr{O}_{E_t}) = 1$." In addition, one needs to require that P avoids any of the nodes in any of the fibers of π.

Example 1.127. A still larger groupoid fibration is $\widetilde{\mathfrak{E}}$, where the fibers are required only to be reduced and irreducible curves of arithmetic genus 1. This groupoid fibration will also include an elliptic curve with additive reduction, i.e. a genus 1 curve with a cusp. (This groupoid fibration is analogous to the stack of degenerate triangles including the one-point triangle; see Exercise 1.59.)

For more details on $\mathfrak{E}$, $\overline{\mathfrak{E}}$, and $\widetilde{\mathfrak{E}}$, see [8].

Example 1.128. Let X be a fixed smooth projective curve over k. A *family of vector bundles of rank r and degree d over X*, parametrized by the k-scheme T, is a vector bundle V of rank r over $X \times T$, such that, for every $t \in T$, the pullback of V to X_t has degree d. A morphism of families of vector bundles from V'/T' to V/T is a pair (f, ϕ) which fits into a cartesian diagram

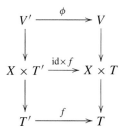

Let us denote this groupoid fibration by $\mathfrak{V}_X^{r,d}$.

Exercise 1.129. Let Γ be a groupoid. The *mass* of Γ is given by

$$\#\Gamma = \sum_{x \in \text{ob } \Gamma/\cong} \frac{1}{\#\text{Aut}(x)}.$$

The sum is taken over the set of isomorphism classes of objects of Γ. Consider Example 1.128 with $k = \mathbb{F}_q$, the finite field with q elements, and $X = \mathbb{P}^1$. Prove that $\#\mathfrak{V}_{\mathbb{P}^1}^{r,d}(\text{Spec } \mathbb{F}_q)$ converges and find its value. □

Representable morphisms

Every k-scheme X defines a groupoid fibration $\underline{X}$ as follows. An $\underline{X}$-family over T is a k-morphism $T \to X$, and pullbacks are defined by composition. This groupoid fibration is special: all fibers $\underline{X}(T)$ are sets (not groupoids), and pullbacks are unique (not unique up to unique isomorphism). We may think of $\underline{X}$ as the functor represented by X.

This construction makes a groupoid fibration out of every scheme, and a morphism of groupoid fibrations out of every morphism of schemes. As one can reconstruct X from $\underline{X}$ (Yoneda's lemma), we lose no information when passing from X to $\underline{X}$, and, in fact, we usually identify X with $\underline{X}$ and omit the underscore from the notation.

If a groupoid fibration $\mathfrak{X}$ is equivalent to $\underline{X}$, for a scheme X, via an equivalence $F : \underline{X} \to \mathfrak{X}$, then X is called the *fine moduli scheme* of $\mathfrak{X}$, and $F(\text{id}_X)$, which is an $\mathfrak{X}$-family parametrized by X, is called the *universal family*. In this case, we also say that $\mathfrak{X}$ is *representable* by the scheme X.

Definition 1.130. A morphism of groupoid fibrations $F : \mathfrak{Y} \to \mathfrak{X}$ is **representable** (more precisely: *representable by schemes*) if, for every $\mathfrak{X}$-family x/T, the groupoid fibration of liftings of x to a $\mathfrak{Y}$-family admits a fine moduli scheme. Thus, there exists a scheme $U \to T$, with a $\mathfrak{Y}$-family y/U, and an isomorphism $\theta : x|_U \to F(y)$ of $\mathfrak{X}$-families over U, such that (U, y, θ) is universal for liftings of x. The universal mapping property can be succinctly specified by saying that the diagram of groupoid fibrations

$$\begin{array}{ccc} U & \xrightarrow{y} & \mathfrak{Y} \\ \downarrow & {\theta}\!\nearrow & \downarrow F \\ T & \xrightarrow{x} & \mathfrak{X} \end{array}$$

is 2-cartesian.

Definition 1.131. A representable morphism of groupoid fibrations is *affine* or *proper* or *smooth* or *flat* or *unramified* or *étale* or *of finite presentation* or *finite* or *an open immersion* or *a closed immersion* (or any other property of morphisms of schemes, stable under base extension) if the morphism $U \to T$ has this property, for all x / T as in Definition 1.130.

Example 1.132. Let $\mathfrak{F}$ be the groupoid fibration of quadruples (T, E, P, s), where (T, E, P) is a family of elliptic curves parametrized by T, and $s : T \to E$ is another section of $E \to T$. Forgetting s defines a morphism of groupoid fibrations $\pi : \mathfrak{F} \to \mathfrak{E}$.

Let (E, P) be an elliptic curve parametrized by T. Then the groupoid fibration of liftings of (E, P) to $\mathfrak{F}$ is represented by $E \to T$ (this is more or less a tautology). Thus π is representable. It is also smooth and proper and of finite presentation. Because for every family of elliptic curves (E, P) the diagram

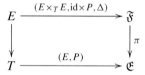

is 2-cartesian, we call $\mathfrak{F} \to \mathfrak{E}$ the *universal elliptic curve*. (The base-point section of $\pi : \mathfrak{F} \to \mathfrak{E}$ is given by $(E, P) \longmapsto (E, P, P)$.)

Similarly, we can define the universal degenerate elliptic curve $\overline{\mathfrak{F}} \to \overline{\mathfrak{E}}$. The morphism $\overline{\mathfrak{F}} \to \overline{\mathfrak{E}}$ is representable, flat, proper, and of finite presentation.

1.3.2 Prestacks

As we have seen, prestacks are groupoid fibrations where isomorphism spaces are well behaved. In the algebraic context, there are several natural conditions that we have to consider.

One of the stronger conditions is the following:

Definition 1.133. The groupoid fibration $\mathfrak{X} \to \mathscr{S}$ has **scheme-representable diagonal** if, for any two objects x / T and y / U, the groupoid fibration of isomorphisms from x to y admits a fine moduli scheme. In other words, there exists a scheme I, with structure maps $I \to T$ and $I \to U$, and an isomorphism of $\mathfrak{X}$-families $\phi : x|_I \to y|_I$, such that (I, ϕ) satisfies the following universal mapping property:

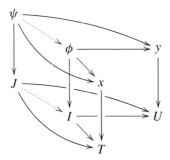

For any scheme J with given morphisms $J \to T$ and $J \to U$, and any isomorphism of $\mathfrak{X}$-families $\psi : x|_J \to y|_J$, there exists a unique morphism of schemes $J \to I$, such that $\phi|_J = \psi$.

The scheme I is called the *scheme of isomorphisms* from x to y, and is also denoted by $\underline{\mathrm{Isom}}(x, y)$. The isomorphism ϕ is called the *universal isomorphism* from x to y.

Exercise 1.134. Prove that there are 2-cartesian diagrams

$$
\begin{array}{ccc}
\underline{\mathrm{Isom}}(x, y) & \longrightarrow & U \\
\downarrow & {}_{\phi /\!\!\!/} & \downarrow {\scriptstyle y} \\
T & \xrightarrow{\;x\;} & \mathfrak{X}
\end{array}
$$

and

$$
\begin{array}{ccc}
\underline{\mathrm{Isom}}(x, y) & \longrightarrow & T \times U \\
\downarrow & {}_{\phi /\!\!\!/} & \downarrow {\scriptstyle x \times y} \\
\mathfrak{X} & \xrightarrow{\;\Delta\;} & \mathfrak{X} \times \mathfrak{X}
\end{array}
$$

If x and y are parametrized by the same scheme T, we can define the scheme $\underline{\mathrm{Isom}}_T(x, y)$, if it exists. It represents, for every $T' \to T$, the isomorphisms from x to y in the fiber $\mathfrak{X}(T)$. Alternatively, it is the pullback

$$
\begin{array}{ccc}
\underline{\mathrm{Isom}}_T(x, y) & \longrightarrow & \underline{\mathrm{Isom}}(x, y) \\
\downarrow & & \downarrow \\
T & \xrightarrow{\;\Delta\;} & T \times T
\end{array}
$$

Prove that $\Delta : \mathfrak{X} \to \mathfrak{X} \times \mathfrak{X}$ is representable by schemes according to Definition 1.130 if and only if, for any two families parametrized by the same scheme T, the groupoid fibration $\underline{\mathrm{Isom}}_T(x, y)$ admits a fine moduli scheme. Prove that this is equivalent to $\mathfrak{X}$ having scheme-representable diagonal according to Definition 1.133. □

We can strengthen the condition by requiring the schemes $\underline{\mathrm{Isom}}(x, y)$ or the morphisms $\underline{\mathrm{Isom}}(x, y) \to T \times U$ to satisfy additional conditions. For example, we can require $\underline{\mathrm{Isom}}(x, y) \to T \times U$ to be an *affine morphism* of schemes, or a *finite* morphism of schemes. This leads to the notion of $\mathfrak{X}$ having *affine diagonal* or *finite diagonal*, respectively. A very common requirement is that the diagonal be of finite presentation.

We can also weaken the condition to require only that $\underline{\mathrm{Isom}}(x, y)$ be an algebraic space.

The weakest possible condition is that $\underline{\mathrm{Isom}}(x, y)$ is only a sheaf in the étale topology. This leads to the notion of **prestack in the étale topology**. This is the "usual" notion of prestack.

Exercise 1.135. Let us prove that $\overline{\mathfrak{E}}$ is a prestack with finite (hence affine) diagonal. We will present a proof, which applies more generally, to explain a commonly used method. For a more direct proof, see Exercise 1.154, which uses Lemma 1.139. Let (T, E, P) and (U, F, Q) be families of degenerate elliptic curves. By passing to the product $T \times U$ we may assume that $T = U$ and that we want to construct the relative isomorphism scheme $\underline{\mathrm{Isom}}_T\big((E, P), (F, Q)\big)$. Because we can glue schemes along open subschemes, the claim that $\underline{\mathrm{Isom}}_T\big((E, P), (F, Q)\big)$ is representable is local in the Zariski topology on T. Families of curves can always be embedded into projective space at least locally, so we may assume that $E \to T$ and $F \to T$ are projective morphisms.

We can quote a general theorem: let $X \to T$ and $Y \to T$ be flat and projective morphisms of schemes. Then $\underline{\mathrm{Isom}}_T(X, Y)$ is represented by a k-scheme that is a (potentially countably infinite) disjoint union of quasi-projective k-schemes. In fact, the scheme $\underline{\mathrm{Isom}}_T(X, Y)$ is an open subscheme of the *Hilbert scheme* of closed subschemes of $X \times_T Y$, via identifying an isomorphism with its graph. For Hilbert schemes, see [13] or [10].

Using this fact, and the fact that $\underline{\mathrm{Isom}}_T\big((E, P), (F, Q)\big)$ is a closed subscheme of $\underline{\mathrm{Isom}}_T(E, F)$, we see that $\mathfrak{E}$ has a scheme-representable diagonal.

In fact, $\mathfrak{E}$ has a *finite* diagonal. To prove this, we can exploit that fact that E and F are curves: in this case, the condition on a subscheme of $E \times F$ to define an isomorphism is a condition on the Hilbert polynomial of the subscheme, because it is just a condition on the degrees. Therefore, our scheme of isomorphisms is projective over T. One checks that fiber-wise there are only finitely many isomorphisms, and then uses the fact that a projective morphism with finite fibers is finite. $\square$

Exercise 1.136. The groupoid fibration $\mathfrak{V}_X^{r,d}$ does not have a finite-type diagonal. For every vector bundle E over X, the automorphism group $\mathrm{Aut}(E)$ is a linear algebraic k-group, but there is no bound on the dimension of $\mathrm{Aut}(E)$, as E varies in $\mathfrak{V}_X^{r,d}(k)$.

On the other hand, $\mathfrak{V}_X^{r,d}$ can be covered by open subfibrations $\mathfrak{V}_X^{r,d,N}$, which are prestacks with affine diagonals of finite presentation. Here, $\mathfrak{V}_X^{r,d,N}$ consists of bundles that are Castelnuovo–Mumford N-regular (see [20]). A family of N-regular bundles E over $X \times T$ admits (at least locally in T) a resolution

$$P_1 \longrightarrow P_0 \longrightarrow E \longrightarrow 0 ,$$

where the P_i are direct sums of $\mathscr{O}(n)$, for $n \ll 0$, and $\mathscr{O}(1)$ is a very ample invertible sheaf on X. If F is another family of N-regular bundles over $X \times T$, we have an exact sequence

$$0 \longrightarrow \pi_* \mathscr{H}om(E, F) \longrightarrow \pi_* \mathscr{H}om(P_0, F) \longrightarrow \pi_* \mathscr{H}om(P_1, F),$$
$$(1.44)$$

where $\pi : X \times T \to T$ is the projection. As $\pi_* \mathscr{H}om(P_i, F)$ commutes with arbitrary base change, and is a vector bundle over T, we see that $\pi_* \mathscr{H}om(E, F)$ is representable by an affine T-scheme, namely the fibered product (1.44). Similarly, $\pi_* \mathscr{H}om(F, E)$, $\pi_* \mathscr{E}nd(E)$, and $\pi_* \mathscr{E}nd(F)$ are affine T-schemes. Finally, $\underline{\mathrm{Isom}}(E, F)$ is a fibered product of affine T-schemes

$$
\begin{array}{ccc}
\underline{\mathrm{Isom}}(E, F) & \longrightarrow & T \\
\downarrow & & \downarrow \\
\pi_* \mathscr{H}om(E, F) \times_T \pi_* \mathscr{H}om(F, E) & \longrightarrow & \pi_* \mathscr{E}nd(E) \times_T \pi_* \mathscr{E}nd(F)
\end{array}
$$

and is therefore an affine T-scheme itself. It is also of finite presentation. □

Example 1.137. Let X be a k-variety, and let G be an algebraic k-group acting on X. Define a groupoid fibration $[X/G]^{\mathrm{pre}}$ as follows: families parametrized by the scheme T are morphisms $x : T \to X$. Morphisms in $[X/G]^{\mathrm{pre}}$ are pairs (f, ϕ), where $f : T' \to T$ is a morphism of parameter schemes, and $\phi : T' \to G$ is a morphism, such that $x' = x(f) \cdot \phi$. Hence the fiber $[X/G]^{\mathrm{pre}}(T)$ is the transformation groupoid of the group $G(T)$ acting on the set $X(T)$. Then $[X/G]^{\mathrm{pre}}$ is a prestack with scheme-representable diagonal, because for $x : T \to X$ and $y : U \to X$ we have that

$$
\begin{array}{ccc}
\underline{\mathrm{Isom}}(x, y) & \longrightarrow & T \times U \\
\downarrow & & \downarrow {\scriptstyle x \times y} \\
X \times G & \xrightarrow{\mathrm{pr} \times \sigma} & X \times X
\end{array}
$$

is a cartesian diagram. We note that the properties of the diagonal of $[X/G]^{\mathrm{pre}}$ are the properties of the morphism $X \times G \to X \times X$.

Versal families and their symmetry groupoids

The definition of versal family uses étale covers. If we were to use only Zariski covers, there would not be enough versal families to make the theory interesting.

Definition 1.138. Suppose that $\mathfrak{X}$ is a groupoid fibration. A **versal family** for $\mathfrak{X}$ is a family x/Γ_0 such that

(i) for every $\mathfrak{X}$-family y/T, there exist étale morphisms $U_i \to T$ whose images cover T, and morphisms of k-schemes $f_i : U_i \to \Gamma_0$ such that $y|_{U_i} \cong f_i^* x$,

(ii) the symmetry groupoid $\Gamma_1 = \underline{\mathrm{Isom}}(x, x)$ of x is representable.

A useful analog of Lemma 1.75 in this context is the following.

Lemma 1.139. *If a groupoid fibration $\mathfrak{X}$ admits a versal family whose symmetry groupoid $\Gamma_1 \rightrightarrows \Gamma_0$ has the property that $\Gamma_1 \to \Gamma_0 \times \Gamma_0$ is affine, then $\mathfrak{X}$ is a prestack with affine diagonal.*

Proof. The proof is analogous to the proof of Lemma 1.75. The morphisms $U_i \to T$ and $V_j \to S$ will be étale, and the morphisms $J_{ij} \to U_i \times V_j$ will be affine. By étale descent of affine schemes, it follows that $I \to T \times S$ is affine.

The theory of descent is about generalizing the construction of schemes by gluing along open subschemes to gluing over an étale cover (or more general types of flat covers), as in this proof. For the result needed here, see Théoreme 2 in [12]. See also [10]. □

Example 1.140. Consider the groupoid fibration of degree 2 unramified covers. A family parametrized by the scheme T is a degree 2 finite étale covering $\widetilde{T} \to T$. The one-point scheme $\operatorname{Spec} k$ parametrizes a trivial family, which is versal. If we were to insist on Zariski open covers in Definition 1.138, this would not be the case.

We adapt Definition 1.86 to the present context.

Definition 1.141. An **algebraic groupoid** $\Gamma_1 \rightrightarrows \Gamma_0$ is a groupoid in $\mathscr{S}$, which means that Γ_0 and Γ_1 are k-schemes and that all structure morphisms are morphisms of k-schemes. We will always assume that our algebraic groupoids also satisfy the following:

(i) the diagonal $s \times t : \Gamma_1 \to \Gamma_0 \times \Gamma_0$ is affine,
(ii) the source and target maps $s, t : \Gamma_1 \to \Gamma_0$ are smooth.

The notion of *Morita equivalence* (see Definition 1.93) carries over *mutatis mutandis*.

Definition 1.142. Let $\Gamma_1 \rightrightarrows \Gamma_0$ be an algebraic groupoid. A Γ-**torsor** over the k-scheme T is a pair (P_0, ϕ), where P_0 is a k-scheme, endowed with a smooth surjective morphism $\pi : P_0 \to T$, and $\phi : P_\bullet \to \Gamma_\bullet$ is a morphism of algebraic groupoids, such that (1.37) is a pullback diagram in $\mathscr{S}$, where $P_\bullet$ is the banal groupoid associated to $\pi : P_0 \to T$. The second axiom in Definition 1.86 is not necessary: every smooth surjective morphism admits étale local sections. (If we were to insist on Zariski local sections, we would get a different notion of torsor.)

If $\Gamma_1 \rightrightarrows \Gamma_0$ is an algebraic group $G \rightrightarrows \mathrm{Spec}\, k$, then a torsor is also called a **principal homogeneous G-bundle** or **G-bundle** for short.

Exercise 1.143. Given an algebraic groupoid $\Gamma_1 \rightrightarrows \Gamma_0$, and a smooth morphism of schemes $U_0 \to \Gamma_0$, the fibered product

$$
\begin{array}{ccc}
U_1 & \longrightarrow & U_0 \times U_0 \\
\downarrow & & \downarrow \\
\Gamma_1 & \longrightarrow & \Gamma_0 \times \Gamma_0
\end{array}
$$

defines an algebraic groupoid $U_1 \rightrightarrows U_0$, the **restriction** of $\Gamma_\bullet$ via the morphism $U_0 \to \Gamma_0$. If $U_0 \to \Gamma_0$ is surjective, $U_\bullet$ is Morita equivalent to $\Gamma_\bullet$. $\qquad\square$

Exercise 1.144. Every algebraic groupoid Γ with Γ_0 quasi-compact is Morita equivalent to an algebraic groupoid Γ', with Γ'_0 and Γ'_1 affine, i.e. an *affine groupoid*. This is proved by restricting Γ via $\coprod U_i \to \Gamma_0$, where U_i is a finite affine open cover of Γ_0 and so $\coprod U_i$ is an affine scheme and $\coprod U_i \to \Gamma_0$ is an étale surjection.

Because of this, we could work entirely with affine schemes and affine groupoids to develop the theory of (quasi-compact) algebraic stacks with affine diagonal. We do not do this because many interesting versal families have non-affine parameter space. $\qquad\square$

Exercise 1.145. Construct the *tautological* Γ-torsor. It is parametrized by Γ_0 and has Γ itself as symmetry groupoid. It is versal for the groupoid fibration of Γ-torsors over $\mathscr{S}$. □

Exercise 1.146. Let G be an algebraic group acting on the scheme X. Then a torsor for the algebraic transformation groupoid $X \times G \rightrightarrows X$ is the same thing as a G-bundle, together with an equivariant morphism to X. □

The analog of the gluing property (Exercise 1.12) in the algebraic context is expressed in terms of the étale topology on $\mathscr{S}$ and gives rise to the notion of *stack in the étale topology*, in analogy to Definition 1.77.

Proposition 1.147. *Let Γ be an algebraic groupoid as in Definition 1.141. Then the groupoid fibration of Γ-torsors is a prestack with affine diagonal, and it satisfies the gluing axiom with respect to the étale topology (hence it is a stack in the étale topology).*

Proof. The part about the affine diagonal follows from Exercise 1.145 and Lemma 1.139, thus ultimately from descent for affine schemes. The gluing axiom is similar to the topological case, and again uses descent for affine schemes. The point is that the scheme $P_0 \to T \times \Gamma_0$, which is to be constructed by gluing, is going to be affine over $T \times \Gamma_0$. □

1.3.3 Algebraic stacks

We only consider algebraic stacks with affine diagonal. Most algebraic stacks that occur in the literature have this property.

The following definition of algebraic stacks avoids explicit reference to Grothendieck topologies, algebraic spaces, or descent theory.

Definition 1.148. A groupoid fibration $\mathfrak{X}$ over $\mathscr{S}$ is an **algebraic stack** if

(i) $\mathfrak{X}$ admits a versal family x/Γ_0 whose symmetry groupoid $\Gamma_1 \rightrightarrows \Gamma_0$ is an algebraic groupoid in the sense of Definition 1.141,

(ii) the tautological morphism of groupoid fibrations

$$\mathfrak{X} \longrightarrow (\Gamma\text{-torsors})$$
$$y \longmapsto \underline{\mathrm{Isom}}(y, x)$$

is an equivalence.

The groupoid fibration of Γ-torsors, for an algebraic groupoid Γ, is an algebraic stack; see Example 1.145.

Every algebraic stack is a prestack with affine diagonal because of Lemma 1.139.

Theorem 1.149. *Suppose a groupoid fibration $\mathfrak{X}$ satisfies (i) in Definition 1.148. If $\mathfrak{X}$ satisfies the gluing axiom with respect to the étale topology in $\mathscr{S}$, the tautological morphism $\mathfrak{X} \to$ (Γ-torsors) is an equivalence, and hence $\mathfrak{X}$ is an algebraic stack.*

Proof. (See also Exercise 1.34.) We have to associate to every Γ-torsor P/T an $\mathfrak{X}$-family over T, whose generalized moduli map is the given torsor P/T. Over P_0 we have an $\mathfrak{X}$-family, and between the two pullbacks to P_1 we have an isomorphism of $\mathfrak{X}$-families, and the cocycle condition is satisfied. In other words, we have smooth gluing data for an $\mathfrak{X}$-family. To obtain étale gluing data for the same family, choose an étale surjection $U_0 \to T$ and a section $\sigma : U_0 \to P_0$. We get an induced morphism of banal groupoids $U_\bullet \to P_\bullet$, via which we can pull back our gluing data.   □

Definition 1.150. Suppose a groupoid fibration $\mathfrak{X}$ satisfies (i) in Definition 1.148. Then $\widetilde{\mathfrak{X}} = $ (Γ-torsors) is the **stack associated to** $\mathfrak{X}$. It is an algebraic stack.

Exercise 1.151. Prove that the tautological functor $\mathfrak{X} \to \widetilde{\mathfrak{X}}$ is fully faithful.
                                                                   □

Example 1.152. Let G be a linear algebraic group acting on a scheme X in such a way that $X \times G \to X \times X$ is affine. The groupoid fibration $[X/G]^{\mathrm{pre}}$ of Example 1.137 satisfies (i) in Definition 1.148. The stack associated to $[X/G]^{\mathrm{pre}}$ is $[X/G]$, the stack of torsors for the algebraic transformation groupoid $X \times G \rightrightarrows X$, Exercise 1.146.

Exercise 1.153. The groupoid fibration $\overline{\mathfrak{E}}$ of degenerate elliptic curves is an algebraic stack. We have already proved that $\overline{\mathfrak{E}}$ is a prestack with affine diagonal in Exercise 1.135. For the fact that $\overline{\mathfrak{E}}$ satisfies the gluing axiom with respect to the étale topology on $\mathscr{S}$, we can quote descent for projective schemes; see [10]. Finally, we need a versal family. The general way to produce versal families again uses Hilbert schemes; see [10] and [13]. For details, and the proof that moduli stacks of curves and marked curves of genus other than 1 are algebraic, see [1].

We can also prove that $\overline{\mathfrak{E}}$ is algebraic directly, avoiding descent theory and the use of Theorem 1.149; see Exercise 1.179.

If $k = \mathbb{C}$, the topological stack underlying the algebraic stack $\overline{\mathfrak{E}}$ is the stack of degenerate lattices of Section 1.2.6. □

Exercise 1.154. Denote the affine plane over k, with the origin removed, by W. Denote the two coordinates by g_2 and g_3. The affine equation

$$y^2 = 4x^3 - g_2 x - g_3$$

with homogenization

$$Y^2 Z = 4X^3 - g_2 X Z^2 - g_3 Z^3$$

defines a family of projective plane curves $E \subset \mathbb{P}^2_W$. Together with the section P at infinity, (E, P) is a family of generalized elliptic curves parametrized by W. Show that it is a versal family for $\overline{\mathfrak{E}}$, and that its symmetry groupoid is the transformation groupoid of the multiplicative group $\mathbb{G}_m$ acting on W with weights 4 and 6. (Exercise 1.178 will be useful to prove that the usual procedure for embedding an abstract genus 1 curve into the plane, and putting it into Weierstrass normal form, works in families, at least locally.)

Conclude that $\overline{\mathfrak{E}} \cong [W/\mathbb{G}_m]$. The quotient stack $[W/\mathbb{G}_m]$ is known as the *weighted projective line* with weights 4 and 6, notation $\mathbb{P}(4, 6)$. □

Given the symmetry groupoid $\Gamma_1 \rightrightarrows \Gamma_0$ of a versal family x of an algebraic stack, we obtain a 2-cartesian diagram

$$
\begin{array}{ccc}
\Gamma_1 & \xrightarrow{t} & \Gamma_0 \\
{\scriptstyle s}\downarrow & \nearrow & \downarrow{\scriptstyle x} \\
\Gamma_0 & \xrightarrow{x} & \mathfrak{X}
\end{array}
$$

The morphism $x : \Gamma_0 \to \mathfrak{X}$ given by the versal family is representable and smooth. We say that $\Gamma_0 \to \mathfrak{X}$ is a **smooth presentation** of $\mathfrak{X}$. The scheme Γ_0 should be thought of as a *smooth cover* of $\mathfrak{X}$. The stack $\mathfrak{X}$ "looks like" Γ_0, locally (in the smooth topology). The fact that $\Gamma_0 \to \mathfrak{X}$ is smooth (which comes from the requirement that s and t be smooth), is essential for "doing geometry" over $\mathfrak{X}$. For example, it makes it possible to decide when $\mathfrak{X}$ is smooth:

Definition 1.155. The algebraic stack $\mathfrak{X}$ is **smooth** (non-singular) if there exists a smooth presentation $\Gamma_0 \to \mathfrak{X}$, where Γ_0 (hence also Γ_1) is non-singular.

This definition is sensible, because for schemes, given a smooth surjective morphism $Y \to X$ where Y is smooth, then X is smooth.

See Exercise 1.180, for another result that requires smoothness (or at least flatness) of the structure morphism $\Gamma_0 \to \mathfrak{X}$ of a presentation.

Definition 1.156. An algebraic stack is called a **separated Deligne–Mumford stack** if its diagonal is finite and unramified.

Of course, $\overline{\mathfrak{E}}$ is a smooth separated Deligne–Mumford stack.

Exercise 1.157. An algebraic groupoid is *étale* if its source and target morphism are étale morphisms of schemes. Prove that every smooth separated Deligne–Mumford stack $\mathfrak{X}$ admits a versal family whose symmetry groupoid is étale by imitating the proof of Proposition 1.110. Let us remark that, by using sheaves of differentials rather than tangent bundles, it can be shown that the assumption that $\mathfrak{X}$ be smooth is not necessary; see [16]. $\qquad\square$

Exercise 1.158. Consider the groupoid fibration of "triangles with centroid at the origin." This is the groupoid fibration of triples $(T', A, \mathscr{L})$, where T'/T is a degree 3 finite étale cover of the parameter scheme T, and $\mathscr{L}$ is a line bundle over the parameter space T. Moreover, $A : T' \to \mathscr{L}$ is a morphism, with the property that, for every geometric point $t \to T$, the three points $A(T'_t) \subset \mathscr{L}_t$ add to zero, and no more than two of them coincide. Prove that this is an algebraic stack in two ways.

(i) Use descent for coherent sheaves ([12], Théoreme 1), to prove that Theorem 1.149 applies. Then prove that the family parametrized by $\mathbb{P}^1$ (with homogeneous coordinates x, y), where $\mathscr{L} = \mathscr{O}(1)$ and A is given by the three sections x, y, $-x - y \in \Gamma\big(\mathbb{P}^1, \mathscr{O}(1)\big)$, is a versal family. Prove also that the symmetry groupoid of this family is given by the standard action of S_3 on $\mathbb{P}^1$; see (1.30). Conclude that this stack of triangles is isomorphic to $[\mathbb{P}^1/S_3]$.

(ii) Prove directly that this stack is isomorphic to $[\mathbb{P}^1/S_3]$ by performing an algebraic analog of Exercise 1.45.

Now assume that $k = \mathbb{C}$ and conclude that the topological stack associated to this algebraic stack is isomorphic to $\mathfrak{L}$. Thus, we have endowed $\mathfrak{L}$ with the structure of a smooth separated Deligne–Mumford stack.

For general k, we will now denote the algebraic stack of triangles by $\mathfrak{L}$. $\qquad\square$

Exercise 1.159. Prove that the stack of triangles $\mathfrak{L}$ from Exercise 1.158 is isomorphic to the stack of triple sections of a rank 1 projective bundle with a marked section, ∞, where no points are allowed to come together. □

Exercise 1.160. Suppose that G is a linear k-group acting on the k-variety X. If $X \times G \to X \times X$ is finite, then $[X/G]$ is a separated Deligne–Mumford stack. If X is smooth, so is the quotient stack $[X/G]$. □

Exercise 1.161. The open substack $\mathfrak{V}_X^{r,d,N} \subset \mathfrak{V}_X^{r,d}$ of N-regular bundles is algebraic: it satisfies the gluing property with respect to the étale topology because of descent for coherent sheaves. It has affine diagonal by Exercise 1.136. So all that is left to do is exhibit a versal family with smooth source and target maps. This is provided by the universal family of the Quot-scheme of quotients of $\mathcal{O}_X(-N)^{\oplus h(N)}$ of Hilbert polynomial h, where h is the Hilbert polynomial of bundles on X of rank r and degree d. The fact that this family is versal follows directly from properties of regularity. Once we restrict to the open subscheme of the Quot-scheme where the quotient is N-regular, the source and target map of the symmetry groupoid become smooth, because then the symmetry groupoid is a transformation groupoid for the group $GL_{h(N)}$. We conclude that $\mathfrak{V}_X^{r,d,N}$ is a quotient stack. □

Exercise 1.162. Prove that any 2-fibered product of algebraic stacks is algebraic. □

1.3.4 The coarse moduli space

Definition 1.163. Let $\mathfrak{X}$ be an algebraic stack. A **coarse moduli scheme** for $\mathfrak{X}$ is a scheme X, together with a morphism $\mathfrak{X} \to X$, which satisfies the following properties:

(i) $\mathfrak{X} \to X$ is universal for morphisms to schemes, in the sense that any morphism $\mathfrak{X} \to Y$, where Y is a scheme, factors uniquely through X:

(ii) for every flat morphism of schemes $Y \to X$, form the 2-fibered product

Then $\mathfrak{Y} \to Y$ satisfies property (i).

A coarse moduli scheme is unique up to unique isomorphism, if it exists. There is no reason why a coarse moduli scheme should exist in general, or why, even if it exists, we should be able to prove anything useful about it.

Example 1.164. Suppose G is a finite group acting on an affine k-scheme of finite type $X = \operatorname{Spec} A$. Then a *universal categorical quotient* (see [19]) exists. It is given by $X = \operatorname{Spec} A^G$. It is a coarse moduli scheme for $[X/G]$. We use the notation X/G for this moduli scheme.

Example 1.165. Suppose $\mathfrak{X}$ is a quotient stack $\mathfrak{X} = [X/G]$, where G is a reductive k-group, and X is a finite type k-scheme. Suppose further that all points of X are semi-stable with respect to some linearization of the G-action. Then the geometric invariant theory quotient $X /\!\!/ G$ is a coarse moduli scheme; see [19].

Example 1.166. Suppose that $\mathfrak{X}$ is a 1-dimensional smooth separated Deligne–Mumford stack. Then we construct a coarse moduli scheme as follows: choose an étale morphism from an affine curve $U_0 \to \mathfrak{X}$ (Exercise 1.157), and let $U_1 \rightrightarrows U_0$ be the corresponding étale groupoid. Restrict this groupoid to the function field L of U_0. The restricted groupoid corresponds to a diagram $L \rightrightarrows B$, where B is a finite étale L-algebra in two ways. Let $K \subset L$ be the equalizer of the two maps $L \rightrightarrows B$, and let $\overline{X}$ be the complete non-singular curve with function field K.

Then, given any étale groupoid presentation $Y_1 \rightrightarrows Y_0$ of $\mathfrak{X}$, both Y_1 and Y_0 are smooth curves (not connected), and there is a unique morphism of groupoids from $Y_1 \rightrightarrows Y_0$ to $\overline{X} \rightrightarrows \overline{X}$, and hence a morphism $\mathfrak{X} \to \overline{X}$. Let $X \subset \overline{X}$ be the image of $\mathfrak{X} \to \overline{X}$, which is an open subcurve of $\overline{X}$. Then $\mathfrak{X} \to X$ is a coarse moduli space.

For higher-dimensional separated Deligne–Mumford stacks, we have to enlarge our class of spaces to include separated algebraic spaces.

Definition 1.167. A **separated algebraic space** is a separated Deligne–Mumford stack for which every object in every fiber $\mathfrak{X}(T)$ over every scheme T is completely asymmetric.

Proposition 1.168. *An algebraic stack is a separated algebraic space if and only if its diagonal is a closed immersion.*

Proof. This follows from the fact that a finite unramified morphism, which is universally injective on points, is a closed immersion. □

Algebraic spaces are a generalization of schemes. They behave a lot like schemes, except that they are not locally affine in the Zariski topology, only in the étale topology.

Definition 1.169. Let $\mathfrak{X}$ be a separated Deligne–Mumford stack. A **coarse moduli space** for $\mathfrak{X}$ is a separated algebraic space X, together with a morphism $\mathfrak{X} \to X$, which satisfies the two properties of Definition 1.163, where Y denotes separated algebraic spaces rather than schemes.

Proposition 1.170. *Every finite type separated Deligne–Mumford stack $\mathfrak{X}$ admits a coarse moduli space X, which is a finite type separated algebraic space. Moreover, if $\bar{k}$ is the algebraic closure of k, then $\mathfrak{X}(\bar{k})/\sim \to X(\bar{k})$ is bijective.*

Proof. We briefly sketch the proof, because the proof shows that $\mathfrak{X}$ is, *étale locally in X*, a quotient stack. We try to imitate the poof of Theorem 1.108, of course, but the main reason why the proof does not carry over is that the Zariski topology on $\Gamma_0 \times \Gamma_0$ is not generated by boxes, like the product topology, which we used twice in the other proof.

Let $\Gamma_1 \rightrightarrows \Gamma_0$ be an étale presentation of $\mathfrak{X}$, and let $P_0 \in \Gamma_0$ be a point with automorphism group G. We pass to a different presentation of $\mathfrak{X}$. In fact, let $\Gamma'_1 \rightrightarrows \Gamma'_0$ be the groupoid of "stars with #$G - 1$ rays" in $\Gamma_1 \rightrightarrows \Gamma_0$. Elements of Γ'_0 are triples (x, ϕ, y), where $x \in \Gamma_0$, $y = (y_g)_{g \in G, g \neq 1}$ is a family of elements of Γ_0, and $\phi = (\phi_g)_{g \in G, g \neq 1}$ is a family of elements of Γ_1, where, for $g \in G$, $g \neq 1$, we have $\phi_g : x \to y_g$ in the groupoid $\Gamma_1 \rightrightarrows \Gamma_0$. Then $\Gamma'_1 \rightrightarrows \Gamma'_0$ is another étale presentation of $\mathfrak{X}$. Replacing $\Gamma_1 \rightrightarrows \Gamma_0$ by $\Gamma'_1 \rightrightarrows \Gamma'_0$, we may assume that there is an embedding $\Gamma_0 \times G \to \Gamma_1$, which identifies $\{P_0\} \times G$ with $G \subset \Gamma_1$, and makes the diagram

$$
\begin{array}{ccc}
\Gamma_0 \times G & \longrightarrow & \Gamma_1 \\
{\scriptstyle \text{pr}} \downarrow & & \downarrow {\scriptstyle s} \\
\Gamma_0 & \xrightarrow{\ \text{id}\ } & \Gamma_0
\end{array}
$$

commute. Replacing Γ_0 by the connected component containing P_0, and passing to the restricted groupoid, we assume that Γ_0 is connected. (Of course, this may result in a presentation for an open substack of $\mathfrak{X}$, which is fine as we need only construct the moduli space locally by the claimed compatibility with flat base change.)

Now, arguing as in the proof of Theorem 1.108, we prove that we obtain a commutative diagram

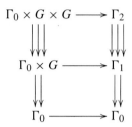

This means that we have constructed a group action of G on Γ_0 and a morphism of groupoids from the transformation groupoid of G on Γ_0 to $\Gamma_1 \rightrightarrows \Gamma_0$. Finally, we restrict to an affine open neighborhood U_0 of P_0 in Γ_0, such that

$$(s \times t)^{-1}\big(\Delta(U_0)\big) \subset \Gamma_0 \times G \subset \Gamma_1 .$$

We obtain an étale morphism of stacks $[U_0/G] \to \mathfrak{X}$. The main point is that this morphism preserves automorphism groups.

Finally, the various coarse moduli spaces U_0/G give an étale presentation for the coarse moduli space of $\mathfrak{X}$. So, if X is the coarse moduli space for $\mathfrak{X}$, we obtain a 2-cartesian diagram

$$
\begin{array}{ccc}
[U_0/G] & \xrightarrow{\text{étale}} & \mathfrak{X} \\
\downarrow & & \downarrow \\
U_0/G & \xrightarrow{\text{étale}} & X
\end{array}
$$

which explains that $\mathfrak{X}$ is, locally in the étale topology on X, a quotient by a finite group.   □

1.3.5 Bundles on stacks

Definition 1.171. A **coherent sheaf** $\mathscr{F}$ over the algebraic stack $\mathfrak{X}$ consists of the following data:

(i) for every $\mathfrak{X}$-family x/T, a coherent sheaf $x^*\mathscr{F}$ over the scheme T,
(ii) for every morphism of $\mathfrak{X}$-families $x'/T' \to x/T$, an isomorphism of coherent sheaves $(x^*\mathscr{F})|_{T'} \to (x')^*\mathscr{F}$.

The isomorphisms in (ii) have to be compatible with each other in the obvious way.

If we think of the family x/T as giving a morphism of stacks $T \xrightarrow{x} \mathfrak{X}$, then $x^*\mathscr{F}$ is the pullback of $\mathscr{F}$ over $\mathfrak{X}$ along x. This explains the notation.

The sections of $x^*\mathscr{F}$ over T are called the *sections of $\mathscr{F}$ over T*. A *global section* of $\mathscr{F}$ is a rule that assigns to every x/T a section of $x^*\mathscr{F}$ over T, compatible with pullbacks. We write $\Gamma(\mathfrak{X}, \mathscr{F})$ for the space of global sections of $\mathscr{F}$.

Example 1.172. The line bundle ω over $\overline{\mathfrak{E}}$ is defined by

$$E^*\omega = P^*\Omega_{E/T},$$

for any family of generalized elliptic curves E/T with identity section P : $T \to E$. Let us denote the dual of ω by t. If $k = \mathbb{C}$, the line bundle over the underlying topological stack of $\overline{\mathfrak{E}}$ corresponding to t is the ambient bundle of the tautological degenerate lattice, denoted $\mathscr{L}$ in Exercise 1.118.

Example 1.173. The structure sheaf $\mathcal{O}_{\mathfrak{X}}$ of any stack $\mathfrak{X}$ is defined by $x^*\mathcal{O}_{\mathfrak{X}} = \mathcal{O}_T$, for every $\mathfrak{X}$-family x/T.

Example 1.174. For any representable morphism $\mathfrak{Y} \to \mathfrak{X}$ of stacks, the sheaf of relative differentials $\Omega_{\mathfrak{Y}/\mathfrak{X}}$ on $\mathfrak{Y}$ is defined by

$$y^*\Omega_{\mathfrak{Y}/\mathfrak{X}} = z^*\Omega_{Z/T},$$

for any $\mathfrak{Y}$-family y/T. Here Z and z are defined by the cartesian diagram

Example 1.175. For any morphism of stacks $F : \mathfrak{X} \to \mathfrak{Y}$, and coherent sheaf $\mathscr{F}$ on $\mathfrak{Y}$, the pullback $F^*\mathscr{F}$ on $\mathfrak{X}$ is defined by

$$x^*F^*\mathscr{F} = F(x)^*\mathscr{F},$$

for any $\mathfrak{X}$-family x/T.

Example 1.176. The line bundle ω on $\overline{\mathfrak{E}}$ can also be defined as $P^*\Omega_{\overline{\mathfrak{F}}/\overline{\mathfrak{E}}}$, where $P : \overline{\mathfrak{E}} \to \overline{\mathfrak{F}}$ is the universal section of the universal generalized elliptic curve.

Example 1.177. For $n > 0$, a vector bundle V_n of rank n on $\overline{\mathfrak{E}}$ is defined by

$$E^* V_n = \pi_* \mathcal{O}_E(nP),$$

for any family of generalized elliptic curves E/T with structure map $\pi : E \to T$ and section $P : T \to E$. To prove this, first note that P defines an effective Cartier divisor on E, so that $\mathcal{O}_E(nP)$ is a well-defined line bundle on E. Then $\pi_* \mathcal{O}_E(nP)$ and $R^1 \pi_* \mathcal{O}_E(nP)$ are coherent sheaves on T. Using cohomology and base change (and Nakayama's lemma), deduce from the fact that $H^1(E_t, \mathcal{O}(nP)) = 0$, for any fiber E_t of E, that $R^1 \pi_* \mathcal{O}_E(nP) = 0$. Then apply cohomology and base change again, to deduce that the formation of $\pi_* \mathcal{O}_E(nP)$ commutes with arbitrary base change. Finally, apply cohomology and base change a third time, to deduce from the fact that $H^0(E_t, \mathcal{O}(nP))$ has dimension n, for every fiber E_t of E, that $\pi_* \mathcal{O}_E(nP)$ is locally free of rank n.

Exercise 1.178. Prove that $V_1 = \mathcal{O}_{\overline{\mathfrak{E}}}$. Prove that, for every $n \geq 2$, there is an exact sequence of vector bundles on $\overline{\mathfrak{E}}$:

$$0 \longrightarrow V_{n-1} \longrightarrow V_n \longrightarrow \mathfrak{t}^{\otimes n} \longrightarrow 0 .$$

Conclude that, if the characteristic of k is neither 2 nor 3, then, for every nowhere-vanishing section θ of $\mathfrak{t}$ over a scheme T, there exist unique sections x of V_2 and y of V_3 over T, mapping to θ^2 and $2\theta^3$, respectively, and satisfying an equation of the form $y^2 = 4x^3 - g_2 x - g_3$ in V_6 over T. Here, g_2 and g_3 are regular functions on T, not vanishing simultaneously.

The induced morphism $\theta \longmapsto (g_2, g_3)$, from the total space of $\mathfrak{t}$ with its zero section removed to $\mathbb{A}^2 \setminus 0$, is $\mathbb{G}_m$-equivariant if $\mathbb{G}_m$ acts with weights 4 and 6 on $\mathbb{A}^2$. Thus, g_2 and g_3 naturally give rise to global sections $g_2 \in \omega^{\otimes 4}$ and $g_3 \in \omega^{\otimes 6}$.

Deduce that every degenerate elliptic curve (E, P) over a parameter scheme T naturally embeds into $\mathbb{P}(\mathcal{O}_T \oplus \omega^{\otimes 2} \oplus \omega^{\otimes 3})$. Recall that $\mathbb{P}(\mathcal{O}_T \oplus \omega^{\otimes 2} \oplus \omega^{\otimes 3}) = \mathrm{Proj}\left(\mathrm{Sym}_{\mathcal{O}_T}(\mathcal{O}_T \oplus \mathfrak{t}^{\otimes 2} \oplus \mathfrak{t}^{\otimes 3})\right)$. There is a natural homomorphism $\mathfrak{t}^{\otimes 6} \to \mathrm{Sym}^3_{\mathcal{O}_T}(\mathcal{O}_T \oplus \mathfrak{t}^{\otimes 2} \oplus \mathfrak{t}^{\otimes 3})$ whose image generates a homogeneous sheaf of ideals which cuts out E_T inside $\mathbb{P}^2(\mathcal{O}_T \oplus \omega^{\otimes 2} \oplus \omega^{\otimes 3})$. The word "natural" in this context means *commutes with pullback of families*. Naturality implies that the construction is universal; in other words, the universal degenerate elliptic curve $\overline{\mathfrak{F}}$ embeds into $\mathbb{P}_{\overline{\mathfrak{E}}}(\mathcal{O}_{\overline{\mathfrak{E}}} \oplus \omega^{\otimes 2} \oplus \omega^{\otimes 3})$. □

Exercise 1.179. Prove that the morphism $\theta \longmapsto (g_2, g_3)$ from the total space of the line bundle $\mathfrak{t}$ over $\overline{\mathfrak{E}}$ minus its zero section to $W = \mathbb{A}^2 \setminus \{(0,0)\}$ is an isomorphism.

Deduce that W is a fine moduli space for the stack of triples (E, P, θ), where (E, P) is a degenerate elliptic curve parametrized by T, say, and θ is a global invertible section of E^*t over T.

The morphism of groupoid fibrations $\overline{\mathfrak{E}} \to [W/\mathbb{G}_m]$ given by (g_2, g_3) is an equivalence. An inverse is given by mapping a line bundle $\mathscr{L}$, with sections $g_2 \in \mathscr{L}^{\otimes 4}$ and $g_3 \in \mathscr{L}^{\otimes 6}$, to the elliptic curve in $\mathbb{P}(\mathcal{O} \oplus \mathscr{L}^{\otimes 2} \oplus \mathscr{L}^{\otimes 3})$ with equation $y^2 = 4x^3 - g_2 x - g_3$.

This gives another proof that $\overline{\mathfrak{E}}$ is the weighted projective line $\overline{\mathfrak{E}} = \mathbb{P}(4, 6)$, avoiding Hilbert schemes (for representability of <u>Isom</u>-spaces) and descent (for the étale gluing property). $\square$

Exercise 1.180. If x/X is a versal family for $\mathfrak{X}$, with symmetry groupoid $X_1 \rightrightarrows X_0$, a coherent sheaf over $\mathfrak{X}$ may be specified by the following data: a coherent sheaf $\mathscr{F}_0$ over X_0, and an isomorphism $\phi : s^*\mathscr{F} \to t^*\mathscr{F}$, such that $\mu^*\phi = p_2^*\phi \circ p_1^*\phi$, where $p_1, \mu, p_2 : X_2 \to X_1$ are the first projection, the groupoid multiplication, and the second projection, respectively. If $\mathscr{F}$ is the corresponding coherent sheaf on $\mathfrak{X}$, prove that we have an exact sequence

$$0 \longrightarrow \Gamma(\mathfrak{X}, \mathscr{F}) \longrightarrow \Gamma(X_0, \mathscr{F}) \xrightarrow{t^* - \phi \circ s^*} \Gamma(X_1, t^*\mathscr{F}) .$$

For example, a coherent sheaf on the quotient stack $[X/G]$ is the same thing as a G-equivariant coherent sheaf on X, and global sections over $[X/G]$ are invariant global sections over X. $\square$

Exercise 1.181. A global section of $\omega^{\otimes n}$ is called a **modular form of weight** n. Prove that the ring of modular forms is a polynomial ring over the field k, generated by $g_2 \in \Gamma(\overline{\mathfrak{E}}, \omega^{\otimes 4})$ and $g_3 \in \Gamma(\overline{\mathfrak{E}}, \omega^{\otimes 6})$. Thus, we have an isomorphism of graded rings

$$\bigoplus_{n=0}^{\infty} \Gamma(\overline{\mathfrak{E}}, \omega^{\otimes n}) = k[g_2, g_3] ,$$

where the degrees of g_2 and g_3 are 4 and 6, respectively. $\square$

Exercise 1.182. Let $X_1 \rightrightarrows X_0$ be an étale presentation of a smooth separated Deligne–Mumford stack $\mathfrak{X}$. We get an induced groupoid by passing to the total spaces of the tangent bundles $TX_1 \rightrightarrows TX_0$. Exercise 1.180 shows that this data gives rise to a vector bundle on $\mathfrak{X}$, the **tangent bundle** of $\mathfrak{X}$, notation $T_{\mathfrak{X}}$. $\square$

Exercise 1.183. Let $\mathfrak{X}$ be an orbifold, i.e. a smooth separated Deligne–Mumford stack whose coarse moduli space is connected, and which has a non-empty open substack which is a scheme. Prove that the frame bundle of $T_{\mathfrak{X}}$ is an algebraic space, and deduce that $\mathfrak{X}$ is a global quotient of an algebraic space by GL_n, where $n = \dim \mathfrak{X}$.   □

Exercise 1.184. Find k, such that $T_{\overline{\mathfrak{C}}} = \omega^{\otimes k}$.   □

1.3.6 Stacky curves: the Riemann–Roch theorem

As an example of algebraic geometry over stacks, we briefly discuss the Riemann–Roch theorem for line bundles over stacky curves. For an account of some of the basics of stacky curves in the analytic category, see [4].

We assume that the ground field k has characteristic 0, to avoid issues with non-separable morphisms and wild ramification. We also assume k to be algebraically closed. A *curve* is a 1-dimensional smooth connected scheme over k. Curves are quasi-projective. A *complete curve* is a projective curve.

Definition 1.185. A **stacky curve** is a 1-dimensional smooth separated Deligne–Mumford stack $\mathfrak{X}$, whose coarse moduli space X (which is a curve) is irreducible. The stacky curve $\mathfrak{X}$ is **complete** if X is complete. An **orbifold curve** is a stacky curve $\mathfrak{X}$ which is generically a scheme, i.e. there is a non-empty open subscheme $U \subset X$ of the coarse moduli space such that $\mathfrak{X} \times_X U \to U$ is an isomorphism.

Every dominant morphism $Y \to \mathfrak{X}$ from a curve to a stacky curve is representable and has a relative sheaf of differentials $\Omega_{Y/\mathfrak{X}}$, and therefore a ramification divisor $R_{Y/\mathfrak{X}}$. The closed points of Y have ramification indices relative $\mathfrak{X}$.

Orbifold curves and root stacks

Definition 1.186. Let X be a curve, let P be a closed point of X, and let $r > 0$ be an integer. The associated **root stack** $\mathfrak{X} = X[\sqrt[r]{P}]$ is defined such that an $\mathfrak{X}$-family parametrized by T consists of

(i) a morphism $T \to X$,

(ii) a line bundle L over T,

(iii) an isomorphism $\phi : L^{\otimes r} \xrightarrow{\sim} \mathcal{O}_X(P)|_T$,

(iv) a section s of L over T, such that $\phi(s^r)$ is the canonical section 1 of $\mathcal{O}_X(P)|_T$.

Isomorphisms of $\mathfrak{X}$-families are given by isomorphisms of line bundles, respecting ψ and s.

To construct a versal family for $X[\sqrt[r]{P}]$, choose an affine open neighborhood U of P in X and a uniformizing parameter π at P, and assume that the order of π at all points of $U - \{P\}$ is 0. Then let $V \to U$ be the Riemann surface of $\sqrt[r]{\pi}$; its affine coordinate ring is $\mathcal{O}(V) = \mathcal{O}(U)[\sqrt[r]{\pi}]$. There is a unique point Q of V lying over P, and the ramification index $e(Q/P)$ is equal to r.

The parameter space of our versal family is $V \sqcup (X - \{P\})$. The line bundle is $\mathcal{O}_V(Q)$ on V and the structure sheaf on $X - \{P\}$. We see that the root stack is isomorphic to X away from P, and so we may as well assume that $X = U$. Then we can take the family parametrized by V alone as a versal family. The symmetry groupoid of this family over V is the transformation groupoid of μ_r, the group or rth roots of unity acting on V by Galois transformations given by the natural action of μ_r on the rth roots of π. We see that $U[\sqrt[r]{P}] \cong [V/\mu_r]$, and that $U = V/\mu_r$.

We conclude that the root stack $\mathfrak{X} = X[\sqrt[r]{P}]$ is an orbifold curve. There is a morphism $\mathfrak{X} \to X$ that makes X the coarse moduli space of $\mathfrak{X}$ and is an isomorphism away from P. The fiber of $\mathfrak{X} \to X$ over P is isomorphic to $B\mu_r$. We can view $\mathfrak{X}$ as obtained by inserting a stacky point of order $1/r$ into the curve X at the point P.

The stack $\mathfrak{X}$ comes with a canonical line bundle over it, and we write it as $\mathcal{O}_{\mathfrak{X}}(1/rP)$. This line bundle has a canonical section, which we write as 1.

Suppose $Y \to X$ is a non-constant morphism of curves. Then there exists a morphism $Y \to X[\sqrt[r]{P}]$ such that

commutes (such a lift is unique up to unique isomorphism) if and only if the ramification indices $e(Q/P)$ of all $Q \in Y$ lying over P are divisible by r. In this case, the ramification index of such a point Q in Y relative to $X[\sqrt[r]{P}]$ is $1/re(Q/P)$. In particular, if the ramification index $e(Q/P)$ is equal to r, the morphism $Y \to X[\sqrt[r]{P}]$ is unramified at Q. Because ramification indices are multiplicative with respect to composition of morphisms, it makes sense to say that $X[\sqrt[r]{P}]$ is ramified of order r over X.

Suppose that $\mathfrak{X} \to X$ is an orbifold curve together with its coarse moduli space, and assume that $\mathfrak{X} \to X$ is an isomorphism away from $P \in X$. We define the **ramification index** of $\mathfrak{X} \to X$ over P to be the integer r such that, for any étale presentation $Y \to \mathfrak{X}$, and any point $Q \in Y$ mapping to P in X, the ramification index $e(Q/P)$ is equal to r. Then there is a canonical X-morphism $\mathfrak{X} \to X[\sqrt[r]{P}]$, because over any such Y there is a canonical rth root of the line bundle $\mathcal{O}_X(P)$. This morphism $\mathfrak{X} \to X[\sqrt[r]{P}]$ is an isomorphism, because any such $Y \to X$ factors (at least locally) through the curve obtained from X by adjoining the rth root of a uniformizing parameter at P.

We conclude that any orbifold curve with only one stacky point is a root stack.

By considering n-tuples of line bundles with sections, we can also glue in stacky points of orders $1/r_1, \ldots, 1/r_n$ at points $P_1, \ldots, P_n$ into X. If $\mathfrak{X}$ is obtained in this way from X, we say that $\mathfrak{X}$ is the **root stack** associated to the effective divisor $\sum_i^n (r_i - 1) P_i$ on X, notation $\mathfrak{X} = X[\sqrt[r_1]{P_1}, \ldots, \sqrt[r_n]{P_n}]$. (The divisor $\sum_i^n (r_i - 1) P_i$ should be thought of as the discriminant of $\mathfrak{X}/X$.) Over this stack we have line bundles $\mathcal{O}_{\mathfrak{X}}(1/r_i P_i)$, each with a canonical section 1. For a (not necessarily effective) divisor $\sum_i s_i P_i$ on X, we can form the tensor product

$$\mathcal{O}_{\mathfrak{X}}\left(\sum_{i=1}^n \frac{s_i}{r_i} P_i \right) = \mathcal{O}_{\mathfrak{X}}\left(\frac{1}{r_1} P_1 \right)^{\otimes s_1} \otimes \cdots \otimes \mathcal{O}_{\mathfrak{X}}\left(\frac{1}{r_n} P_n \right)^{\otimes s_n}.$$

In this way, any $\mathbb{Q}$-divisor $\sum_i q_i P_i$ on X with $q_i \in \mathbb{Q}$, such that $r_i q_i \in \mathbb{Z}$ for all i, can be thought of as a proper (i.e. integral) divisor on $\mathfrak{X}$. All divisors on $\mathfrak{X}$ come about in this way. Therefore, we get a canonical bijection between divisors on $\mathfrak{X}$ and $\mathbb{Q}$-divisors on X, which become integral when multiplied by the r_i.

We have proved the following theorem.

Theorem 1.187. *Let $\mathfrak{X}$ be an orbifold curve with coarse moduli curve X.*

(i) *There is a unique effective divisor $\sum_{i=1}^n (r_i - 1) P_i$ on X such that $\mathfrak{X} \cong X[\sqrt[r_1]{P_1}, \ldots, \sqrt[r_n]{P_n}]$.*

(ii) *The ramification index $e(P)$ of $\mathfrak{X}$ at $P \in X$ is equal to 1, unless $P = P_i$, for some $i = 1, \ldots, n$, in which case it is $e(P) = r_i$.*

(iii) *Every divisor on $\mathfrak{X}$ is the pullback from X of a unique $\mathbb{Q}$-divisor $\sum_{j=1}^m q_j P_j$ on X, such that $e(P_j)q_j \in \mathbb{Z}$, for all $j = 1, \ldots, m$.*

We will always identify every divisor on $\mathfrak{X}$ with the corresponding $\mathbb{Q}$-divisor on X.

Theorem 1.188. *Let $\mathfrak{X}$ be an orbifold curve with coarse moduli curve X. Denote by π the structure morphism $\pi : \mathfrak{X} \to X$. Let D be a divisor on $\mathfrak{X}$, also considered as a $\mathbb{Q}$-divisor on X. Then*

$$\pi_* \mathscr{O}_{\mathfrak{X}}(D) = \mathscr{O}_X(\lfloor D \rfloor),$$

and $R^i \pi_* \mathscr{O}_{\mathfrak{X}}(D) = 0$, for all $i > 0$. Here, $\lfloor D \rfloor$ is the integral divisor on X, obtained by rounding down all coefficients of D. In particular, we have

$$H^i\big(\mathfrak{X}, \mathscr{O}_{\mathfrak{X}}(D)\big) = H^i\big(X, \mathscr{O}_X(\lfloor D \rfloor)\big),$$

for all $i \geq 0$.

Proof. The claim about π_* is easily deduced from the fact that $\mathfrak{X}$ is a root stack. The claim about $R^i\pi_*$, for $i > 0$, requires some basic cohomology theory for stacks; see, for example, [3]. The main point is that group cohomology of a finite group with coefficients in k vanishes. □

Corollary 1.189. (Orbifold Riemann–Roch) *Assume that $\mathfrak{X}$ is a complete orbifold curve. Let g be the genus of the coarse moduli curve X. We have*

$$\chi\big(\mathfrak{X}, \mathscr{O}_{\mathfrak{X}}(D)\big) = \deg\lfloor D \rfloor + 1 - g$$

and

$$\dim \Gamma\big(\mathfrak{X}, \mathscr{O}_{\mathfrak{X}}(D)\big) = \deg\lfloor D \rfloor + 1 - g,$$

if $\lfloor D \rfloor$ is non-special, in particular if $\deg\lfloor D \rfloor > 2g - 2$.

Example 1.190. Consider, for example, the stack $\mathfrak{W}$ of triangles with the Weierstrass compactification, and its canonical line bundle $\mathscr{L}$ (which contains the universal triangle). This line bundle comes with two sections $g_2 \in \mathscr{L}^{\otimes 2}$ and $g_3 \in \mathscr{L}^{\otimes 3}$. These sections do not vanish simultaneously. The quotient g_3/g_2 is then a meromorphic section of $\mathscr{L}$. Using the coordinate j on the coarse moduli space $\mathbb{P}^1$ of $\mathfrak{W}$, we see that the divisor of zeroes of g_3/g_2 is equal to

$$D = \frac{1}{2}(1728) - \frac{1}{3}(0)$$

because g_2 vanishes to order 3 at $j = 0$ and g_3 vanishes to order 2 at $j = 1728$. We conclude that

$$\mathscr{L} \cong \mathscr{O}_{\mathfrak{W}}\Big(\frac{1}{2}(1728) - \frac{1}{3}(0)\Big),$$

and $\deg \mathscr{L} = 1/6$. We conclude that all $\mathscr{L}^{\otimes n}$, for $n \geq 0$, are non-special, and hence

$$\dim \Gamma(\mathfrak{W}, \mathscr{L}^{\otimes n}) = \Big\lfloor \frac{n}{2} \Big\rfloor + \Big\lfloor \frac{-n}{3} \Big\rfloor + 1 = \begin{cases} \lfloor \frac{n}{6} \rfloor & \text{if } n \equiv 1 \mod 6 \\ \lfloor \frac{n}{6} \rfloor + 1 & \text{otherwise,} \end{cases}$$

for all $n \geq 0$.

Example 1.191. Now consider the stack $\mathfrak{L}$ of triangles, Exercise 1.158, with its tautological line bundle $\mathscr{L}$. The canonical morphism $(g_2, g_3) : \mathfrak{L} \to \mathfrak{W}$ is given by mapping the universal triple section of $\mathscr{L}$ to its second symmetric polynomial $g_2 \in \mathscr{L}^{\otimes 2}$ and its third symmetric polynomial $g_3 \in \mathscr{L}^{\otimes 3}$. Hence $(g_2, g_3)^* \mathscr{L} = \mathscr{L}$. Therefore, we have $\Gamma(\mathfrak{L}, \mathscr{L}) = \Gamma(\mathfrak{W}, \mathscr{L})$. Note that $\mathscr{O}_{\mathfrak{L}}(\frac{1}{2}\infty) \cong \mathscr{O}_{\mathfrak{L}}(\frac{1}{2}1728)$.

Stacky curves

Definition 1.192. For a finite group G, a G-**gerbe** is a morphism of algebraic stacks $\mathfrak{X} \to \mathfrak{Y}$ such that there exists a smooth presentation $Y \to \mathfrak{Y}$, such that the pullback $\mathfrak{X} \times_{\mathfrak{Y}} Y \to Y$ is isomorphic to $Y \times BG$.

Remark 1.193. Usually, the term G-gerbe means something stronger, namely a G-gerbe together with a trivialization of an associated $\mathrm{Out}(G)$-torsor, the *band* of G. For details, see, for example, [6].

Example 1.194. The morphism $\overline{\mathfrak{E}} \to \mathfrak{W}$ (see Exercise 1.117) is described algebraically as follows: a line bundle ω, with sections $g_2 \in \omega^{\otimes 4}$ and $g_3 \in \omega^{\otimes 6}$ (Exercise 1.179), is mapped to the line bundle $\mathscr{L} = \omega^{\otimes 2}$, with the same sections $g_2 \in \mathscr{L}^{\otimes 2}$ and $g_3 \in \mathscr{L}^{\otimes 3}$. Alternatively, it is given by the morphism of transformation groupoids $W \times \mathbb{G}_m \to W \times \mathbb{G}_m$, $(g_2, g_3, u) \mapsto (g_2, g_3, u^2)$.

The morphism $\overline{\mathfrak{E}} \to \mathfrak{W}$ is a $\mathbb{Z}_2$-gerbe. In fact, one way to think of $\overline{\mathfrak{E}}$ is as the stack of square roots $\omega^{\otimes 2} = \mathscr{L}$ of the tautological line bundle $\mathscr{L}$ on $\mathfrak{W}$. Pulling back via the smooth presentation $W \to \mathfrak{W}$, this turns into the stack of square roots of the trivial bundle on W. This is $W \times B\mathbb{Z}_2$.

Let $\mathfrak{X}$ be a stacky curve and let $X_1 \rightrightarrows X_0$ be an étale groupoid presentation of $\mathfrak{X}$. Then all connected components of both X_1 and X_0 are curves. Let $\overline{X}_1$ be the normalization of the image of the morphism $X_1 \to X_0 \times X_0$. Then the connected components of $\overline{X}_1$ are also curves. Moreover, we have a factorization $X_1 \to \overline{X}_1 \to X_0 \times X_0$, where $X_1 \to \overline{X}_1$ is finite and hence surjective. By the functorial properties of normalization, $\overline{X}_1 \rightrightarrows X_0$ is an algebraic groupoid. Moreover, source and target maps of this groupoid are unramified and hence étale. We have a morphism of groupoids $X_1 \to \overline{X}_1$, which is also unramified and hence étale, and the kernel of this morphism is a finite étale group scheme $G \to X_0$. In fact, we have what is known as a *central extension of groupoids*:

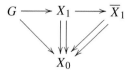

All three groupoids in this sequence have the same space of objects. Because $\mathfrak{X}$ is connected, it follows that G is a twisted form of a single finite group G_0. Let $\overline{\mathfrak{X}}$ be the stack associated to $\overline{X}_1 \rightrightarrows X_0$; it is an orbifold curve. The morphism of groupoids $X_1 \to \overline{X}_1$ defines a morphism of stacks $\mathfrak{X} \to \overline{\mathfrak{X}}$, and this morphism is a G_0-gerbe. We have proved the following theorem.

Theorem 1.195. *Every stacky curve $\mathfrak{X}$ is isomorphic to a gerbe over an orbifold curve $\overline{\mathfrak{X}}$, hence a gerbe over a root stack. The orbifold $\overline{\mathfrak{X}}$ is uniquely determined by $\mathfrak{X}$, and is called the* **underlying orbifold** *of $\mathfrak{X}$.*

Theorem 1.196. *Let $\mathfrak{X}$ be a stacky curve and let $\pi : \mathfrak{X} \to \overline{\mathfrak{X}}$ be the morphism to the underlying orbifold curve. If $\mathscr{L}$ is a line bundle on $\mathfrak{X}$, either $\mathscr{L}$ comes from a line bundle $\overline{\mathscr{L}}$ on $\overline{\mathfrak{X}}$ via pullback along π, or $\pi_*\mathscr{L} = 0$. Moreover, $R^i\pi_*\mathscr{L} = 0$, for all $i > 0$. In the first case, we have, for all $i \geq 0$,*

$$H^i(\mathfrak{X}, \mathscr{L}) = H^i(\overline{\mathfrak{X}}, \overline{\mathscr{L}}) \, ;$$

in the second case, we have, for all $i \geq 0$,

$$H^i(\mathfrak{X}, \mathscr{L}) = 0 \, .$$

Proof. The first claim about π_* can be checked generically, where it follows from the fact that a 1-dimensional representation of a finite group is either trivial or has no invariant subspace. The rest is not hard using some basic cohomology theory of stacks. Just as for Theorem 1.188, the main point is that group cohomology of a finite group with coefficients in k vanishes. □

Together with Corollary 1.189, Theorem 1.196 determines the cohomology of stacky curves with values in line bundles, i.e. the stacky Riemann–Roch theorem. It shows that the result is not much different than the one for orbifold curves, so instead of formulating the theorem we finish with an example.

Example 1.197. Consider the bundle ω of modular forms over $\overline{\mathfrak{E}}$, the stack of generalized elliptic curves. The corresponding orbifold curve is $\mathfrak{W}$, the stack of triangles in the Weierstrass compactification. Let $\mathscr{L}$ denote the tautological bundle on $\mathfrak{W}$. If n is even, $\omega^{\otimes n} = \pi^* \mathscr{L}^{\otimes \frac{n}{2}}$; if n is odd, $\pi_* \omega^{\otimes n} = 0$. It follows that, for all $n \geq 0$,

$$H^i(\overline{\mathfrak{E}}, \omega^{\otimes n}) = 0, \qquad \text{for all } i > 0,$$

$$\dim \Gamma(\overline{\mathfrak{E}}, \omega^{\otimes n}) = \begin{cases} 0 & \text{if } n \text{ is odd} \\ \lfloor \frac{n}{12} \rfloor & \text{if } n \equiv 2 \mod 12 \\ \lfloor \frac{n}{12} \rfloor + 1 & \text{otherwise}. \end{cases}$$

Note that these dimensions agree with the dimensions that one can read off from Exercise 1.181

References

[1] Enrico Arbarello, Maurizio Cornalba, and Phillip A. Griffiths, with a contribution by Joseph Daniel Harris. *Geometry of Algebraic Curves, Volume II*. Grundlehren der mathematischen Wissenschaften, vol. 268. Springer, Heidelberg, 2011.

[2] M. Artin. Versal deformations and algebraic stacks. *Invent. Math.*, 27:165–189, 1974.

[3] K. Behrend. Cohomology of stacks. In *Intersection Theory and Moduli*, ICTP Lect. Notes, XIX, pp. 249–294 (electronic). Abdus Salam International Centre for Theoretical Physics, Trieste, 2004.

[4] Kai Behrend and Behrang Noohi. Uniformization of Deligne-Mumford curves. *J. Reine Angew. Math.*, 599:111–153, 2006.

[5] N. Bourbaki. *Éléments de Mathématique. Topologie Générale*. Hermann, Paris, 1971, chaps. 1–4.

[6] J.-L. Brylinski. *Loop Spaces, Characteristic Classes and Geometric Quantization*. Progress in Mathematics, vol. 107. Birkhäuser, Boston, MA, 1993.

[7] J. H. Conway. The orbifold notation for surface groups. In *Groups, Combinatorics and geometry (Durham, 1990)*, M. Liebeck and J. Saxl, eds. London Mathematical Society Lecture Notes Series vol. 165. Cambridge University Press, Cambridge, 1992, pp. 438–447.

[8] P. Deligne. Courbes elliptiques: formulaire d'après J. Tate. In *Modular Functions of One Variable, IV (Proc. Internat. Summer School, Univ. Antwerp, Antwerp, 1972)*, B. J. Birch and W. Kuyk, eds. Lecture Notes in Mathematics, vol. 476. Springer, Berlin, 1975, pp. 53–73.

[9] P. Deligne and D. Mumford. The irreducibility of the space of curves of given genus. *Inst. Hautes Études Sci. Publ. Math.*, 36:75–109, 1969.

[10] Barbara Fantechi, Lothar Göttsche, Luc Illusie, Steven L. Kleiman, Nitin Nitsure, and Angelo Vistoli. *Fundamental Algebraic Geometry*. Mathematical Surveys and Monographs, vol. 123. American Mathematical Society, Providence, RI, 2005. [Grothendieck's FGA explained.]

[11] Jean Giraud. *Cohomologie non Abélienne*. Die Grundlehren der mathematischen Wissenschaften in Einzeldarstellungen, vol. 179. Springer, Berlin, 1971.

[12] Alexander Grothendieck. Technique de descente et théorèmes d'existence en géometrie algébrique. I. Généralités. Descente par morphismes fidèlement plats. In *Séminaire N. Bourbaki, Vol. 5*, Exp. no. 190. Soc. Math. France, Paris, 1995, pp. 299–327.

[13] Alexander Grothendieck. Techniques de construction et théorèmes d'existence en géométrie algébrique. IV. Les schémas de Hilbert. In *Séminaire N. Bourbaki, Vol. 6*, Exp. no. 221. Soc. Math. France, Paris, 1995, pp. 249–270.

[14] Richard Hain. Lectures on moduli spaces of elliptic curves. In *Transformation Groups and Moduli Spaces of Curves*. Adv. Lect. Math. (ALM), vol. 16. International Press, Somerville, MA, 2011, pp. 95–166.

[15] R. Hartshorne. *Algebraic Geometry*. Graduate Texts in Mathematics no. 52. Springer-Verlag, New York, 1977.

[16] Gérard Laumon and Laurent Moret-Bailly. *Champs Algébriques*. Ergebnisse der Mathematik und ihrer Grenzgebiete. 3. Folge, vol. 39. Springer-Verlag, Berlin, 2000.

[17] Kirill C. H. Mackenzie. *General Theory of Lie Groupoids and Lie Algebroids*. London Mathematical Society Lecture Note Series, vol. 213. Cambridge University Press, Cambridge, 2005.

[18] I. Moerdijk and J. Mrčun. *Introduction to Foliations and Lie Groupoids*. Cambridge Studies in Advanced Mathematics, vol. 91. Cambridge University Press, Cambridge, 2003.

[19] D. Mumford, J. Fogarty, and F. Kirwan. *Geometric Invariant Theory* 3rd edn. Ergebnisse der Mathematik und ihrer Grenzgebiete (2), vol. 34. Springer-Verlag, Berlin, 1994.

[20] David Mumford, with a section by G. M. Bergman. *Lectures on Curves on an Algebraic Surface*. Annals of Mathematics Studies, no. 59. Princeton University Press, Princeton, NJ, 1966.

[21] B. Noohi. Foundations of topological stacks I. arXiv:math/0503247 [math.AG].

2

BPS states and the $P = W$ conjecture

W.-Y. Chuang

National Taiwan University

D.-E. Diaconescu

Rutgers University

and

G. Pan

Rutgers University

Abstract

A string theoretic framework is presented for the work of Hausel and Rodriguez-Vilegas as well as de Cataldo, Hausel and Migliorini on the cohomology of character varieties. The central element of this construction is an identification of the cohomology of the Hitchin moduli space with BPS states in a local Calabi–Yau threefold. This is a summary of several talks given during the Moduli Space Program 2011 at the Isaac Newton Institute.

2.1 Introduction

Consider an M-theory compactification on a smooth projective Calabi–Yau threefold Y. M2-branes wrapping holomorphic curves in Y yield supersymmetric BPS states in the five-dimensional effective action. These particles are electrically charged under the low energy $U(1)$ gauge fields. The lattice of electric charges is naturally identified with the second homology lattice $H_2(Y, \mathbb{Z})$. Quantum states of massive particles in five dimensions also form multiplets of the little group $SU(2)_L \times SU(2)_R \subset Spin(4, 1)$, which is the stabilizer of the time direction in $\mathbb{R}^5$. The unitary irreducible representations of $SU(2)_L \times SU(2)_R$ may be labeled by pairs of half-integers

Moduli Spaces, eds. L. Brambila-Paz, O. García-Prada, P. Newstead and R. Thomas. Published by Cambridge University Press. © Cambridge University Press 2014.

$(j_L, j_R) \in \left(\frac{1}{2}\mathbb{Z}\right)^2$, which are the left, respectively right, moving spin quantum numbers. In conclusion, the space of five-dimensional BPS states admits a direct sum decomposition

$$\mathcal{H}_{BPS}(Y) \simeq \bigoplus_{\beta \in H_2(Y,\mathbb{Z})} \bigoplus_{j_L, j_R \in \frac{1}{2}\mathbb{Z}} \mathcal{H}_{BPS}(Y, \beta, j_L, j_R).$$

The refined Gopakumar–Vafa invariants are the BPS degeneracies

$$N(Y, \beta, j_L, j_R) = \dim \mathcal{H}_{BPS}(Y, \beta, j_L, j_R).$$

The unrefined invariants are BPS indices,

$$N(Y, \beta, j_L) = \sum_{j_R \in \frac{1}{2}\mathbb{Z}} (-1)^{2j_R+1} (2j_R + 1) N(Y, \beta, j_L, j_R).$$

String theory arguments [18] imply that BPS states should be identified with cohomology classes of moduli spaces of stable pure dimension sheaves on Y. More specifically, let $\mathcal{M}(Y, \beta, n)$ be the moduli space of slope (semi)stable pure dimension-one sheaves F on Y with numerical invariants

$$\mathrm{ch}_2(F) = \beta, \qquad \chi(F) = n.$$

Suppose furthermore that (β, n) are primitive, such that there are no strictly semistable points. If $\mathcal{M}(Y, \beta, n)$ is smooth, the BPS states are in one-to-one correspondence with cohomology classes of the moduli space. In this case, there is a geometric construction of the expected $SL(2)_L \times SL(2)_R$ action on the BPS Hilbert space if in addition there is a natural map $h : \mathcal{M}(Y, \beta, n) \to \mathcal{B}$ to a smooth projective variety $\mathcal{B}$ whose generic fibers are smooth Abelian varieties. Then [22] the action follows from decomposition theorem [2, 9] as well as from the relative hard Lefschetz theorem [10]. In particular, the action of the positive roots J_L^+, J_R^+ is given by the cup product with a relative ample class ω_h, respectively the pull back of an ample class $\omega_{\mathcal{B}}$, on the base. One then obtains a decomposition

$$H^*(\mathcal{M}(Y, \beta, n)) \simeq \bigoplus_{(j_L, j_R) \in \mathbb{Z}^2} R(j_L, j_R)^{\oplus d(j_L, j_R)},$$

where $R(j_L, j_R)$ is the irreducible representation of $SL(2)_L \times SL(2)_R$ with highest weight (j_L, j_R). A priori, the multiplicities $d(j_L, j_R)$ should depend on n for a fixed curve class β. As no such dependence is observed in the low energy theory, one is led to further conjecture that the $d(j_L, j_R)$ are in fact independent of n, as long as the numerical invariants (β, n) are primitive. Granting this additional conjecture, the refined BPS invariants are given by $N(Y, \beta, j_L, j_R) = d(j_L, j_R)$.

In more general situations, no rigorous mathematical construction of a BPS cohomology theory is known. There is, however, a rigorous construction of unrefined GV invariants via stable pairs [44, 45], which will be briefly reviewed shortly. It is worth noting that the BPS cohomology theory would have to detect the scheme structure and the obstruction theory of the moduli space as is the case in [44, 45].

Concrete examples where the moduli space $\mathcal{M}(Y, \beta, n)$ is smooth are usually encountered in local models, in which case Y is a noncompact threefold.

Example 2.1. Let S be a smooth Fano surface and let Y be the total space of the canonical bundle K_S; $\pi : Y \to S$ is the natural projection and $\sigma : S \to Y$ is the zero section. Let $\mathcal{O}_S(1)$ be a very ample line bundle on S. For any coherent sheaf F on Y with proper support, define the Hilbert polynomial of F by

$$P_F(m) = \chi(F \otimes_Y \pi^* \mathcal{O}_S(m)).$$

For sheaves F with one dimensional support,

$$P_F(m) = m r_F + n_F, \qquad r_F, n_F \in \mathbb{Z}, \ r_F > 0.$$

Such a sheaf will be called (semi)stable if

$$r_F n_{F'} \ (\leq) \ r_{F'} n_F$$

for any proper nontrivial subsheaf $0 \subset F' \subset F$. Note that if (r_F, n_F) are coprime, any semistable sheaf with numerical invariants (r_F, n_F) must be stable. Moreover, [23, Lem. 7.1] proves that any stable sheaf F on Y with Hilbert polynomial $P_F(m) = m r_F + n_F$, with $r_F > 0$, must be the extension by zero, $F = \sigma_* E$, of a stable sheaf E on S. Furthermore, [23, Lem. 7.2] proves that $\mathrm{Ext}_S^2(E, E) = 0$.

Let D be an effective divisor on S of degree $d > 0$ with respect to the polarization $\mathcal{O}_S(1)$. Let $n \in \mathbb{Z}$ such that (d, n) are coprime. Let $\mathcal{M}^s(Y, D, n)$ be the moduli space of stable dimension-one sheaves F on Y with numerical invariants

$$\mathrm{ch}_2(F) = \sigma_*(\mathrm{ch}_1(\mathcal{O}_D)), \qquad \chi(F) = n, \qquad n \in \mathbb{Z},$$

and let $\mathcal{M}^s(S, D, n)$ be the moduli space of stable dimension-one sheaves E on S with numerical invariants

$$\mathrm{ch}_1(E) = \mathrm{ch}_1(\mathcal{O}_D), \qquad \chi(E) = n.$$

Then [23, Lem. 7.1] implies that there is an isomorphism $\mathcal{M}^s(Y, D, n) \simeq \mathcal{M}^s(S, D, n)$. The vanishing result in [23, Lem. 7.2] implies that $\mathcal{M}^s(S, D, n)$

is smooth according to [24, Thm. 4.5.1]. Moreover, there is a well-defined morphism $\mathcal{M}^s(S, D, n) \to |D|$ sending E to its determinant [50, Prop. 3.0.2]. Therefore, $\mathcal{M}^s(Y, D, n)$ is smooth projective and equipped with a morphism $h : \mathcal{M}^s(Y, r, d) \to |D|$.

Example 2.2. Let X be a smooth projective curve and D an effective divisor on X, possibly trivial. Let Y be the total space of the rank-two bundle $\mathcal{O}_X(-D) \oplus K_X(D)$. Note that $H_2(Y) \simeq \mathbb{Z}$ is generated by the class σ of the zero section. Let $\mathcal{O}_X(1)$ be a very ample line bundle on X. For any dimension-one sheaf F on Y with proper support, define

$$P_F(m) = \chi(F \otimes_Y \pi^* \mathcal{O}_X(1)) = mr_F + n_F,$$

where $\pi : Y \to X$ is the natural projection. Then F is (semi)stable if

$$r_F n_{F'} \ (\leq) \ r_{F'} n_F$$

for any proper nontrivial subsheaf $0 \subset F' \subset F$.

Let (d, n) be a pair of coprime integers, with $d > 0$. Then there is a quasi-projective moduli space $\mathcal{M}(Y, d, n)$ of stable dimension-one sheaves F on Y with proper support and numerical invariants

$$\mathrm{ch}_2(F) = d\sigma, \qquad \chi(F) = n.$$

Let $\mathcal{H}_r^e(X)$ be the moduli space of rank $r \geq 1$, degree $e \in \mathbb{Z}$ stable Hitchin pairs on X. Then it is easy to prove the following statements:

(a) if $D = 0$ and $(d, n) = 1$, there is an isomorphism

$$\mathcal{M}(Y, d, n) \simeq \mathcal{H}_d^{n+d(g-1)}(X) \times \mathbb{C};$$

(b) if $D \neq 0$ and $(d, n) = 1$, there is an isomorphism

$$\mathcal{M}(Y, d, n) \simeq \mathcal{H}_d^{n+d(g-1)}(X).$$

The proof is analogous to the proof of [12, Thm 1.9], the details being omitted.

As discussed, unrefined GV numbers can be defined via Donaldson–Thomas [36] or stable pair invariants [44]. For smooth projective Calabi–Yau threefolds, such invariants are defined by integration of virtual cycles on a component of the Hilbert scheme of curves, respectively the stable pair moduli space. When Y is a non-compact Calabi–Yau threefold, as in Example 2.2, one has to employ equivariant virtual integration as in [4] and [43] because the moduli spaces are noncompact. The torus action used in this construction is a fiberwise action on Y with weights $+1$, -1 on the direct summands $K_X(D)$, $\mathcal{O}_X(-D)$ leaving the zero section pointwise fixed. Compactness of the fixed loci in Donaldson–Thomas theory was proven in [43], and for stable pair theory in [12]. Moreover, the equivalence between reduced Donaldson–Thomas theory and stable pair theory has been proven in [3] for smooth projective Calabi–Yau threefolds. Certain versions of this result were also proven in [47] and [48]. For the quasi-projective varieties in Example 2.2, this equivalence follows, in principle combining the results of [4], [43], and [37, Sect. 5]. To explain this briefly, recall that the local GW, respectively DT, theory of curves has been computed in [4] and [43] using degenerations of Y to normal crossing divisors, where each component is a rank-two bundle over $\mathbb{P}^1$ and each component intersects at most three others along common fibers. Therefore in order to prove their equivalence it suffices to prove equivalence of the resulting relative local theories. The same strategy will lead to a proof of DT/stable pair correspondence using the results of [37, Sect. 5] to prove the equivalence of relative GW and stable pair theories. The details have not been fully worked out anywhere in the literature, but this result will be assumed in this paper.

In a IIA compactification on Y, $\mathcal{Z}_{DT}(Y, q, Q)$ is the generating function for the degeneracies of BPS states corresponding to bound states of one D6-brane and arbitrary D2-D0 brane configurations on Y [11, Sect. 6]. According to [18] and [13], M-theory/IIA duality yields an alternative expression for this generating function in terms of the five-dimensional BPS indices $N(Y, \beta, j_L)$. Then

$$\mathcal{Z}_{DT}(Y, q, Q) = \exp\left(F_{GV}(Y, q, Q)\right), \qquad (2.1)$$

where

$$F_{GV}(Y, q, Q) =$$

$$\sum_{k \geq 1} \sum_{\beta \in H_2(Y), \beta \neq 0} \sum_{j_L \in \frac{1}{2}\mathbb{Z}} \frac{Q^{k\beta}}{k} (-1)^{2j_L} N(Y, \beta, j_L) \frac{q^{-2kj_L} + \cdots + q^{2kj_L}}{(q^{k/2} - q^{-k/2})^2}.$$

$$(2.2)$$

Relation (2.1) can either be inferred from [18], relying on the GW/DT correspondence conjectured in [36], or directly derived on physical grounds from Type IIA/M-theory duality [13]. Note that the generating function $F_{GV}(Y, q, Q)$ in (2.1) may be rewritten in the form [30]

$$F_{GV}(Y, q, Q) = \sum_{g \geq 0} \sum_{\beta \neq 0} n_{g,\beta} u^{2g-2} \sum_{k \geq 1} \frac{1}{k} \left(\frac{\sin(ku/2)}{u/2}\right)^{2g-2} Q^{k\beta}, \qquad (2.3)$$

where $q = -e^{-iu}$ and

$$N(Y, \beta, j_L) = \sum_{g \geq 2j_L} \binom{2g}{q + 2j_L} n_{g,\beta}.$$

In the mathematics literature, relation (2.1) with $F_{GV}(Y, q, Q)$ of the form (2.3), where $n_{g,\beta} \in \mathbb{Z}$, is known as the strong rationality conjecture [44]. It was proven for irreducible curve classes on smooth projective Calabi–Yau threefolds in [45] and for general curve classes in [49] with a technical caveat concerning holomorphic Chern–Simons functions for perverse coherent sheaves. In all these cases the proof does not provide a cohomological interpretation of the invariants $n_{g,\beta}$.

According to [28], a similar relation is expected to hold between refined stable pair invariants and the GV numbers $N(Y, \beta, j_L, j_R)$. As explained in [14], refined stable pair invariants are obtained as a specialization of the virtual motivic invariants of Kontsevich and Soibelman [33]. Then one expects [28] a relation of the form

$$\mathcal{Z}_{DT,Y}(q, Q, y) = \exp\left(F_{GV,Y}(q, Q, y)\right), \tag{2.4}$$

where

$$F_{GV,Y}(q, Q, y) = \sum_{k \geq 1} \sum_{\beta \in H_2(Y), \beta \neq 0} \sum_{j_L, j_R \in \frac{1}{2}\mathbb{Z}} \frac{Q^{k\beta}}{k}(-1)^{2j_L + 2j_R} N(Y, \beta, j_L, j_R)$$

$$\times q^{-k} \frac{(q^{-2kj_L} + \cdots + q^{2kj_L})(y^{-2kj_R} + \cdots + y^{2kj_R})}{(1 - (qy)^{-k})(1 - (qy^{-1})^{-k})}. \tag{2.5}$$

Expression (2.5) was written in [28] in different variables, $(q^{-1}y, q^{-1}y^{-1})$.

The main goal of this chapter is to point out that the refined GV expansion (2.4) for a local curve geometry is related via a simple change of variables to the Hausel–Rodriguez-Villegas formula for character varieties. There a few conjectural steps involved in this identification. First, it relies on an explicit conjectural formula for the refined stable pair theory of a local curve derived in Section (2.3) from geometric engineering and instanton sums. In fact, it is expected that a rigorous construction of motivic stable pair theory of local curves should be possible following the program of Kontsevich and Soibelman [33]. A conjectural motivic formula generalizing equation (2.12) has recently been written down by Mozgovoy [41]. Second, as explained in detail in Section (2.4), the refined GV invariants of the local curve are in fact perverse Betti numbers of the Hitchin moduli space. Therefore, the conversion of the HRV formula into a refined GV expansion relies on the identification between the weight filtration on the cohomology of character varieties and the perverse filtration on the cohomology of the Hitchin system conjectured by de Cataldo, Hausel and Migliorini [8]. This will be referred to as the $P = W$ conjecture. The connection found here provides independent physics-based evidence for this conjecture. Finally, note that further evidence for all the claims of the present paper comes from the recent rigorous results of [41], [38], and [39]. In [41] it is rigorously proven that the refined theory of the local curve implies the HRV conjecture for the Poincaré polynomial of the Hitchin system via motivic wallcrossing, while [38], and [39] prove expansion formulas analogous to (2.4) for families of irreducible reduced plane curves.

Acknowledgements We are very grateful to Tamas Hausel and Fernando Rodriguez-Villegas for illuminating discussions on their work. We would also like to thank Jim Bryan, Ugo Bruzzo, Ron Donagi, Oscar García-Prada, Lothar Göttsche, Jochen Heinloth, Dominic Joyce, Ludmil Katzarkov, Bumsig Kim, Melissa Liu, Davesh Maulik, Greg Moore, Sergey Mozgovoy, Kentaro Nagao, Alexei Oblomkov, Rahul Pandharipande, Tony Pantev, Vivek Shende, Artan Sheshmani, Alexander Schmitt, Jacopo Stoppa, Balazs Szendroi, Andras Szenes, Michael Thaddeus, Richard Thomas, and Zhiwei Yun for very helpful conversations. D.-E.D. would like to thank the organizers of the Moduli Space Program 2011 at the Isaac Newton Institute for the partial support during completion of this work, as well as a very stimulating mathematical environment. The work of D.-E.D. was also supported in part by NSF grant PHY-0854757-2009. The work of W.-Y.C. was supported by NSC grant 99-2115-M-002-013-MY2 and the Golden-Jade fellowship of the Kenda Foundation.

2.2 Hausel–Rodriguez-Villegas formula and $P = W$

Let X be a smooth projective curve over $\mathbb{C}$ of genus $g \geq 1$, and let $p \in X$ be an arbitrary closed point. Let $\gamma_p \in \pi_1(X \setminus \{p\})$ be the natural generator associated to p. For any coprime integers $r \in \mathbb{Z}_{\geq 1}$, $e \in \mathbb{Z}$, the character variety $\mathcal{C}_r^e(X)$ is the moduli space of representations

$$\phi : \pi_1(X \setminus \{p\}) \to GL(r, \mathbb{C}), \qquad \phi(\gamma_p) = e^{2i\pi e/r} I_r,$$

modulo conjugation. $\mathcal{C}_r^e(X)$ is a smooth quasi-projective variety, and its rational cohomology $H^*(\mathcal{C}_r^e(X))$ carries a weight filtration

$$W_0^k \subset \cdots W_i^k \subset \cdots \subset W_{2k}^k = H^k(\mathcal{C}_r^n(X)). \tag{2.6}$$

According to [19], $W_{2i}^k = W_{2i+1}^k$ for all $i = 0, \ldots, 2k$, hence one can define the mixed Poincaré polynomial

$$W(\mathcal{C}_r^e(X), z, t) = \sum_{i,k} \dim(W_i^k / W_{i-1}^k) t^k z^{i/2}. \tag{2.7}$$

Moreover, it was proven in [19] that $W(\mathcal{C}_r^e(X), z, t)$ is independent of e for fixed r, with (r, e) coprime. Therefore it will be denoted in the following by $W_r(z, t)$. Obviously $W_r(1, t)$ is the usual Poincaré polynomial. Note that one can equally well use compactly supported cohomology in (2.7), which is related to cohomology without support condition by Poincaré duality [19],

$$H_c^k(\mathcal{C}_r^n(X)) \times H^{2d-k}(\mathcal{C}_r^n(X)) \to \mathbb{C}.$$

The difference would be an irrelevant overall monomial factor. Using number theoretic considerations, Hausel and Rodriguez-Villegas [19] derive a conjectural formula for the mixed Poincaré polynomials $W_r(z, t)$ as follows.

2.2.1 Hausel–Rodriguez-Villegas formula

The conjecture formulated in [19] expresses the generating function

$$F_{HRV}(z, t, T) = \sum_{r,k \geq 1} B_r(z^k, t^k) W_r(z^k, t^k) \frac{T^{kr}}{k},$$

$$B_r(z, t) = \frac{(zt^2)^{(1-g)r(r-1)}}{(1-z)(1-zt^2)},$$

as

$$F_{HRV}(z, t, T) = \ln Z_{HRV}(z, t, T) \tag{2.8}$$

where $Z_{HRV}(z, t, T)$ is a sum of rational functions associated to Young diagrams. Given a Young diagram μ

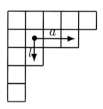

let μ_i be the length of the ith row, $|\mu|$ the total number of boxes of μ, and μ^t the transpose of μ. For any box $\square = (i, j) \in \mu$, let

$$a(\square) = \mu_i - j, \qquad l(\square) = \mu_j^t - i, \qquad h(\square) = a(\square) + l(\square) + 1$$

be the arm, leg, and hook length, respectively. Then

$$\mathcal{Z}_{HRV}(z, t, T) = \sum_\mu \mathcal{H}_g^\mu(z, t) T^{|\mu|},$$

where

$$\mathcal{H}_g^\mu(z, t) = \prod_{\Box \in \mu} \frac{(zt^2)^{l(\Box)(2-2g)}(1 - z^{h(\Box)} t^{2l(\Box)+1})^{2g}}{(1 - z^{h(\Box)} t^{2l(\Box)+2})(1 - z^{h(\Box)} t^{2l(\Box)})}.$$

The main observation in this paper is that equation (2.8) can be identified with the expansion of the refined Donaldson–Thomas series of a certain Calabi–Yau threefold in terms of numbers of BPS states.

2.2.2 Hitchin system and $P = W$

Let $\mathcal{H}_r^e(X)$ be the moduli space of stable Higgs bundles (E, Φ) on X, where Φ is a Higgs field with coefficients in K_X. For coprime (r, e) this is a smooth quasi-projective variety equipped with a projective Hitchin map

$$h : \mathcal{H}_r^e(X) \to \mathcal{B}$$

to the affine variety

$$\mathcal{B} = \oplus_{i=1}^r H^0(K_X^{\otimes i}).$$

The decomposition of the derived direct image $Rh_*\underline{\mathbb{Q}}$ into perverse sheaves yields [9, 8] a perverse filtration

$$0 = P_0^k \subset P_1^k \subset \cdots \subset P_k^k = H^k(\mathcal{H}_r^e(X))$$

on cohomology. Following the construction in [8, Sect. 1.4.1], let $H^k(\mathcal{B}, Rh_*\underline{\mathbb{Q}})$ denote the kth hypercohomology group and ${}^\mathrm{p}\tau_{\leq p} Rh_*\underline{\mathbb{Q}}$ denote the truncations of $Rh_*\underline{\mathbb{Q}}$. Then set

$$P_p H^k(\mathcal{B}, Rh_*\underline{\mathbb{Q}}) = \mathrm{Im}\big(H^k(\mathcal{B}, {}^\mathrm{p}\tau_{\leq p} Rh_*\underline{\mathbb{Q}}) \to H^k(\mathcal{B}, {}^\mathrm{p}\tau_{\leq p} Rh_*\underline{\mathbb{Q}})\big)$$

and

$$P_p^k = P_p H^{k-d}(\mathcal{B}, Rh_*\underline{\mathbb{Q}}[d]),$$

where $d = \dim_\mathbb{C} \mathcal{B}$.

It is well known that $\mathcal{C}_r^e(X)$ and $\mathcal{H}_r^e(X)$ are identical as smooth real manifolds. This result is due to [20] and [15] for rank $r = 2$, and to [7] and [46] for general $r \geq 2$. Therefore there is a natural identification $H^*(\mathcal{C}_r^e(X)) = H^*(\mathcal{H}_r^e(X))$. Then it is conjectured in [8] that the two filtrations W_j^k, P_j^k coincide,

$$W_{2j}^k = P_j^k$$

for all k, j. This is proven in [8] for Hitchin systems of rank $r = 2$.

For future reference, note that a relative ample class ω with respect to h yields a hard Lefschetz isomorphism [10]:

$$\omega^l : Gr^P_{d-l} H^k (\mathcal{H}^e_r(X)) \xrightarrow{\sim} Gr^P_{d+l} H^{k+2l} (\mathcal{H}^e_r(X)).$$

This is known as the relative hard Lefschetz theorem.

Note that granting the $P = W$ conjecture, equation (2.8) yields explicit formulas for the perverse Poincaré polynomial of the Hitchin moduli space. In particular, by specialization to $z = 1$ it determines the Poincaré polynomial of the Hitchin moduli space of any rank $r \geq 1$.

2.3 Refined stable pair invariants of local curves

Let Y be the total space of the rank-two bundle $\mathcal{O}_X(-D) \oplus K_X(D)$, where D is an effective divisor of degree $p \geq 0$ on X, as in Example 2.2. Note that $H_2(Y) \simeq \mathbb{Z}$ is generated by the class σ of the zero section. Following [44], stable pairs on Y are two term complexes $P = (\mathcal{O}_Y \xrightarrow{s} F)$ where F is a pure dimension-one sheaf and s is a generically surjective section. Since Y is noncompact in the present case, it will also be required that F have compact support, which must be necessarily a finite cover of X. The numerical invariants of F will be

$$\mathrm{ch}_2(F) = d\sigma, \qquad \chi(F) = n.$$

Then, according to [44], there is a quasi-projective fine moduli space $\mathcal{P}(Y, d, n)$ of pairs of type (d, n) equipped with a symmetric perfect obstruction theory. The moduli space also carries a torus action induced by the $\mathbb{C}^\times$ action on Y which scales $\mathcal{O}_X(-D)$, $K_X(D)$ with weights $-1, 1$. Virtual numbers of stable pairs can be defined by equivariant virtual integration by analogy with [4] and [43]. On smooth projective Calabi–Yau threefolds, the virtual number of pairs is equal to the Euler characteristic of the moduli space weighted by the Behrend function [1]. The analogous relation,

$$P(\beta, n) = \chi^B(\mathcal{P}(Y, d, n)),$$

for equivariant residual invariants of local curves follows from [12, Thm. 1.9] and [6, Lem. 3.1]. Let

$$Z_{PT}(Y, q, Q) = 1 + \sum_{d \geq 1} \sum_{n \in \mathbb{Z}} P(d, n) Q^d q^n.$$

Applying the motivic Donaldson–Thomas formalism of Kontsevich and Soibelman, one obtains a refinement $P^{ref}(d, n, y)$ of stable pair invariants modulo foundational issues. The $P^{ref}(d, n, y)$ are Laurent polynomials of the formal variable y with integral coefficients. In a string theory compactification on Y, these coefficients are numbers of D6-D2-D0 bound states with given four-dimensional spin quantum number. The resulting generating series will be denoted by $Z^{ref}_{PT}(Y, q, Q, y)$.

2.3.1 TQFT formalism

A TQFT formalism for unrefined Donaldson–Thomas theory of a local curve has been developed in [43], in parallel with a similar construction [4] in Gromov–Witten theory. Very briefly, the final result is that the generating series of local invariants is obtained by gluing vertices corresponding to a pair-of-pants decomposition of the Riemann surface X. Each such vertex is a rational function $P_{\mu_i}(q)$ labeled by three partitions μ_i, $i = 1, 2, 3$, corresponding to the three boundary components. In the equivariant Calabi–Yau case, a non-trivial result is obtained only for identical partitions, $\mu_i = \mu$, $i = 1, 2, 3$, in which case

$$P_\mu(q) = \prod_{\square \in \mu}(q^{h(\square)/2} - q^{-h(\square)/2}).$$

Then the generating function is given by

$$Z_{DT}(Y, q, Q) = \sum_\mu (-1)^{p|\mu|} q^{-(g-1-p)\kappa(\mu)}(P_\mu(q))^{2g-2} Q^{|\mu|}, \qquad (2.9)$$

where

$$\kappa(\mu) = \sum_{\square \in \mu}(i(\square) - j(\square)).$$

2.3.2 Refined invariants from instanton sums

Although the refined stable pair invariants are not rigorously constructed for higher genus local curves, string duality leads to an explicit conjectural formula for the series $Z^{ref}_{PT}(Y, q, Q, y)$. This follows using geometric engineering [31, 40, 25, 34] of supersymmetric five-dimensional gauge theories.

For completeness, geometric engineering is a correspondence between local Calabi–Yau threefolds and five-dimensional gauge theories with eight supercharges. Such gauge theories are classified by triples (G, R, p), where G is a compact semisimple Lie group and R is a unitary representation of G, and $p \in \mathbb{Z}$. Physically R encodes the matter content of the theory, and p is the level of a five-dimensional Chern–Simons term. For certain triples (G, R, p) (but not all), there exists a noncompact smooth Calabi–Yau threefold $Y_{(G,R,p)}$ such that the gauge theory specified by (G, R, p) is the extreme infrared limit of M-theory in the presence of a gravitational background specified by $Y_{(G,R,p)}$. Many such examples are known [31, 29], but the list is not exhaustive, and there is no known necessary and sufficient condition on (G, R, p) guaranteeing the existence of $Y_{(G,R)}$.

For example, if $G = SU(N)$, $N \geq 2$, R is the zero representation, and $p = 0$, the corresponding threefold $Y_{(SU(N),0,0)}$ is constructed as follows. Let Y be the total space of the rank-two bundle $\mathcal{O}_{\mathbb{P}^1} \oplus \mathcal{O}_{\mathbb{P}^1}(-2)$ and let μ_N be the multiplicative group of Nth roots of unity. There is a fiberwise action $\mu_N \times Y \to Y$ where the generator $\eta = e^{2i\pi/N}$ acts by multiplication by (η, η^{-1}) on the two summands. The quotient Y/μ_N is a singular toric variety. Then $Y_{(SU(N),0,0)}$ is the unique toric crepant resolution of Y/μ_N.

Moreover, suppose Y is replaced in the preceding paragraph by a rank-two bundle of the form $\mathcal{O}_X(-D) \oplus \mathcal{O}_X(K_X + D)$, with X a curve of genus $g \geq 1$, as in Example 2.2. Then there exists a corresponding gauge theory, and it has gauge group $G = SU(N)$, matter content $R = ad(G)^{\oplus g}$, and level $p = \deg(D)$, where $ad(G)$ denotes the adjoint representation.

An important mathematical prediction of this correspondence is an identification between a generating function of stable pair invariants of $Y_{(G,R,p)}$ and the five-dimensional equivariant instanton sum of the gauge theory (G, R, p) defined by Nekrasov in [42]. Some care is needed in formulating a precise relation; since $Y_{(G,R,p)}$ are noncompact, the stable pair invariants must be defined as residual equivariant invariants with respect to a torus action. In addition, this identification also involves a nontrivial change of formal variables, which is known in many examples, but has no general prescription.

Therefore a more precise formulation of this conjecture would state that there exists a torus action on $Y_{(G,R,p)}$ such that the residual equivariant stable pair theory is well defined, and its generating function equals the equivariant instanton sum of the gauge theory (G, R, p) up to change of variables. Such statements have been formulated and proven in many examples where $Y_{(G,R,p)}$ is a toric Calabi–Yau threefold [16, 26, 27, 17, 21, 32, 35, 28]. Furthermore, a refined version of the geometric engineering conjecture is available due to the work of [28], where it has been checked for $SU(N)$ with $N = 2, 3$ and $(R, p) = (0, 0)$.

In the present case, the geometric engineering conjecture yields [5] an explicit prediction for the residual stable pair theory of the threefolds Y in Example 2.2. Because of a subtlety of physical nature, this case was treated in [5] as a limit of $SU(2)$ gauge theory with $R = ad(G)^{\oplus g}$ and level

$p = \deg(D)$. Omitting the computations, which are given in detail in [5, Sect. 3], note that the final result can be presented in terms of equivariant K-theoretic invariants of the Hilbert scheme of points in $\mathbb{C}^2$ as follows.

Let $\mathcal{H}ilb^k(\mathbb{C}^2)$ denote the Hilbert scheme of length $k \geq 1$ zero-dimensional subschemes of $\mathbb{C}^2$. It is smooth, quasi-projective, and carries a $\mathbf{G} = \mathbb{C}^\times \times \mathbb{C}^\times$-action induced by the natural scaling action on $\mathbb{C}^2$. Let $R_\mathbf{G}$ denote the representation ring of $\mathbf{G}$, and let $q_1, q_2 : \mathbf{G} \to \mathbb{C}$ be the characters defined by

$$q(t_1, t_2) = t_1, \qquad q_2(t_1, t_2) = t_2.$$

Also, let ch $: R_\mathbf{G} \to \mathbb{Z}[q_1, q_2]$ denote the canonical ring isomorphism assigning to any representation R the character ch(R).

Now let $\mathcal{V}_k$ denote the tautological vector bundle on the Hilbert scheme whose fiber at a point $[Z]$ is the space of global sections $H^0(\mathcal{O}_Z)$. For each pair of integers $(g, p) \in \mathbb{Z}^2$, $g \geq 0$, $p \geq 0$, let

$$\mathcal{E}_k^{g,p} = T^*\mathcal{H}ilb^k(\mathbb{C}^2)^{\oplus g} \otimes \det(\mathcal{V}_k)^{1-g-p}.$$

By construction, $\mathcal{E}_k^{g,p}$ has a natural $\mathbf{G}$-equivariant structure which yields a linear $\mathbf{G}$-action on the sheaf cohomology groups $H^i(\wedge^j \mathcal{E}_k^{g,p})$ of its exterior powers. Moreover, as observed, for example, in [35], although these spaces are infinite dimensional, each irreducible representation of $\mathbf{G}$ has finite multiplicity in the decomposition of $H^i(\wedge^j \mathcal{E}_k^{g,p})$. Therefore one can formally define the equivariant $\chi_{\tilde{y}}$-genus of $\mathcal{E}_k^{(g,p)}$,

$$\chi_{\tilde{y}}(\mathcal{E}_k^{(g,p)}) = \sum_{i,j}(-\tilde{y})^j(-1)^i \operatorname{ch} H^i(\wedge^j \mathcal{E}_k^{(g,p)}),$$

as an element of $\mathbb{Z}[[q_1, q_2]]$. The equivariant K-theoretic partition function is defined by

$$Z_{inst}(q_1, q_2, \tilde{Q}, \tilde{y}) = \sum_{k \geq 0} \chi_{\tilde{y}}(\mathcal{E}_k^{g,p})\tilde{Q}^k. \tag{2.10}$$

A fixed point theorem gives an explicit formula for $Z_{inst}(q_1, q_2, \tilde{Q}, \tilde{y})$ as a sum over partitions:

$$Z_{inst}(q_1, q_2, \tilde{Q}, \tilde{y}) = \sum_\mu \prod_{\square \in \mu}(q_1^{-l(\square)}q_2^{-a(\square)})^{g-1+p}$$

$$\times \frac{(1 - \tilde{y}q_1^{-l(\square)}q_2^{a(\square)+1})^g(1 - \tilde{y}q_1^{l(\square)+1}q_2^{-a(\square)})^g}{(1 - q_1^{-l(\square)}q_2^{a(\square)+1})(1 - q_1^{l(\square)+1}q_2^{-a(\square)})} \tilde{Q}^{|\mu|}.$$

The resulting conjectural expression for the refined stable pair partition function is then [5]

$$Z_{PT}^{ref}(Y, q, Q, y) = Z_{inst}(q^{-1}y, qy, (-1)^{g-1}y^{2-g}Q, y^{-1}). \tag{2.11}$$

A straightforward computation shows that

$$Z_{PT}^{ref}(Y, q, Q, y) = \sum_{\mu} \Omega_{g,p}^{\mu}(q, y) Q^{|\mu|}, \tag{2.12}$$

where

$$\Omega_{g}^{\mu}(q, y) = (-1)^{p|\mu|} \prod_{\square \in \mu} \left[\left(q^{l(\square)-a(\square)} y^{-(l(\square)+a(\square))} \right)^{p} (qy^{-1})^{(2l(\square)+1)(g-1)} \right.$$
$$\left. \times \frac{(1 - q^{-h(\square)} y^{l(\square)-a(\square)})^{2g}}{(1 - q^{-h(\square)} y^{l(\square)-a(\square)-1})(1 - q^{-h(\square)} y^{l(\square)-a(\square)+1})} \right].$$

The change of variables in (2.11) does not have a conceptual derivation. This conjecture is supported by extensive numerical computations involving wall-crossing for refined invariants in [5]. Further supporting evidence for formula (2.11) is obtained by comparison with the unrefined TQFT formula (2.9) for local curves. Specializing the right-hand side of (2.11) at $y = 1$, one obtains

$$Z_{PT}^{ref}(Y, q, Q, 1) = \sum_{\mu} Q^{|\mu|} \prod_{\square \in \mu} (-1)^{p|\mu|} q^{(g-1+p)(l(\square)-a(\square))}$$
$$\times (q^{h(\square)/2} - q^{-h(\square)/2})^{2g-2}.$$

Agreement with (2.9) follows from the identity

$$\sum_{\square \in \mu} (l(\square) - a(\square)) = \sum_{\square \in \mu} (j(\square) - i(\square)) = -\kappa(\mu).$$

Finally, note that expression (2.11) with $p = 0$ is related to the left-hand side of the HRV formula by

$$Z_{HRV}(z, t, T) = Z_{PT}^{ref}(Y, (zt)^{-1}, (zt^2)^{g-1}T, t). \tag{2.13}$$

2.4 HRV formula as a refined GV expansion

This section spells out in detail the construction of refined GV invariants of a threefold Y, as in Example 2.2, with $p = \deg(D) = 0$, in terms of the perverse filtration on the cohomology of the Hitchin moduli space. In this case, the generic fibers and the base of the Hitchin map $h : \mathcal{H}_r^e(X) \to \mathcal{B}$ have equal complex dimension d. Using the conjectural formula (2.11), it will be shown that equation (2.4) yields the HRV formula by a monomial change of variables. As observed in Example 2.2, the moduli space of slope stable pure dimension-one sheaves F on Y with compact support and numerical invariants,

$$ch_2(F) = r\sigma, \qquad \chi(F) = n,$$

is isomorphic to $\mathbb{C} \times \mathcal{H}_r^{n+r(g-1)}(X)$ provided that $(r, n) = 1$. Therefore, following the general arguments in the Introduction (Section 2.1), one should be able to define refined GV invariants using the decomposition theorem for the Hitchin map $h : \mathcal{H}_r^e(X) \to \mathcal{B}$, $e = n + r(g - 1)$. However, since the base of the Hitchin fibration is a linear space, there will not exist an $SL(2)_L \times SL(2)_R$ action on cohomology as required by M-theory. In this situation one can only define an $SL(2)_L \times \mathbb{C}_R^\times$-action, where $\mathbb{C}_R^\times$ can be thought of as a Cartan subgroup of $SL(2)_R$. This action can be explicitly described in terms of the perverse sheaf filtration constructed in [8, Sect. 1.4], which was briefly reviewed in Section 2.2.2.

Note that, given a relative ample class ω for the Hitchin map, there is a preferred splitting

$$H^k(\mathcal{H}_r^e(X)) \simeq \bigoplus_p Gr_p H^k(\mathcal{H}_r^e(X)) \tag{2.14}$$

of the perverse sheaf filtration presented in detail in [8, Sect. 1.4.2, 1.4.3]. Moreover, the relative Lefschetz isomorphism

$$\omega^l : Gr_{d-l}^P H^k(\mathcal{H}_r^e(X)) \xrightarrow{\sim} Gr_{d+l}^P H^{k+2l}(\mathcal{H}_r^e(X))$$

yields a decomposition

$$Gr_p^P H^k(\mathcal{H}_r^e(X)) \simeq \bigoplus_{i+2j=p} Q^{i,j;k}, \qquad Q^{i,j;k} = \omega_h^j Q^{i,0;k-2j},$$

where

$$Q^{i,0;k} = \mathrm{Ker}\big(\omega_h^{d-i+1} : Gr_i^P H^k(\mathcal{H}_r^e(X)) \to Gr_{2d-i+1}^P H^{k+2(d-i+1)}(\mathcal{H}_r^e(X))\big)$$

for all $0 \leq i \leq d$. Let $Q^{i,j} = \bigoplus_{k \geq 0} Q^{i,j;k}$. By construction, for fixed $0 \leq i \leq d$, there is an isomorphism

$$\bigoplus_{j=0}^{d-i} Q^{i,j} \simeq R_{(d-i)/2}^{\oplus \dim(Q^{i,0})},$$

where R_{j_L} is the irreducible representation of $SL(2)_L$ with spin $j_L \in \frac{1}{2}\mathbb{Z}$. The generator J_L^+ is represented by cup-product with ω, and $Q^{i,j}$ is the eigenspace of the Cartan generator J_L^3 with eigenvalue $j - (d-i)/2$. Note that cup product with ω preserves the grading $k - d - 2j$ and therefore one can define an extra $\mathbb{C}^\times$-action on $H^*(\mathcal{H}_r^e(X))$ that scales $Q^{i,j;k}$ with weight $d + 2j - k$. This torus action will be denoted by $\mathbb{C}_R^\times \times H^*(\mathcal{H}_r^e(X)) \to H^*(\mathcal{H}_r^e(X))$. Note also that

$$d + 2j - k \geq -d$$

because $j \geq 0$ and $k \leq -2d$.

In conclusion, in the present local curve geometry the $SL(2)_L \times SL(2)_R$ action on the cohomology of the moduli space of D2-D0 branes is replaced by an $SL(2)_L \times \mathbb{C}_R^\times$ action. This is certainly puzzling from a physical perspective because the BPS states are expected to form five-dimensional spin multiplets. The absence of a manifest $SL(2)_R$ symmetry of the local BPS spectrum is due to noncompactness of the moduli space. This is simply a symptom of the fact that there is no well-defined physical decoupling limit associated to a local higher genus curve, as considered here, in M-theory. In principle, in order to obtain a physically sensible theory, one would have to construct a Calabi–Yau threefold $\overline{Y}$ containing a curve X with infinitesimal neighborhood isomorphic to Y so that the moduli space $\mathcal{M}_{\overline{Y}}(r[X], n)$ is compact and there is an embedding $H^*(\mathcal{M}_Y(r, n)) \subset H^*(\mathcal{M}_{\overline{Y}}(r[X], n))$. The cohomology classes in the complement would then provide the missing components of the five-dimensional spin multiplets. Such a construction seems to be very difficult, and it is not in fact needed for the purpose of the present paper.

Given the $SL(2)_L \times \mathbb{C}_R^\times$ action described in the preceding paragraph, one can define the following local version of the refined Gopakumar–Vafa expansion (2.5):

$$F_{GV,Y}(q, Q, y) = \sum_{k\geq 1} \sum_{r\geq 1} \sum_{j_L=0}^{d/2} \sum_{l\geq -d} \frac{Q^{kr}}{k}(-1)^{2j_L+l} N_r((j_L, l))$$
$$\times \frac{q^{-k}(q^{-2kj_L} + \cdots + q^{2kj_L})y^{kl}}{(1 - (qy)^{-k})(1 - (qy^{-1})^{-k})}, \tag{2.15}$$

where

$$N_r(j_L, l) = \dim(Q^{d-2j_L, 0; d+l}).$$

The same change of variables as in equation (2.13) yields

$$F_{GV,Y}((zt)^{-1}, (zt^2)^{g-1}T, t) = \sum_{k\geq 1} \sum_{r\geq 1} \frac{T^{kr}}{k} B_r(z^k, t^k) P_r(z^k, t^k), \tag{2.16}$$

where $B_r(z, t)$ is defined above equation (2.8), and

$$P_r(z, t) = \sum_{j=0}^{d} \sum_{l\geq 0}(-1)^{j+l} N_r((j-d)/2, l-d) \, t^l \, (1 + \cdots + (zt)^{2j}).$$

It is clear that the change of variables

$$(q, Q, y) = ((zt)^{-1}, (zt^2)^{g-1}T, t)$$

identifies the HRV formula (2.8) with the refined GV expansion (2.4) for a local curve, provided that

$$P_r(z, t) = W_r(z, t). \tag{2.17}$$

However, given the cohomological definition of the refined GV invariants $N_r(j_L, l)$, relation (2.16) follows from the $P = W$ conjecture of [8]. This provides a string theoretic explanation, as well as strong evidence for this conjecture.

References

[1] K. Behrend. Donaldson-Thomas type invariants via microlocal geometry. *Ann. of Math. (2)*, 170(3):1307–1338, 2009.

[2] A. A. Beĭlinson, J. Bernstein, and P. Deligne. Faisceaux pervers. Analysis and topology on singular spaces, I (Luminy, 1981). *Astérisque*, 100:5–171, 1982.

[3] T. Bridgeland. Hall algebras and curve-counting invariants. *J. Amer. Math. Soc.*, 24(4):969–998, 2011.

[4] J. Bryan and R. Pandharipande. The local Gromov-Witten theory of curves. *J. Amer. Math. Soc.*, 21(1):101–136 (electronic), 2008. With an appendix by Bryan, C. Faber, A. Okounkov, and Pandharipande.

[5] W.-Y. Chuang, D.-E. Diaconescu, and G. Pan. Wallcrossing and cohomology of the moduli space of Hitchin pairs. *Commun. Num. Theor. Phys.*, 5:1–56, 2011.

[6] W.-Y. Chuang, D.-E. Diaconescu, and G. Pan. Chamber structure and wall-crossing in the ADHM theory of curves II. *J. Geom. Phys.*, 62(2):548–561, 2012.

[7] K. Corlette. Flat G-bundles with canonical metrics. *J. Differ. Geom.*, 28(3):361–382, 1988.

[8] M. de Cataldo, T. Hausel, and L. Migliorini. Topology of Hitchin systems and Hodge theory of character varieties. *Ann. of Math. (2)*, 175(3):1329–1407, 2012.

[9] M. A. A. de Cataldo and L. Migliorini. The decomposition theorem, perverse sheaves and the topology of algebraic maps. *Bull. Amer. Math. Soc. (N.S.)*, 46(4):535–633, 2009.

[10] M. A. A. de Cataldo and L. Migliorini. The perverse filtration and the Lefschetz hyperplane theorem. *Ann. of Math. (2)*, 171(3):2089–2113, 2010.

[11] F. Denef and G. W. Moore. Split states, entropy enigmas, holes and halos. *JHEP*, 1111:129, 2011.

[12] D. E. Diaconescu. Moduli of ADHM sheaves and local Donaldson-Thomas theory. *J. Geom. Phys.*, 64(4):763–799.

[13] R. Dijkgraaf, C. Vafa, and E. Verlinde. M-theory and a topological string duality, hep-th/0602087, 2006.

[14] T. Dimofte and S. Gukov. Refined, motivic, and quantum. *Lett. Math. Phys.*, 91:1, 2010.

[15] S. K. Donaldson. Twisted harmonic maps and the self-duality equations. *Proc. London Math. Soc. (3)*, 55(1):127–131, 1987.

[16] T. Eguchi and H. Kanno. Five-dimensional gauge theories and local mirror symmetry. *Nucl. Phys.*, B586:331–345, 2000.

[17] T. Eguchi and H. Kanno. Topological strings and Nekrasov's formulas. *JHEP*, 12:006, 2003.

[18] R. Gopakumar and C. Vafa. M theory and topological strings II. arXiv:9812127, 1998.

[19] T. Hausel and F. Rodriguez-Villegas. Mixed Hodge polynomials of character varieties. *Invent. Math.*, 174(3):555–624, 2008. With an appendix by Nicholas M. Katz.

[20] N. J. Hitchin. The self-duality equations on a Riemann surface. *Proc. London Math. Soc. (3)*, 55(1):59–126, 1987.

[21] T. J. Hollowood, A. Iqbal, and C. Vafa. Matrix models, geometric engineering and elliptic genera. *JHEP*, 03:069, 2008.

[22] S. Hosono, M.-H. Saito, and A. Takahashi. Relative Lefschetz action and BPS state counting. *Internat. Math. Res. Not.* 2001 (15):783–816, 2001.

[23] Z. Hua. Chern-Simons functions on toric Calabi-Yau threefolds and Donaldson-Thomas theory. arXiv:1103.1921, 2011.

[24] D. Huybrechts and M. Lehn. *The Geometry of Moduli Spaces of Sheaves*. Aspects of Mathematics, E31. Braunschweig: Friedrich Vieweg & Sohn, 1997.

[25] K. A. Intriligator, D. R. Morrison, and N. Seiberg. Five-dimensional supersymmetric gauge theories and degenerations of Calabi-Yau spaces. *Nucl. Phys.*, B497:56–100, 1997.

[26] A. Iqbal and A.-K. Kashani-Poor. Instanton counting and Chern-Simons theory. *Adv. Theor. Math. Phys.*, 7:457–497, 2004.

[27] A. Iqbal and A.-K. Kashani-Poor. SU(N) geometries and topological string amplitudes. *Adv. Theor. Math. Phys.*, 10:1–32, 2006.

[28] A. Iqbal, C. Kozcaz, and C. Vafa. The refined topological vertex. *JHEP*, 10:069, 2009.

[29] S. Katz, P. Mayr, and C. Vafa. Mirror symmetry and exact solution of 4-D N=2 gauge theories: 1. *Adv. Theor. Math. Phys.*, 1:53–114, 1998.

[30] S. H. Katz, A. Klemm, and C. Vafa. M-theory, topological strings and spinning black holes. *Adv. Theor. Math. Phys.*, 3:1445–1537, 1999.

[31] S. H. Katz and C. Vafa. Geometric engineering of N=1 quantum field theories. *Nucl. Phys.*, B497:196–204, 1997.

[32] Y. Konishi. Topological strings, instantons and asymptotic forms of Gopakumar-Vafa invariants. hep-th/0312090, 2003.

[33] M. Kontsevich and Y. Soibelman. Stability structures, Donaldson-Thomas invariants and cluster transformations. arXiv.org:0811.2435, 2008.

[34] A. E. Lawrence and N. Nekrasov. Instanton sums and five-dimensional gauge theories. *Nucl. Phys.*, B513:239–265, 1998.

[35] J. Li, K. Liu, and J. Zhou. Topological string partition functions as equivariant indices. *Asian J. Math.*, 10(1):81–114, 2006.

[36] D. Maulik, N. Nekrasov, A. Okounkov, and R. Pandharipande. Gromov-Witten theory and Donaldson-Thomas theory. I. *Compos. Math.*, 142(5):1263–1285, 2006.

[37] D. Maulik, R. Pandharipande, and R. Thomas. Curves on K3 surfaces and modular forms. *J. Topology*, 3(4):937–996, 2010.

[38] D. Maulik and Z. Yun. Macdonald formula for curves with planar singularities. arXiv:1107.2175.

[39] L. Migliorini and V. Shende. A support theorem for Hilbert schemes of planar curves. arXiv:1107.2355, 2011.

[40] D. R. Morrison and N. Seiberg. Extremal transitions and five-dimensional supersymmetric field theories. *Nucl. Phys.*, B483:229–247, 1997.

[41] S. Mozgovoy. Solution of the motivic ADHM recursion formula. *Math. Res. Not.*, 2012(18):4218–4244, 2012.

[42] N. A. Nekrasov. Seiberg-Witten prepotential from instanton counting. *Adv. Theor. Math. Phys.*, 7:831–864, 2004.

[43] A. Okounkov and R. Pandharipande. The local Donaldson-Thomas theory of curves. *Geom. Topol.*, 14:1503–1567, 2010.

[44] R. Pandharipande and R. P. Thomas. Curve counting via stable pairs in the derived category. *Invent. Math.*, 178(2):407–447, 2009.

[45] R. Pandharipande and R. P. Thomas. Stable pairs and BPS invariants. *J. Amer. Math. Soc.*, 23(1):267–297, 2010.

[46] C. T. Simpson. Higgs bundles and local systems. *Inst. Hautes Études Sci. Publ. Math.*, 75(1):5–95, 1992.

[47] J. Stoppa and R. P. Thomas. Hilbert schemes and stable pairs: GIT and derived category wall crossings. *Bull. Soc. Math. France*, 139(3):297–339, 2011.

[48] Y. Toda. Curve counting theories via stable objects I. DT/PT correspondence. *J. Amer. Math. Soc.*, 23(4):1119–1157, 2010.

[49] Y. Toda. Generating functions of stable pair invariants via wall-crossings in derived categories. In *New Developments in Algebraic Geometry, Integrable Systems and Mirror Symmetry (RIMS, Kyoto, 2008)*, M.-H. Saito, S. Hosono, and K. Yoshioka, eds. Advanced Studies in Pure Mathematics 59. Tokyo: Mathematical Society of Japan, 2010, pp. 389–434.

[50] Y. Yuan. Determinant line bundles on moduli spaces of pure sheaves on rational surfaces and strange duality. *Asian J. Math.*, 16(3):451–478, 2012.

3

Representations of surface groups and Higgs bundles

Peter B. Gothen[1]

Universidade do Porto

3.1 Introduction

In this chapter we give an introduction to Higgs bundles and their application to the study of surface group representations. This is based on two fundamental theorems. The first is the theorem of Corlette and Donaldson on the existence of harmonic metrics in flat bundles, which we treat in Lecture 1 (Section 3.2), after explaining some preliminaries on surface group representations, character varieties and flat bundles. The second is the Hitchin–Kobayashi correspondence for Higgs bundles, which goes back to the work of Hitchin and Simpson; this is the main topic of Lecture 2 (Section 3.3). Together, these two results allow the character variety for representations of the fundamental group of a Riemann surface in a Lie group G to be identified with a moduli space of holomorphic objects, known as G-Higgs bundles. Finally, in Lecture 3 (Section 3.4), we show how the $\mathbb{C}^*$-action on the moduli space G-Higgs bundles can be used to study its topological properties, thus giving information about the corresponding character variety.

Owing to lack of time and expertise, we do not treat many other important aspects of the theory of surface group representations, such as the approach using bounded cohomology (e.g. [9, 10]), higher Teichmüller theory (e.g. [17]), or ideas related to geometric structures on surfaces (e.g. [28]). We also do not touch on the relation of Higgs bundle moduli with mirror symmetry and the Geometric Langlands Programme (e.g. [33], [39]).

[1] Member of VBAC (Vector Bundles on Algebraic Curves). Partially supported by the FCT (Portugal) with EU (COMPETE) and national funds through the projects PTDC/MAT/099275/2008 and PTDC/MAT/098770/2008, and through Centro de Matemática da Universidade do Porto (PEst-C/MAT/UI0144/2011).

Moduli Spaces, eds. L. Brambila-Paz, O. García-Prada, P. Newstead and R. Thomas. Published by Cambridge University Press. © Cambridge University Press 2014.

In keeping with the lectures, we do not give proofs of most results. For more details and full proofs, we refer to the literature. Some references that the reader may find useful are the papers of Hitchin [35, 37], García-Prada [19], Goldman [25, 28, 27] and also [3], [4], [21], [20] and [6].

Notation Our notation is mostly standard. Smooth p-forms are denoted by Ω^p and smooth (p, q)-forms by $\Omega^{p,q}$. We shall occasionally confuse vector bundles and locally free sheaves.

Acknowledgments It is a pleasure to thank the organizers of the Newton Institute Semester on Moduli Spaces, and especially Leticia Brambila-Paz, for the invitation to lecture in the School on Moduli Spaces and for making it such a pleasant and stimulating event. I would also like to thank the participants in the course for their interest and for making the tutorials a fun and rewarding experience. It is impossible to mention all the mathematicians to whom I am indebted and who have generously shared their insights on the topics of these lectures over the years, without whom these notes would not exist. But I would like to express my special gratitude to Bill Goldman, Nigel Hitchin and my collaborators Ignasi Mundet i Riera, Steve Bradlow and Oscar García-Prada.

3.2 Lecture 1: Character varieties for surface groups and harmonic maps

In this lecture we give some basic definitions and properties of character varieties for representations of surface groups. We then explain the theorem of Corlette and Donaldson on the existence of harmonic maps in flat bundles, which is one of the two central results in the non-abelian Hodge theory correspondence (the other one being the Hitchin–Kobayashi correspondence, which will be treated in Lecture 2).

3.2.1 Surface group representations and character varieties

More details on the following can be found in, e.g. [25].

Let Σ be a closed oriented surface of genus g. The fundamental group of Σ has the standard presentation

$$\pi_1 \Sigma = \langle a_1, b_1, \ldots, a_g, b_g \mid \prod [a_i, b_i] = 1 \rangle, \tag{3.1}$$

where $[a_i, b_i] = a_i b_i a_i^{-1} b_i^{-1}$ is the commutator.

Let G be a real reductive Lie group. We denote its Lie algebra by $\mathfrak{g} = \mathrm{Lie}(G)$. Though not strictly necessary for everything that follows, we shall assume that G is connected. We shall also fix a non-degenerate quadratic form on G, invariant under the adjoint action of G (when G is semisimple, the Killing form or a multiple thereof will do).

By definition, a *representation* of $\pi_1\Sigma$ in G is a homomorphism $\rho\colon \pi_1\Sigma \to G$. Let $\mathrm{Ad}\colon G \to \mathrm{Aut}(\mathfrak{g})$ be the adjoint representation of G on its Lie algebra $\mathfrak{g}$. We say that ρ is *reductive*[2] if the composition

$$\mathrm{Ad}\circ\rho\colon \pi_1\Sigma \to \mathrm{Aut}(\mathfrak{g})$$

is completely reducible. Denote by $\mathrm{Hom}^{\mathrm{red}}(\pi_1\Sigma, G) \subset \mathrm{Hom}(\pi_1\Sigma, G)$ the subset of reductive representations.

Definition 3.1. The *character variety* for representations of $\pi_1\Sigma$ in G is

$$\mathcal{M}^{\mathrm{B}}(\Sigma, G) = \mathrm{Hom}^{\mathrm{red}}(\pi_1\Sigma, G)/G,$$

where the G-action is by simultaneous conjugation:

$$(g\cdot\rho)(x) = g\rho(x)g^{-1}.$$

The character variety is also known as the *Betti moduli space* (in Simpson's language [48]).

Note that, using the presentation (3.1), a representation ρ is given by a $2g$-tuple of elements in G satisfying the relation. Hence we get an inclusion $\mathrm{Hom}(\pi_1\Sigma, G) \hookrightarrow G^{2g}$, which endows $\mathrm{Hom}(\pi_1\Sigma, G)$ with a natural topology.[3] However, it turns out that the quotient space $\mathrm{Hom}(\pi_1, G)/G$ is not in general Hausdorff. The restriction to reductive representations remedies this problem.

We also remark that, if G is a complex reductive algebraic group, the character variety can be constructed as an affine GIT quotient (this is classical; a nice exposition is contained in [11, §3.1]).

3.2.2 Review of connections and curvature in principal bundles

Recall that a (smooth) *principal G-bundle* on Σ is a smooth fiber bundle $\pi\colon E \to \Sigma$ with a G-action (normally taken to be on the right) which is free and transitive on each fiber. Moreover, E is required to admit G-equivariant

---

[2] When G is algebraic, an alternative equivalent definition is to ask for the Zariski closure of $\rho(\Sigma) \subset G$ to be reductive.

[3] This is in fact the same as the compact-open topology on the mapping space $\mathrm{Hom}(\pi_1\Sigma, G)$, where we give $\pi_1\Sigma$ the discrete topology.

local trivializations $E_{|U} \cong U \times G$ over small open sets $U \subset \Sigma$ (where G acts by right multiplication on the second factor of the product $U \times G$). Note that the fiber E_x over any $x \in \Sigma$ is a G-torsor, so, choosing an element $e \in E_x$, we get a canonical identification $E_x \cong G$.

Example 3.2.
(1) The *frame bundle* of a rank n complex vector bundle $V \to \Sigma$ is a principal $GL(n, \mathbb{C})$-bundle, which has fibers

$$E_x = \{e \colon \mathbb{C}^n \overset{\cong}{\to} V_x \mid e \text{ is a linear isomorphism}\}.$$

(2) The universal covering $\tilde{\Sigma} \to \Sigma$ is a principal $\pi_1 \Sigma$-bundle over Σ. In this case the action is on the left.

Whenever we have a principal G-bundle $E \to \Sigma$ and a smooth G-space V (i.e. V is a smooth manifold on which G acts by smooth maps), we obtain a fiber bundle $E(V)$ with fibers modeled on V by taking the quotient of $E \times V$ under the diagonal G-action:

$$E(V) = E \times_G V \to \Sigma.$$

In particular, if V is a vector space with a linear G-action, we obtain a vector bundle $E(V)$ with fibers modeled on V. An important instance of this construction is when $V = \mathfrak{g}$ is acted on by G via the adjoint action. The resulting vector bundle $\text{Ad } E := E(\mathfrak{g})$ is then known as the *adjoint bundle* of E.

There is a bijective correspondence between sections $s \colon \Sigma \to E(V)$ of the bundle $\pi \colon E(V) \to \Sigma$ and G-equivariant maps $\tilde{s} \colon E \to V$, given by

$$s(x) = [e, \tilde{s}(e)]$$

for $e \in E(V)_x = \pi^{-1}(x)$ and $x \in \Sigma$. Similarly, a G-equivariant differential p-form $\alpha \in \Omega^p(E, V)$ descends to an $E(V)$-valued p-form $\tilde{\alpha} \in \Omega^p(\Sigma, E(V))$ if and only if it is *tensorial*, i.e. it vanishes on the vertical tangent spaces $T_e^v E = T_e E_x$ to E.

A *connection* in a principal G-bundle $E \to \Sigma$ is given by a smooth G-invariant Lie-algebra-valued 1-form $A \in \Omega^1(E, \mathfrak{g})$, which restricts to the identity on the vertical tangent spaces $T_e^v E$ under the natural identification $T_e^v E \cong \mathfrak{g}$ given by the choice of $e \in E_x$. Equivalently, a connection corresponds to the choice of a horizontal complement $T_e^h E = \text{Ker}(A(e) \colon T_e E \to \mathfrak{g})$ to $T_e^v E$ in each $T_e E$. Moreover, the G-invariance means that these complements correspond under the G-action. The difference of two connections is a tensorial form, so it follows that the space $\mathcal{A}$ of connections on E is an affine space modeled on $\Omega^1(\Sigma, \text{Ad } E)$.

Given a connection A in a principal bundle E, we obtain a covariant derivative

$$d_A : \Omega^0(\Sigma, E(V)) \to \Omega^1(\Sigma, E(V))$$

on sections in any associated vector bundle $E(V)$ as follows. Let $s \in \Omega^0(\Sigma, E(V))$ and let $\tilde{s} : E \to V$ be the corresponding G-equivariant map as above. Then we define a tensorial 1-form $\widetilde{d_A(s)}$ on E by composing $d\tilde{s}$ with the projection $TE \to T^h E$ defined by A, and let $d_A(s) \in \Omega^1(\Sigma, E(V))$ be the corresponding $E(V)$-valued 1-form.

Given a connection in E, the horizontal subspaces define a G-invariant distribution on the total space of E. The obstruction to integrability of this horizontal distribution is given by the *curvature*

$$F(A) = dA + \tfrac{1}{2}[A, A] \in \Omega^2(E, \mathfrak{g})$$

of the connection A, where the bracket $[A, A]$ is defined by combining the wedge product on forms with the Lie bracket on $\mathfrak{g}$. One checks that $F(A)$ is in fact a tensorial form and therefore descends to a 2-form on Σ with values in the adjoint bundle, which we denote by the same symbol,

$$F(A) \in \Omega^2(\Sigma, E(\mathfrak{g})).$$

A connection A is *flat* if $F(A) = 0$. A principal G-bundle $E \to \Sigma$ with a flat connection is called a *flat bundle*. Equivalently, a flat bundle is one for which the structure group G is discrete. The Frobenius theorem has the following immediate consequence.

Proposition 3.3. *Let $E \to \Sigma$ be a flat bundle and let $e_0 \in E_{x_0}$ for some $x_0 \in \Sigma$. Then, for any sufficiently small neighborhood $U \subset \Sigma$, there is a unique section $s \in \Omega^0(U, E_{|U})$ such that $d_A(s) = 0$ and $s(x_0) = e_0$.*

3.2.3 Surface group representations and flat bundles

Given a G-bundle E on Σ with a connection A, it follows from the existence and uniqueness theorem for ordinary differential equations that we can lift any loop γ in Σ to a covariantly constant loop in E (i.e. one whose tangent vectors are horizontal for the connection). In this way we obtain a well-defined parallel transport $E_x \to E_x$, which is given by multiplication by a unique group element, the *holonomy of A along γ*, denoted by $h_A(\gamma) \in G$. Moreover, if the connection A is flat, it follows from Proposition 3.3 that the holonomy only depends on the homotopy class of γ and thus we obtain the *holonomy representation* of $\pi_1 \Sigma$:

$$\rho_A \colon \pi_1 \Sigma \to G \tag{3.2}$$

defined by $\rho_A([\gamma]) = h_A(\gamma)$. We say that a flat connection A is *reductive* if its holonomy representation is a reductive representation of $\pi_1 \Sigma$ in G.

On the other hand, let $\rho \colon \pi_1 \Sigma \to G$ be a representation. We can then define a principal G-bundle E_ρ by taking the quotient

$$E_\rho = \widetilde{\Sigma} \times_{\pi_1 \Sigma} G,$$

where $\pi_1 \Sigma$ acts on the universal cover $\widetilde{\Sigma} \to \Sigma$ by deck transformations and on G by left multiplication via ρ. Moreover, since $\widetilde{\Sigma} \to \Sigma$ is a covering, there is a natural choice of horizontal subspaces in E_ρ. Therefore this bundle has a naturally defined connection which is evidently flat.

One sees that these two constructions are inverses of each other. Next we introduce the natural equivalence relation on (flat) connections, and promote this correspondence to a bijection between equivalence classes of flat connections and points in the character variety.

3.2.4 Flat bundles and gauge equivalence

The *gauge group*[4] is the automorphism group

$$\mathcal{G} = \Omega^0(\Sigma, \mathrm{Aut}(E)),$$

where $\mathrm{Aut}(E) = E \times_{\mathrm{Ad}} G \to \Sigma$ is the bundle of automorphisms of E. The gauge group acts on the space of connections $\mathcal{A}_E$ via

$$g \cdot A = gAg^{-1} + gdg^{-1}.$$

Moreover, the corresponding action on the curvature is given by

$$F(g \cdot A) = gF(A)g^{-1} \tag{3.3}$$

and hence $\mathcal{G}$ preserves the subspace of flat connections on E.

Recall that principal G-bundles on Σ are classified (up to smooth isomorphism) by a characteristic class

$$c(E) \in H^2(\Sigma, \pi_1 G) \cong \pi_1 G.$$

Here we are using the fact that G is connected and that Σ is a closed oriented surface. Fix $d \in \pi_1 G$ and let $E \to \Sigma$ be a principal G-bundle with $c(E) = d$. We can then consider the quotient space

$$\mathcal{M}_d^{\mathrm{dR}}(\Sigma, G) = \{ A \in \mathcal{A} \mid F(A) = 0 \text{ and } A \text{ is reductive} \}/\mathcal{G},$$

which is known as the *de Rham moduli space* (recall that $\mathcal{A}$ denotes the space of connections).

---

[4] This is the mathematician's definition. To a physicist, the gauge group is the structure group G.

Proposition 3.4. *If flat connections* B_i *correspond to representations* $\rho_i \colon \pi_1 \Sigma \to G$ *for* $i = 1, 2$, *then* B_1 *and* B_2 *are gauge equivalent if and only if there is a* $g \in G$ *such that* $\rho_1 = g\rho_2 g^{-1}$.

This proposition implies that there is a bijection

$$\mathcal{M}_d^{\mathrm{dR}}(\Sigma, G) \cong \mathcal{M}_d^{\mathrm{B}}(\Sigma, G), \tag{3.4}$$

where we denote by $\mathcal{M}_d^{\mathrm{B}}(\Sigma, G) \subset \mathcal{M}^{\mathrm{B}}(\Sigma, G)$ the subspace of representations with characteristic class d.

3.2.5 Harmonic metrics in flat bundles

Let $G' \subset G$ be a Lie subgroup. Recall that a *reduction of structure group* in a principal G-bundle $E \to \Sigma$ to $G' \subset G$ is a section

$$h \colon \Sigma \to E/G'$$

of the bundle $E/G' = E \times_G (G/G')$, picking out a G'-orbit in each fiber E_x.

Let us now fix a maximal compact subgroup $H \subset G$. This choice, together with the invariant inner product on $\mathfrak{g}$, gives rise to a *Cartan decomposition*:

$$\mathfrak{g} = \mathfrak{h} + \mathfrak{m}, \tag{3.5}$$

where $\mathfrak{h}$ is the Lie algebra of H and $\mathfrak{m}$ is its orthogonal complement.

Definition 3.5. A *metric* in a principal G-bundle $E \to \Sigma$ is a reduction of structure group to $H \subset G$.

If $E_\rho = \widetilde{\Sigma} \times_\rho G$ is a flat bundle, we have

$$E/H = \widetilde{\Sigma} \times_\rho (G/H),$$

and hence a metric h in E corresponds to a $\pi_1 \Sigma$-equivariant map

$$\widetilde{h} \colon \widetilde{\Sigma} \to G/H.$$

The energy of the metric h is essentially the integral over Σ of the norm squared of the derivative of h. In the following we make this concept precise. To start with we give Σ a Riemannian metric and note that G/H is a Riemannian manifold. Hence we can calculate the norm $|D\widetilde{h}(\widetilde{x})|$ at any point $\widetilde{x} \in \widetilde{X}$. Furthermore, as the group G acts on G/H by isometries, the derivative of $\widetilde{h}$ satisfies

$$|D\widetilde{h}(\widetilde{x})| = |D\widetilde{h}(\gamma \cdot \widetilde{x})|$$

for any $\gamma \in \pi_1 \Sigma$. Alternatively, we may proceed as follows. Let $T^v E \to \Sigma$ be the vertical tangent bundle of E. The fact that E is flat means that there is a natural projection $p \colon TE \to T^v E$ and we can define the vertical part of the derivative of h as the composition

$$Dh = p \circ dh \colon T\Sigma \to TE \to T^v E.$$

Clearly we have $|Dh(x)| = |\widetilde{Dh}(\tilde{x})|$ for any $\tilde{x} \in E_x$.

Definition 3.6. Let Σ be a closed oriented surface with a Riemannian metric and let $E \to \Sigma$ be a flat principal G-bundle. The *energy* of a metric h in E is given by

$$\mathcal{E}(h) = \int_\Sigma |Dh|^2 \mathrm{vol}.$$

Remark 3.7. Recall that on a surface the integral of a 1-form is conformally invariant. Hence it suffices to give Σ a conformal structure in order to make the energy functional well defined.

Definition 3.8. A metric h in a flat G-bundle $E \to \Sigma$ is *harmonic* if it is a critical point of the energy functional.

Next we want to reformulate this in terms of connections. Let $i \colon E_H \to E$ be the principal H-bundle obtained by the reduction of structure group defined by the metric h, and denote the flat connection on E by $B \in \Omega^1(E, \mathfrak{g})$. Using the Cartan decomposition (3.5) we can then write

$$i^* B = A + \psi, \tag{3.6}$$

where $A \in \Omega^1(E_H, \mathfrak{h})$ defines a connection E_H and $\psi \in \Omega^1(E_H, \mathfrak{m})$ is a tensorial 1-form, which therefore descends to a section, denoted by the same symbol

$$\psi \in \Omega^1(E_H(\mathfrak{m})).$$

Note that we have a canonical identification

$$E_H(\mathfrak{m}) \cong T^v E,$$

and that under this identification we have

$$\psi = Dh,$$

as is easily checked. To calculate the critical points of the energy functional, take a deformation of the metric h of the form

$$h_t = \exp(t \cdot s)h \in \Omega^0(\Sigma, E/H)$$

for $s \in \Omega^0(\Sigma, E_H(\mathfrak{m}))$. One then calculates

$$\frac{d}{dt}(\mathcal{E}(h_t))_{|t=0} = \langle \psi, d_A s \rangle,$$

from which we deduce the following.

Proposition 3.9. *Let h be a metric in a flat bundle $E \to \Sigma$ and let (A, ψ) be defined by (3.6). Then h is harmonic if and only if*

$$d_A^* \psi = 0.$$

3.2.6 The Corlette–Donaldson theorem

The following result was proved independently by Donaldson [15] (for $G = \mathrm{SL}(2, \mathbb{C})$) and Corlette [13] (for more general groups and base manifolds of dimension higher than two); see also Labourie [41]. The idea of the proof is to adapt the proof of Eells–Sampson on the existence of harmonic maps into negatively curved target manifolds to the present "twisted situation."

Theorem 3.10. *A flat bundle $E \to \Sigma$ corresponding to a representation $\rho\colon \pi_1 \Sigma \to G$ admits a harmonic metric if and only if ρ is reductive.*

In terms of the pair (A, θ) given by (3.6), the flatness condition on B becomes

$$\begin{aligned} F(A) + \tfrac{1}{2}[\theta, \theta] &= 0, \\ d_A \theta &= 0, \end{aligned} \tag{3.7}$$

as can be seen by considering the $\mathfrak{h}$- and $\mathfrak{m}$-valued parts of the equation $F(B) = 0$ separately. This motivates the following definition.

Definition 3.11. Let $E_H \to X$ be a principal H-bundle on X, let A be a connection on E_H, and let $\theta \in \Omega^1(X, E_H(\mathfrak{m}))$. The triple (E_H, A, θ) is called a *harmonic bundle* if the equations

$$F(A) + \tfrac{1}{2}[\theta, \theta] = 0, \tag{3.8}$$
$$d_A \theta = 0, \tag{3.9}$$
$$d_A^* \theta = 0 \tag{3.10}$$

are satisfied.

Next we want to obtain a statement at the level of moduli spaces (analogous to (3.4)). Fix a reduction $E_H \hookrightarrow E$ and consider the gauge groups

$$\mathcal{H} = \mathrm{Aut}(F_H) = \Omega^0(\Sigma, E_H \times_{\mathrm{Ad}} H),$$
$$\mathcal{G} = \mathrm{Aut}(E) = \Omega^0(\Sigma, E \times_{\mathrm{Ad}} G).$$

Then Theorem 3.10 can equivalently be formulated as saying that, for any flat reductive connection B in E, there is a gauge transformation $g \in \mathcal{G}$ such that, writing $g \cdot B = A + \psi$, the triple (E_H, A, ψ) is a harmonic bundle.

Let $d \in \pi_1 H$, and fix E_H with $c(E_H) = d$. The *moduli space of harmonic bundles of topological class d* is given by

$$\mathcal{M}_d^{\mathrm{Har}}(\Sigma, G) = \{(A, \theta) \mid (3.8)\text{–}(3.10) \text{ hold}\}/\mathcal{H}.$$

Now Theorem 3.10 can be complemented by a suitable uniqueness statement (analogous to Proposition 3.4), which allows us to obtain a bijective correspondence

$$\mathcal{M}_d^{\mathrm{dR}}(\Sigma, G) \cong \mathcal{M}_d^{\mathrm{Har}}(\Sigma, G). \tag{3.11}$$

3.3 Lecture 2: *G*-Higgs bundles and the Hitchin–Kobayashi correspondence

In Lecture 1, Section 3.2, we saw that any reductive surface group representation gives rise to an essentially unique harmonic metric in the associated flat bundle. In this lecture, we reinterpret this in holomorphic terms, introducing G-Higgs bundles. Moreover, we explain the Hitchin–Kobayashi correspondence for these.

Recall from Remark 3.7 that we equipped the surface Σ with a conformal class of metrics. This is equivalent to having defined a Riemann surface, which we shall henceforth denote by $X = (\Sigma, J)$.

3.3.1 Lie theoretic preliminaries

Let $H^{\mathbb{C}}$ be the complexification of the maximal compact subgroup $H \subseteq G$, and let $\mathfrak{h}^{\mathbb{C}}$ and $\mathfrak{g}^{\mathbb{C}}$ be the complexifications of the Lie algebras $\mathfrak{h}$ and $\mathfrak{g}$, respectively. In particular, $\mathfrak{h}^{\mathbb{C}} = \mathrm{Lie}(H^{\mathbb{C}})$. However, we do not need to assume the existence of a complexification of the Lie group G.

The Cartan decomposition (3.5) complexifies to

$$\mathfrak{g}^{\mathbb{C}} = \mathfrak{h}^{\mathbb{C}} + \mathfrak{m}^{\mathbb{C}}; \tag{3.12}$$

note that this is a direct sum of vector spaces, but not of Lie algebras. In fact, we have

$$[\mathfrak{h}^{\mathbb{C}}, \mathfrak{h}^{\mathbb{C}}] \subseteq \mathfrak{h}^{\mathbb{C}}, \quad [\mathfrak{h}^{\mathbb{C}}, \mathfrak{m}^{\mathbb{C}}] \subseteq \mathfrak{m}^{\mathbb{C}}, \quad [\mathfrak{m}^{\mathbb{C}}, \mathfrak{m}^{\mathbb{C}}] \subseteq \mathfrak{h}^{\mathbb{C}}.$$

Moreover, we have the $\mathbb{C}$-linear Cartan involution

$$\theta : \mathfrak{g}^{\mathbb{C}} \to \mathfrak{g}^{\mathbb{C}},$$

whose ± 1-eigenspace decomposition is (3.12), the real structure (i.e. $\mathbb{C}$-antilinear involution) corresponding to $\mathfrak{g} \subset \mathfrak{g}^{\mathbb{C}}$,

$$\sigma : \mathfrak{g}^{\mathbb{C}} \to \mathfrak{g}^{\mathbb{C}},$$

and the compact real structure,

$$\widetilde{\tau} = \theta \circ \sigma : \mathfrak{g}^{\mathbb{C}} \to \mathfrak{g}^{\mathbb{C}}.$$

The $+1$-eigenspace of $\widetilde{\tau}$ is a maximal compact subalgebra of $\mathfrak{g}^{\mathbb{C}}$ whose intersection with $\mathfrak{h}^{\mathbb{C}}$ is $\mathfrak{h}$.

We shall also need the *isotropy representation* of $H^{\mathbb{C}}$ on $\mathfrak{m}^{\mathbb{C}}$,

$$\iota : H^{\mathbb{C}} \to \mathrm{GL}(\mathfrak{m}^{\mathbb{C}}), \tag{3.13}$$

which is induced by the complexification of the adjoint action of H on $\mathfrak{g}$.

3.3.2 The Hitchin equations

We extend $\widetilde{\tau}$ to

$$\tau : \Omega^1(X, E(\mathfrak{m}^{\mathbb{C}})) \to \Omega^1(X, E(\mathfrak{m}^{\mathbb{C}}))$$

by combining it with conjugation on the form component. Locally,

$$\tau(\omega \otimes a) := \bar{\omega} \otimes \tau(a)$$

for a complex 1-form ω on X and a section a of $E(\mathfrak{m}^{\mathbb{C}})$. There is an isomorphism

$$\Omega^1(E(\mathfrak{m})) \to \Omega^{1,0}(X, E(\mathfrak{m}^{\mathbb{C}})),$$
$$\theta \longmapsto \frac{\theta - iJ\theta}{2}, \tag{3.14}$$

where J is the complex structure on the tangent bundle of X. The inverse is given by

$$\theta = \varphi - \tau(\varphi). \tag{3.15}$$

This is entirely analogous to the way in which we can write the connection $A \in \Omega^1(E_H, \mathfrak{h})$,

$$A = A^{1,0} + A^{0,1}, \tag{3.16}$$

with $A^{p,q} \in \Omega^{p,q}(E_H(\mathfrak{m}^{\mathbb{C}}))$.

Remark 3.12. Note that

$$E_H(\mathfrak{m}^{\mathbb{C}}) = E_{H^{\mathbb{C}}}(\mathfrak{m}^{\mathbb{C}}),$$

where $E_{H^{\mathbb{C}}} = E_H \times_H H^{\mathbb{C}}$ is the principal $H^{\mathbb{C}}$-bundle obtained by extension of the structure group.

The bijective correspondence $A \leftrightarrow A^{0,1}$ gives us a bijective correspondence between connections A on E_H and holomorphic structures on $E_{H^{\mathbb{C}}}$ (the integrability condition is automatically satisfied because $\dim_{\mathbb{C}} X = 1$).

Correspondingly, for any complex representation V of $H^{\mathbb{C}}$, the vector bundle $E_{H^{\mathbb{C}}}(V)$ becomes a holomorphic bundle, and the covariant derivative on sections of $E_{H^{\mathbb{C}}}(V)$ given by A decomposes as $d_A = \bar{\partial}_A + \partial_A$, where

$$\bar{\partial}_A \colon \Omega^0(X, E_{H^{\mathbb{C}}}(V)) \to \Omega^{0,1}(X, E_{H^{\mathbb{C}}}(V)).$$

The holomorphic sections of $E_{H^{\mathbb{C}}}(V)$ are just the ones which are in the kernel of $\bar{\partial}_A$.

With all this notation in place, one sees, using the Kähler identities, that the harmonic bundle equations (3.8)–(3.10) are equivalent to the *Hitchin equations*:

$$F(A) - [\varphi, \tau(\varphi)] = 0, \tag{3.17}$$
$$\bar{\partial}_A \varphi = 0. \tag{3.18}$$

Thus we have a canonical identification

$$\mathcal{M}_d^{\mathrm{Har}}(X, G) = \mathcal{M}_d^{\mathrm{Hit}}(X, G), \tag{3.19}$$

where we have introduced the moduli space

$$\mathcal{M}_d^{\mathrm{Hit}}(X, G) = \{(A, \varphi) \mid (3.17)\text{–}(3.18) \text{ hold}\}/\mathcal{H}$$

of solutions to the Hitchin equations. This gauge theoretic point of view allows one to give the moduli space $\mathcal{M}_d^{\mathrm{Hit}}(X, G)$ a Kähler structure. While the metric depends on the choice of conformal structure on Σ, the Kähler form is independent of this choice, and in fact coincides with Goldman's symplectic form [29].

3.3.3 *G*-Higgs bundles, stability and the Hitchin–Kobayashi correspondence

The second Hitchin equation (3.18) says that Φ is holomorphic with respect to the structure of a holomorphic bundle. Write $K = T^*X^{\mathbb{C}}$ for the holomorphic cotangent bundle, or *canonical bundle*, of X and H^0 for holomorphic sections. We have thus reached the conclusion that the harmonic bundle gives rise to a holomorphic object, a so-called *G*-Higgs bundle, defined as follows.

Definition 3.13. A *G-Higgs bundle* on X is a pair (E, φ), where $E \to X$ is a holomorphic principal $H^{\mathbb{C}}$-bundle and $\varphi \in H^0(X, E(\mathfrak{m}^{\mathbb{C}}) \otimes K)$.

When G is a complex group, we have that $H^{\mathbb{C}} = G$ and the Cartan decomposition $\mathfrak{g}^{\mathbb{C}} = \mathfrak{g} + i\mathfrak{g}^{\mathbb{C}}$. Hence a *G*-Higgs bundle is a pair (E, φ), where E is a holomorphic principal *G*-bundle and $\varphi \in H^0(X, E(\mathfrak{g}) \otimes K)$. Note that $E(\mathfrak{g}^{\mathbb{C}}) = \operatorname{Ad} E$ is just the adjoint bundle of E.

Another particular case is when $G = H$ is a compact group. Then we have $\varphi = 0$, so a *G*-Higgs bundle is just a holomorphic principal bundle, and the Hitchin equations simply say that $F(A) = 0$.

In the following we give some examples of *G*-Higgs bundles for specific groups.

Example 3.14. Let $G = \mathrm{SU}(n, \mathbb{C})$. Then a *G*-Higgs bundle is just a holomorphic vector bundle $V \to X$ with trivial determinant.

Example 3.15. Let $G = \mathrm{SL}(n, \mathbb{C})$. Then a *G*-Higgs bundle is a pair (V, φ), where $V \to X$ is a holomorphic vector bundle with trivial determinant and $\varphi \in H^0(X, \operatorname{End}_0(E) \otimes K)$ (where $\operatorname{End}_0(E)$ is the subspace of traceless endomorphisms).

Example 3.16. Let $G = \mathrm{SU}(p, q)$. Then a *G*-Higgs bundle is a triple (V, W, φ), where V and W are holomorphic vector bundles on X of rank p and q, respectively, satisfying $\det(V) \otimes \det(W) \cong \mathcal{O}$, and

$$\varphi = (\beta, \gamma) \in H^0(X, \operatorname{Hom}(W, V) \otimes K) \oplus H^0(X, \operatorname{Hom}(V, W) \otimes K).$$

Example 3.17. Let $G = \mathrm{Sp}(2n, \mathbb{R})$. Then a *G*-Higgs bundle is a pair (V, φ), where V is a holomorphic vector bundle on X of rank n and

$$\varphi = (\beta, \gamma) \in H^0(X, S^2V \otimes K) \oplus H^0(X, S^2V^* \otimes K).$$

If $G = \mathrm{SU}(n)$, we are thus in the presence of a complex vector bundle with a flat unitary connection. Such a bundle turns out to be *polystable*. The Narasimhan–Seshadri theorem [43] says, conversely, that if a holomorphic vector bundle is polystable then it admits a metric such that the unique unitary connection compatible with the holomorphic structure is (projectively) flat. There is an analogous statement for other compact G, due to Ramanathan [45]).

These results generalize to G-Higgs bundles. The appropriate stability condition is a bit involved to state in general. However, in the case of Higgs vector bundles it is simply the following. Recall that the *slope* of a vector bundle $E \to X$ is $\mu(E) = \deg(E)/\mathrm{rk}(E)$. Also, we say that a subbundle $F \subset E$ is φ-*invariant* if $\varphi(F) \subset F \otimes K$.

Definition 3.18. A Higgs vector bundle (E, φ) is *semistable* if

$$\mu(F) \leqslant \mu(E)$$

for any φ-invariant subbundle $F \subset E$, and *stable* if, moreover, strict inequality holds whenever F is proper and non-zero. A Higgs vector bundle (E, φ) is *polystable* if it is isomorphic to a direct sum of stable Higgs bundles, all of the same slope.

The stability conditions for G-Higgs bundles can be obtained as a special case of a general stability condition for *pairs*, and we refer the reader to [21] for the detailed formulation. It is worth noting that poly- and semistability of a G-Higgs bundle (E, φ) are equivalent to poly- and semistability of the Higgs vector bundle $(E(\mathfrak{g}^{\mathbb{C}}), \mathrm{ad}(\varphi))$.

The general Hitchin–Kobayashi correspondence for principal pairs [38, 7] now has as a consequence the following Hitchin–Kobayashi correspondence for G-Higgs bundles. (See [21] for the full extension to polystable pairs, as well as a detailed analysis of the case of G-Higgs bundles.)

Theorem 3.19. *Assume that G is semisimple. A G-Higgs bundle (E, φ) is polystable if and only if it admits a reduction of structure group to the maximal compact $H \subset H^{\mathbb{C}}$, unique up to isomorphism of H-bundles, such that the following holds: denoting by A the unique H-connection compatible with the reduction and by $\bar{\partial}_A$ the $\bar{\partial}$-operator induced from the holomorphic structure, the pair (A, φ) satisfies the Hitchin equations (3.17) and (3.18).*

In the context of Higgs bundles, the Hitchin–Kobayashi correspondence goes back to the work of Hitchin [35] and Simpson [47].

Remark 3.20. The assumption that G is semisimple is not essential. Indeed, as can be expected from the situation for usual G-bundles, for reductive G an analogous statement holds if one adds a suitable central term to the right-hand side of the first of the Hitchin equations. For the correspondence with representations, one must then consider homomorphisms of a central extension of the fundamental group. We refer to [21] for more details on this.

Just as for vector bundles, stability of G-Higgs bundles has a dual importance. Namely, apart from its role in the Hitchin–Kobayashi correspondence, it is also the appropriate notion for constructing moduli spaces using GIT. The constructions by Schmitt [46] are in fact sufficiently general to cover many cases of G-Higgs bundles. Thus we have yet another moduli space at our disposal, namely the moduli space

$$\mathcal{M}_d^{\mathrm{Dol}}(X, G)$$

of semistable G-Higgs bundles of topological class $d \in \pi_1 H$. Alternatively, this moduli space can be constructed using a Kuranishi slice method. From this point of view, we fix a principal $H^{\mathbb{C}}$-bundle $E \to \Sigma$ and consider the complex configuration space

$$\mathcal{C}^{\mathbb{C}} = \{(A^{0,1}, \varphi) \mid \bar{\partial}_A \varphi = 0\}.$$

The complex gauge group $\mathcal{H}^{\mathbb{C}}$ acts naturally on this space and on the subspace $\mathcal{C}^{\mathbb{C}}_{\mathrm{polystable}}$ of pairs $(A^{0,1}, \varphi)$ that define the structure of a polystable G-Higgs bundle on E. The moduli space is then

$$\mathcal{M}_d^{\mathrm{Dol}}(X, G) = \mathcal{C}^{\mathbb{C}}_{\mathrm{polystable}}/\mathcal{H}^{\mathbb{C}}.$$

Either way, Theorem 3.19 implies that we have an identification

$$\mathcal{M}_d^{\mathrm{Dol}}(X, G) \cong \mathcal{M}_d^{\mathrm{Hit}}(X, G). \tag{3.20}$$

Putting this together with the previous identifications (3.4), (3.11) and (3.19), we finally obtain the *non-abelian Hodge theorem*.

Theorem 3.21. *Let X be a closed Riemann surface of genus g. Then there is a homeomorphism*

$$\mathcal{M}_d^{\mathrm{B}}(X, G) \cong \mathcal{M}_d^{\mathrm{Dol}}(X, G).$$

Remark 3.22. The fact that the identification of Theorem 3.21 is a homeomorphism is not too hard to see, but more is true: outside of the singular loci, the identification is in fact an analytic isomorphism. On the other hand, it is definitely not algebraic. In this respect, it is instructive to consider the example $G = \mathbb{C}^*$.

3.3.4 The Hitchin map

We recall the definition of the Hitchin map, which plays a central role in the theory of Higgs bundles.

Take a basis $\{p_1, \ldots, p_r\}$ for the invariant polynomials on the Lie algebra $\mathfrak{g}^{\mathbb{C}}$ and let $d_i = \deg(p_i)$. Given a G-Higgs bundle (E, φ), evaluating p_i on φ gives a section $p_i(\varphi) \in H^0(X, K^{d_i})$. The *Hitchin map* is defined to be

$$
\begin{aligned}
h \colon \mathcal{M}_d^{\mathrm{Dol}} &\to B, \\
(E, \varphi) &\longmapsto (p_1(\varphi), \ldots, p_r(\varphi)),
\end{aligned}
\tag{3.21}
$$

where the *Hitchin base* is

$$
B := \bigoplus H^0(X, K^{d_i}).
$$

The Hitchin map is proper and, for G complex, defines an algebraically completely integrable system known as the Hitchin system [36].

3.3.5 The moduli space of $\mathrm{SU}(p, q)$-Higgs bundles

We end this section by illustrating how the Higgs bundle point of view allows for easy proofs of strong results by proving the Milnor–Wood inequality for $\mathrm{SU}(p, q)$-Higgs bundles and discussing a closely related rigidity result.

Recall that an $\mathrm{SU}(p, q)$-Higgs bundle is a quadruple (V, W, β, γ), where V and W are vector bundles on X of rank p and q, respectively, satisfying $\det(V) \otimes \det(W) \cong \mathcal{O}$, and where $\beta \in H^0(X, \mathrm{Hom}(W, V) \otimes) K$ and $\gamma \in H^0(X, \mathrm{Hom}(V, W) \otimes K)$. The topological classification of such bundles is given by $\deg(V) = -\deg(W) \in \mathbb{Z}$. Denote by $\mathcal{M}_d$ the moduli space of $\mathrm{SU}(p, q)$-Higgs bundles with $\deg(V) = d$.

For the case $p = q = 1$, we have $\mathrm{SU}(1, 1) = \mathrm{SL}(2, \mathbb{R})$, and the degree d is just the Euler class of the corresponding flat $\mathrm{SL}(2, \mathbb{R})$-bundle. In 1957 Milnor [42] proved that it satisfies the bound

$$
|d| \leqslant g - 1.
$$

Much more generally, whenever G is non-compact of Hermitian type, one can define an integer invariant, the *Toledo invariant*, of representations $\rho \colon \pi_1 \Sigma \to G$, and there is a bound on the Toledo invariant, usually known as a *Milnor–Wood inequality*. In various degrees of generality, this is due to, among others, Domic and Toledo [14], Dupont [16], Toledo [49, 50] and Turaev [51]. For the case of $G = \mathrm{SU}(p, q)$, the Milnor–Wood inequality is

$$
|d| \leqslant \min\{p, q\}(g - 1).
\tag{3.22}
$$

A proof of this Milnor–Wood inequality using Higgs bundles is very easy to give. For this it is convenient (though not essential) to pass through the usual Higgs vector bundles: since $SU(p, q)$ is a subgroup of $SL(p + q, \mathbb{C})$, to any $SU(p, q)$-Higgs bundle we can associate an $SL(p + q, \mathbb{C})$-Higgs bundle

$$(E, \Phi) = \left(V \oplus W, \begin{pmatrix} 0 & \beta \\ \gamma & 0 \end{pmatrix} \right).$$

Now, if (V, W, β, γ) is polystable, then so is (E, Φ) (this follows immediately from the fact that a solution to the $SU(p, q)$-Hitchin equations on (V, W, β, γ) induces a solution on (E, Φ)). Let $N \subset V$ be the kernel of $\gamma : V \to W \otimes K$, viewed as a subbundle, and let $I \otimes K \subset W \otimes K$ be the subbundle obtained by saturating the image subsheaf. Thus, γ induces a bundle map of maximal rank $\bar{\gamma} : V/N \to I \otimes K$, from which we deduce that

$$\deg(N) - \deg(V) + \deg(I) + (2g - 2)\operatorname{rk}(\gamma) \geqslant 0 \tag{3.23}$$

with equality if and only if $\bar{\gamma}$ is an isomorphism. Moreover, the subbundles $N \subset E$ and $V \oplus I \subset E$ are Φ-invariant, so polystability of (E, Φ) implies that

$$\deg(N) \leqslant 0, \tag{3.24}$$

$$\deg(V) + \deg(I) \leqslant 0. \tag{3.25}$$

Putting together equations (3.23)–(3.25), we obtain

$$\deg(V) \leqslant \operatorname{rk}(\gamma)(g - 1) \tag{3.26}$$

from which the Milnor–Wood inequality (3.22) is immediate for $d \geqslant 0$. When $d \leqslant 0$ a similar argument involving β instead of γ gives the result.

But our arguments in fact give more information for the case when equality holds in (3.22). Assume for definiteness that $d \geqslant 0$ and that $p \leqslant q$. Then, if equality holds in (3.22), we conclude immediately that $\operatorname{rk}(\gamma) = p$ and that $\gamma : V \to I \otimes K$ is an isomorphism. Hence, by polystability of (E, Φ), there is a decomposition $W = I \oplus Q$ and $\beta_{|Q} = 0$. In other words, the $SU(p, q)$-Higgs bundle (V, W, β, γ) decomposes into the $U(p, p)$-Higgs bundle (V, I, β, γ) and the $U(q - p)$-Higgs bundle Q.

From the point of view of representations of the fundamental group, this can be viewed as a rigidity result, which was first proved by Toledo [50] for $p = 1$ and by Hernández [34] for $p = 2$. A more general result, valid in the context of arbitrary groups of Hermitian type, has been proved [10, 8]. From the point of view of Higgs bundles, the results for $U(p, q)$ appeared in [3], and a survey of the situation for other classical groups can be found in [4], while the Ph.D. thesis of Rubio [44] treats the question for general groups using a general Lie theoretic approach.

3.4 Lecture 3: Morse–Bott theory of the moduli space of G-Higgs bundles

In this final lecture we consider the $\mathbb{C}^*$-action on the moduli space of G-Higgs bundles and explain how to use it to study its topology. We consider the Dolbeault moduli space and occasionally use the identification with the gauge theory moduli space of solutions to Hitchin's equation. For simplicity, we denote it simply by $\mathcal{M}_d$. Again, though not strictly necessary, we assume that G is semisimple. To start we need to review some of the deformation theory of G-Higgs bundles.

3.4.1 Simple and infinitesimally simple G-Higgs bundles

Let E be a principal $H^{\mathbb{C}}$-bundle on X. An *automorphism* of E is an equivariant holomorphic bundle map $g\colon E \to E$ which admits a holomorphic inverse. We denote the group of automorphisms of E by $\mathrm{Aut}(E)$. Equivalently, we may define $\mathrm{Aut}(E)$ to be the space of holomorphic sections of the bundle of automorphisms $E \times_{\mathrm{Ad}} G \to X$. Let (E, φ) be a G-Higgs bundle. We denote by $\mathrm{Aut}(E)$ the group of automorphisms of (E, φ):

$$\mathrm{Aut}(E, \varphi) = \{g \in \mathrm{Aut}(E) \mid \mathrm{Ad}(g)(\varphi) = \varphi\}.$$

We also introduce the *infinitesimal automorphism space* (which, at least formally, is the Lie algebra of the automorphism group), defining

$$\mathrm{aut}(E, \varphi) = \{Y \in H^0(X, E(\mathfrak{h}^{\mathbb{C}})) \mid [Y, \varphi] = 0\}.$$

A G-Higgs bundle (E, φ) is *simple* if its automorphism group is the smallest possible, i.e.

$$\mathrm{Aut}(E, \varphi) = Z(H^{\mathbb{C}}) \cap \mathrm{Ker}(\iota).$$

Also, we say that (E, φ) is *infinitesimally simple* if

$$\mathrm{aut}(E, \varphi) = Z(\mathfrak{h}^{\mathbb{C}}) \cap \mathrm{Ker}(d\iota).$$

Note that, for Higgs vector bundles, these two notions are equivalent. This is, however, not true in general, as Example 3.24 shows.

The following result is the G-Higgs bundle version of the well-known fact that a stable vector bundle has only scalar automorphisms.

Proposition 3.23. *Let (E, φ) be a stable G-Higgs bundle. Then it is infinitesimally simple.*

Example 3.24. Let M_1 and M_2 be line bundles on X with $M_i^2 = K$ and $M_1 \neq M_2$. Define $V = M_1 \oplus M_2$, and let $\beta = 0 \in H^0(X, S^2 V \otimes K)$ and $\gamma = \left(\begin{smallmatrix} 1 & 0 \\ 0 & 1 \end{smallmatrix}\right) \in H^0(X, S^2 V^* \otimes K)$. Then it is easy to see that the $Sp(2, \mathbb{R})$-Higgs bundle (V, β, γ) is stable and hence infinitesimally simple. However, it is not simple since it has the automorphism $\left(\begin{smallmatrix} -1 & 0 \\ 0 & 1 \end{smallmatrix}\right)$.

3.4.2 Deformation theory of G-Higgs bundles

Next we outline the deformation theory of G-Higgs bundles. A useful reference for the following material is by Biswas and Ramanan [2].

Definition 3.25. The *deformation complex* of a G-Higgs bundle (E, φ) is the complex of sheaves

$$C^\bullet(E, \varphi) \colon E(\mathfrak{h}^{\mathbb{C}}) \xrightarrow{[-, \varphi]} E(\mathfrak{m}^{\mathbb{C}}) \otimes K.$$

The deformation theory of a G-Higgs bundle (E, φ) is governed by the hypercohomology groups of the deformation complex. Thus, we have the following standard results.

Proposition 3.26. *Let (E, φ) be a G-Higgs bundle.*

(1) There is a canonical identification between the space of infinitesimal deformations of (E, φ) and the hypercohomology group

$$\mathbb{H}^1(C^\bullet(E, \varphi)).$$

(2) There is a long exact sequence

$$0 \to \mathbb{H}^0(C^\bullet(E, \varphi)) \to H^0(E(\mathfrak{h}^{\mathbb{C}})) \xrightarrow{[-, \varphi]} H^0(E(\mathfrak{m}^{\mathbb{C}}) \otimes K)$$
$$\to \mathbb{H}^1(C^\bullet(E, \varphi)) \to H^1(E(\mathfrak{h}^{\mathbb{C}})) \xrightarrow{[-, \varphi]} H^1(E(\mathfrak{m}^{\mathbb{C}}) \otimes K)$$
$$\to \mathbb{H}^2(C^\bullet(E, \varphi)) \to 0.$$

Note that the long exact sequence in Proposition 3.26 immediately implies that there is a canonical identification

$$\mathrm{aut}(E, \Phi) = \mathbb{H}^0(C^\bullet(E, \varphi)).$$

One way of proving the following result is to consider the Kuranishi slice method for constructing the moduli space mentioned in Section 3.3.3.

Proposition 3.27. *Assume that (E, φ) is a stable and simple G-Higgs bundle and that the vanishing $\mathbb{H}^2(C^\bullet(E, \varphi)) = 0$ holds. Then (E, φ) represents a smooth point of the moduli space $\mathcal{M}_d$.*

Remark 3.28. If G is a reductive group that is not necessarily semisimple, one should consider the *reduced deformation complex*, obtained by dividing out by $Z(\mathfrak{g}^\mathbb{C})$; equivalently, this is the deformation complex of the PG-Higgs bundle obtained from the G-Higgs bundle.

3.4.3 The $\mathbb{C}^*$-action and topology of moduli spaces

In order to avoid the problems arising from the presence of singularities, throughout this section we make the assumption that we are in a situation where the moduli space $\mathcal{M}_d$ is smooth.

It is a very important feature of the moduli space of Higgs bundles that it admits an action of the multiplicative group of non-zero complex numbers:

$$\begin{aligned} \mathbb{C}^* \times \mathcal{M}_d &\to \mathcal{M}_d \\ (z, (E, \varphi)) &\mapsto (E, z\varphi). \end{aligned} \tag{3.27}$$

There are two distinct ways of using this action to obtain topological information about the moduli space, as we now explain. However, a theorem of Kirwan ensures that they give essentially equivalent information.

We start by considering a Morse theoretic point of view. For this we use the identification between the Dolbeault moduli space and the moduli space of solutions to the Hitchin equations (3.17) and (3.18) (referred to in Theorem 3.19). Observe that the subgroup $S^1 \subset \mathbb{C}^*$ acts on the moduli space, of solutions to the Hitchin equations. With respect to the (symplectic) Kähler form on the moduli space this action is Hamiltonian and it has a moment map (up to some fixed scaling) given by

$$\begin{aligned} f \colon \mathcal{M}_d &\to \mathbb{R}, \\ (A, \varphi) &\mapsto \|\varphi\|^2 := \int_X |\varphi|^2 \mathrm{vol}. \end{aligned}$$

Hitchin [35] showed, using Uhlenbeck's weak compactness theorem, that f is a proper map. Moreover, it follows from a theorem of Frankel [18] that f is a perfect Bott–Morse function. That $f \colon M \to \mathbb{R}$ is *Bott–Morse* means that its critical points form smooth (connected) submanifolds $N_\lambda \subseteq M$ such that the Hessian of f is non-degenerate along the normal bundle to N_λ in M. That f is *perfect* means that the Poincaré polynomial

$$P_t(M) := \sum t^i \dim H^i(M, \mathbb{Q})$$

can be determined as

$$P_t(M) = \sum_\lambda t^{\text{Index}(N_\lambda)} P_t(N_\lambda); \qquad (3.28)$$

here $\text{Index}(N_\lambda)$ is the *index* of the critical submanifold N_λ, i.e. the real rank of the subbundle of the normal bundle on which the Hessian of f is negative definite.

The condition for f to be a moment map for the Hamiltonian S^1-action on $\mathcal{M}_d$ is

$$\text{grad } f = i\xi,$$

where ξ is the vector field generating the S^1-action. In particular, the critical submanifolds of f are just the components of the fixed locus of the S^1-action. Moreover, if we denote by N_λ^+ the stable manifold of N_λ, we obtain a *Morse stratification*

$$\mathcal{M}_d = \bigcup_\lambda N_\lambda^+.$$

Note that the fact that f is proper and bounded below guarantees that every point in $\mathcal{M}_d$ belongs to one of the N_λ^+.

The more algebraic point of view comes about by looking at the full $\mathbb{C}^*$-action on $\mathcal{M}_d^{\text{Dol}}$. It is a general result of Białynicki-Birula [1] that there is an algebraic stratification defined as follows: let $\{\widetilde{N}_\lambda\}$ be the components of the fixed locus and define

$$\widetilde{N}_\lambda^+ = \{m \in \mathcal{M}_d \mid \lim_{z \to 0} z \cdot m \in \widetilde{N}_\lambda\}.$$

Then the *Białynicki-Birula stratification* is given by

$$\mathcal{M}_d = \bigcup_\lambda \widetilde{N}_\lambda^+.$$

It is perhaps not immediately clear that every point in $\mathcal{M}_d$ lies in one of the $\widetilde{N}_\lambda^+$. It follows, however, from the properness and equivariance (with respect to the suitable weighted $\mathbb{C}^*$-action on the Hitchin base B) of the Hitchin map (3.21).

The whole picture fits into the general setup of $\mathbb{C}^*$-actions on Kähler manifolds arising from Hamiltonian circle actions. In particular, it follows from the results of Kirwan [40] that the Morse and Białynicki-Birula stratifications coincide.

From either point of view, one can now obtain topological information on the moduli space, as pioneered by Hitchin [35] in his calculation of the Poincaré polynomial of the moduli space of rank 2 Higgs bundles. In general, the success of this approach depends crucially on having a good understanding of the topology of the fixed loci N_λ. We remark that the role played by the underlying geometric decomposition of the moduli space is perhaps best brought out by studying in the first place the class of the spaces under study in the K-theory of varieties, and then obtaining from this information such as Hodge and Poincaré polynomials. For examples of this point of view we refer to [12] or [23].

3.4.4 Calculation of Morse indices

Let us consider the fixed points of the circle action on $\mathcal{M}_d$. For simplicity, we start out with an ordinary Higgs vector bundle (E, φ), where E is a vector bundle and $\varphi \in H^0(X, \mathrm{End}(E) \otimes K)$.

The following is easily proved (see [35] or [48]).

Proposition 3.29. *The Higgs bundle (E, φ) is a fixed point of the circle action on $\mathcal{M}_d^{\mathrm{Dol}}$ if and only if it is a* Hodge bundle, *i.e. there is a decomposition*

$$E = E_0 \oplus \ldots E_p$$

and, with respect to this decomposition, φ has weight 1, by which we mean that $\varphi(E_k) \subseteq E_{k+1} \otimes K$.

The basic idea is that the weight k subbundle $E_k \subset E$ is the ik-eigenbundle of the infinitesimal automorphism $\psi = \lim_{\theta \to 0} g(\theta)$ counteracting the circle action, where

$$(E, e^{i\theta}\varphi) = g(\theta) \cdot (E, \varphi).$$

For G-Higgs bundles in general, the simplest procedure is to work out the shape of the Hodge bundles (fixed under the circle action) in each individual case. Note that if the G-Higgs bundle (E, φ) is fixed, then so is the adjoint Higgs vector bundle $(E(\mathfrak{g}^\mathbb{C}), \mathrm{ad}(\varphi))$, and therefore it is a Hodge bundle. Moreover, since the infinitesimal automorphism ψ lies in $E(\mathfrak{h})$, the decomposition of $E(\mathfrak{g}^\mathbb{C})$ in eigenbundles is compatible with the decomposition $E(\mathfrak{g}^\mathbb{C}) = E(\mathfrak{h}^\mathbb{C}) \oplus E(\mathfrak{m}^\mathbb{C})$. It follows that there are decompositions

$$E(\mathfrak{h}^\mathbb{C}) = \bigoplus E(\mathfrak{h}^\mathbb{C})_k,$$
$$E(\mathfrak{m}^\mathbb{C}) = \bigoplus E(\mathfrak{m}^\mathbb{C})_k,$$

and that with respect to these we have

$$\mathrm{ad}(\varphi)\colon E(\mathfrak{h}^{\mathbb{C}})_k \to E(\mathfrak{m}^{\mathbb{C}})_{k+1} \otimes K,$$
$$\mathrm{ad}(\varphi)\colon E(\mathfrak{m}^{\mathbb{C}})_k \to E(\mathfrak{h}^{\mathbb{C}})_{k+1} \otimes K.$$

In particular, the deformation complex of (E, φ) decomposes as

$$C^{\bullet}(E, \varphi) = \bigoplus C_k^{\bullet}(E, \varphi),$$

where the weight k piece of the deformation complex is given by

$$C_k^{\bullet}(E, \varphi)\colon E(\mathfrak{h}^{\mathbb{C}})_k \xrightarrow{[-,\varphi]} E(\mathfrak{m}^{\mathbb{C}})_{k+1} \otimes K. \qquad (3.29)$$

An easy calculation (see, for example, [22] for the case of ordinary parabolic Higgs bundles, which is essentially the same as the present one) now shows the following.

Proposition 3.30. *Let (E, φ) be a stable G-Higgs bundle which is fixed under the circle action and represents a smooth point of the moduli space. With the notation introduced above, we have*

$$\dim N_\lambda^+ = \dim \mathbb{H}^1(C_{\leqslant 0}^{\bullet}), \qquad (3.30)$$
$$\dim N_\lambda^- = \dim \mathbb{H}^1(C_{\geqslant 0}^{\bullet}), \qquad (3.31)$$
$$\dim N_\lambda = \dim \mathbb{H}^1(C_0^{\bullet}). \qquad (3.32)$$

Hence, the Morse index of the critical submanifold N_λ is

$$\mathrm{index}(N_\lambda) = 2\dim \mathbb{H}^1(C_{>0}^{\bullet}).$$

Bott–Morse theory shows that the number of connected components of the moduli space equals that of the subspace of local minima of f. Thus, for the determination of this most basic of topological invariants, it is important to have a convenient criterion for the Morse index to be zero. This is provided by the following result ([3, Prop. 4.14]; see [5, Lem. 3.11] for a corrected proof).

Proposition 3.31. *Let (E, φ) represent a critical point of f. Then (E, φ) represents a local minimum of f if and only if the map*

$$[-, \varphi]\colon E(\mathfrak{h}^{\mathbb{C}}) \to E(\mathfrak{m}^{\mathbb{C}}) \otimes K$$

is an isomorphism for all $k > 0$.

3.4.5 The moduli space of Sp($2n$, $\mathbb{R}$)-Higgs bundles

We end by illustrating how the ideas explained in this section work, by considering the case $G = \text{Sp}(2n, \mathbb{R})$.

Recall that an Sp($2n$, $\mathbb{R}$)-Higgs bundle is a triple (V, β, γ), where $V \to X$ is a rank n vector bundle, $\beta \in H^0(X, S^2V) \otimes K$ and $\gamma \in H^0(X, S^2V^*)$. The topological classification of such bundles is given by $\deg(V) \in \mathbb{Z}$. Denote by $\mathcal{M}_d$ the moduli space of Sp($2n$, $\mathbb{R}$)-Higgs bundles with $\deg(V) = d$.

In the following we outline the application of the Morse theoretic point of view for determining the number of connected components of $\mathcal{M}_d$. We should point out that $\mathcal{M}_d$ is not a smooth variety, so that care must be taken in dealing with singularities in applying the theory. We shall ignore this issue for reasons of space, and in order to bring out more clearly the main ideas. We refer to [20] for full details.

Note that an Sp($2n$, $\mathbb{R}$)-Higgs bundle is in particular an SU(n, n)-Higgs bundle. Hence we have from (3.22) that the Milnor–Wood inequality

$$|d| \leqslant n(g - 1) \tag{3.33}$$

holds. Say that an Sp($2n$, $\mathbb{R}$)-Higgs bundle is *maximal* if equality holds.

Note that taking V to its dual and interchanging β and γ defines an isomorphism $\mathcal{M}_d \cong \mathcal{M}_{-d}$. Hence we shall assume, without loss of generality, that $d \geqslant 0$ for the remainder of this section.

Denote by $N_0 \subset \mathcal{M}_d$ the subspace of local minima of f. In the non-maximal case, Proposition 3.31 leads to the following result.

Proposition 3.32. *Assume that $0 < d < n(g - 1)$. Then the subspace of local minima $N_0 \subset \mathcal{M}_d$ consists of all (V, β, γ) with $\beta = 0$. If $d = 0$, the subspace of local minima $N_0 \subset \mathcal{M}_0$ consists of all (V, β, γ) with $\beta = 0$ and $\gamma = 0$.*

Thus, for $d = 0$, the subspace N_0 can be identified with the moduli space of polystable vector bundles of degree 0. Since this moduli space is known to be connected, we conclude that $\mathcal{M}_0$ is also connected.

For $0 < d < n(g - 1)$, the moduli space $\mathcal{M}_d$ is known to be connected only for $n = 1$ (by the results of Goldman [26], re-proved by Hitchin [35] using Higgs bundles) and for $n = 2$ by García-Prada and Mundet [24] (see also [31]). However, for $n \geqslant 3$, the connectedness of N_0 – and hence $\mathcal{M}_d$ – appears to be difficult to establish.

On the other hand, when $d = n(g - 1)$ is maximal, the complete answer is known from the work of Goldman and Hitchin [26, 35] when $n = 1$, from [30] when $n = 2$, and from [20] when $n \geqslant 3$. It is as follows.

Theorem 3.33. *Let* $\mathcal{M}_{\max}$ *be the moduli space of* $\mathrm{Sp}(2n, \mathbb{R})$-*Higgs bundles* (V, β, γ) *with* $\deg(V) = n(g - 1)$. *Then*

(1) $\#\pi_0\mathcal{M}_{\max} = 2^{2g}$ *for* $n = 1$,
(2) $\#\pi_0\mathcal{M}_{\max} = 3 \cdot 2^{2g} + 2g - 4$ *for* $n = 2$,
(3) $\#\pi_0\mathcal{M}_{\max} = 3 \cdot 2^{2g}$ *for* $n \geqslant 3$.

We end by briefly explaining how this result comes about. Hitchin [37] showed that whenever G is a split real form, the moduli space of G-Higgs bundles has a distinguished component, now known as the *Hitchin component*, which can be concisely described in terms of representations of the fundamental group: it consists of G-Higgs bundles corresponding to representations which factor through a Fuchsian representation of the fundamental group in $\mathrm{SL}(2, \mathbb{R})$, where $\mathrm{SL}(2, \mathbb{R}) \hookrightarrow G$ is embedded as a so-called principal three-dimensional subgroup (when $G = \mathrm{Sp}(2n, \mathbb{R})$ this is just the irreducible representation of $\mathrm{SL}(2n, \mathbb{R})$ on $\mathbb{R}^{2n}$). For every n, the moduli space $\mathcal{M}_{\max}$ has 2^{2g} Hitchin components, which, however, become identified when projected to the moduli space for the projective group $\mathrm{PSp}(2n, \mathbb{R})$.

To explain the appearance of the remaining components, recall the argument used to prove the Milnor–Wood inequality (3.22) in Section 3.3.5. This shows that, for a maximal $\mathrm{Sp}(2n, \mathbb{R})$-Higgs bundle (V, β, γ) (with $d \geqslant 0$), we have an isomorphism

$$\gamma: V \to V^* \otimes K.$$

Hence, because γ is symmetric, V admits a K-valued everywhere non-degenerate quadratic form. Defining $W = V \otimes K^{-n/2}$ and $Q = \gamma \otimes 1_{K^{-n/2}}$, we obtain an $\mathrm{O}(n, \mathbb{C})$-bundle (W, Q), meaning that we obtain new topological invariants defined by the Stiefel–Whitney classes w_1 and w_2 of (W, Q). These then give rise to new subspaces $\mathcal{M}_{w_1,w_2}$, and, using the Morse theoretic approach, one shows that they are in fact connected components. When $n = 2$, even more components appear because, when $w_1 = 0$, there is a reduction to the circle $\mathrm{SO}(2, \mathbb{C}) \subset \mathrm{O}(2, \mathbb{C})$, and this give rise to an integer invariant because $\mathrm{SO}(2) = S^1$.

Remark 3.34. These new invariants have been studied (and generalized) from the point of view of surface group representations in the work of Guichard and Wienhard [32].

Remark 3.35. Let (W, Q) be the $\mathrm{O}(n, \mathbb{C})$-bundle arising from a maximal $\mathrm{Sp}(2n, \mathbb{R})$-Higgs bundle and define $\theta = (\beta \otimes 1_{K^{n/2}}) \circ Q: W \to W \otimes K^2$. Then $((W, Q), \theta)$ is a $\mathrm{GL}(n, \mathbb{R})$-Higgs bundle, except for the fact the twisting is by the square of the canonical bundle rather than the canonical bundle itself. This observation is the beginning of an interesting story known as the "Cayley correspondence"; for more on this, we refer the reader to [4] and [44].

References

[1] A. Białynicki-Birula, Some theorems on actions of algebraic groups, *Ann. of Math. (2)* **98** (1973), 480–497. MR 0366940 (51 #3186).

[2] I. Biswas and S. Ramanan, An infinitesimal study of the moduli of Hitchin pairs, *J. London Math. Soc. (2)* **49** (1994), 219–231.

[3] S. B. Bradlow, O. García-Prada, and P. B. Gothen, Surface group representations and U(p, q)-Higgs bundles, *J. Differential Geom.* **64** (2003), 111–170.

[4] S. B. Bradlow, O. García-Prada, and P. B. Gothen, *Maximal surface group representations in isometry groups of classical Hermitian symmetric spaces*, Geometriae Dedicata **122** (2006), 185–213.

[5] S. B. Bradlow, O. García-Prada, and P. B. Gothen, Homotopy groups of moduli spaces of representations, *Topology* **47** (2008), 203–224.

[6] S. B. Bradlow, O. García-Prada, and P. B. Gothen, Deformations of maximal representations in Sp(4, $\mathbb{R}$), *Quart. J. Math.* **63** (2012), 795–843.

[7] S. B. Bradlow, O. García-Prada, and I. Mundet i Riera, Relative Hitchin-Kobayashi correspondences for principal pairs, *Quart. J. Math.* **54** (2003), 171–208.

[8] M. Burger, A. Iozzi, and A. Wienhard, Surface group representations with maximal Toledo invariant, *C. R. Math. Acad. Sci. Paris* **336** (2003), no. 5, 387–390.

[9] M. Burger, A. Iozzi, and A. Wienhard, Higher Teichmüller spaces: from SL(2, $\mathbb{R}$) to other Lie groups, In *Handbook of Teichmüller Theory*, A. Papadopoulos, ed. Strasbourg: IRMA arXiv: 1004.2894 [math.GT] (in press).

[10] M. Burger, A. Iozzi, and A. Wienhard, Surface group representations with maximal Toledo invariant, *Ann. of Math. (2)* **172** (2010), no. 1, 517–566.

[11] A. Casimiro and C. Florentino, Stability of affine g-varieties and irreducibility in reductive groups, *Int. J. Math* **23** (2012).

[12] W.-Y. Chuang, D.-E. Diaconescu, and G. Pan, Wallcrossing and cohomology of the moduli space of Hitchin pairs, *Commun. Number Theor. Phys.* **5** (2011), 1–56.

[13] K. Corlette, Flat G-bundles with canonical metrics, *J. Differential Geom.* **28** (1988), 361–382.

[14] A. Domic and D. Toledo, The Gromov norm of the Kaehler class of symmetric domains, *Math. Ann.* **276** (1987), 425–432.

[15] S. K. Donaldson, Twisted harmonic maps and the self-duality equations, *Proc. London Math. Soc. (3)* **55** (1987), 127–131.

[16] J. L. Dupont, *Bounds for Characteristic Numbers of Flat Bundles*, Springer LNM vol. 763(1978). Berlin: Springer, pp. 109–119.

[17] V. V. Fock and A. B. Goncharov, Moduli spaces of local systems and higher Teichmuller theory, *Publ. Math. Inst. Hautes Études Sci.* **103** (2006), 1–211.

[18] T. Frankel, Fixed points and torsion on Kähler manifolds, *Ann. of Math. (2)* **70** (1959), 1–8.

[19] O. García-Prada, Higgs bundles and surface group representations. In *Moduli Spaces and Vector Bundles*, L. Brambila-Paz, S. B. Bradlow, O. Gracía-Prada, and S. Ramanan, eds. London Mathematical Society Lecture Note Series 359. Cambridge: Cambridge University Press (2009), pp. 265–310.

[20] O. García-Prada, P. B. Gothen, and I. Mundet i Riera, Higgs bundles and surface group representations in the real symplectic group, *J. Topology* **6** (2013), 64–118.

[21] O. García-Prada, P. B. Gothen, and I. Mundet-Riera, *The Hitchin-Kobayashi correspondence, Higgs pairs and surface group representations*, preprint, 2012, `arXiv:0909.4487v3 [math.AG]`.

[22] O. García-Prada, P. B. Gothen, and V. Muñoz, Betti numbers of the moduli space of rank 3 parabolic Higgs bundles, *Mem. Am. Math. Soc.* **187** (2007), no. 879.

[23] O. García-Prada, J. Heinloth, and A. Schmitt, On the motives of moduli of chains and Higgs bundles, *J. Europ. Math. Soc.*, `arXiv:1104.5558v1 [math.AG]` (in press).

[24] O. García-Prada and I. Mundet i Riera, Representations of the fundamental group of a closed oriented surface in Sp(4, $\mathbb{R}$), *Topology* **43** (2004), 831–855.

[25] W. M. Goldman, *Representations of Fundamental Groups of Surfaces*, Springer LNM 1167. 1985, Berlin: Springer, pp. 95–117.

[26] W. M. Goldman, Topological components of spaces of representations, *Invent. Math.* **93** (1988), 557–607.

[27] W. M. Goldman, Higgs bundles and geometric structures on surfaces. In *The Many Facets of Geometry*, O. García-Prada, J. P. Bourguignon, and S. Salamon, eds. Oxford: Oxford University Press, 2010, pp. 129–163.

[28] W. M. Goldman, Locally homogeneous geometric manifolds, in *Proceedings of the International Congress of Mathematicians 2010* R. Bhatia, A. Pal, G. Rangarajan, V. Srinivas, and M. Vanninathan, eds. Singapore: World Scientific (2011).

[29] William M. Goldman, The symplectic nature of fundamental groups of surfaces, *Adv. in Math.* **54** (1984), 200–225.

[30] P. B. Gothen, Components of spaces of representations and stable triples, *Topology* **40** (2001), 823–850.

[31] P. B. Gothen and A. Oliveira, Rank two quadratic pairs and surface group representations, *Geometriae Dedicata* (2012), first published online: 16 March 2012. doi:10.1007/s10711-012-9709-1.

[32] O. Guichard and A. Wienhard, Topological invariants of Anosov representations, *J. Topol.* **3** (2010), no. 3, 578–642.

[33] T. Hausel, Global topology of the Hitchin system. In *Handbook of Moduli: Vol. II* G. Farkas and I. Morrison, eds. Advanced Lectures in Mathematics 25. Somerville, MA: International Press, 2013, pp. 29–70.

[34] L. Hernández, Maximal representations of surface groups in bounded symmetric domains, *Trans. Amer. Math. Soc.* **324** (1991), 405–420.

[35] N. J. Hitchin, The self-duality equations on a Riemann surface, *Proc. London Math. Soc. (3)* **55** (1987), 59–126.

[36] N. J. Hitchin, Stable bundles and integrable systems, *Duke Math. J.* **54** (1987), 91–114.

[37] N. J. Hitchin, Lie groups and Teichmüller space, *Topology* **31** (1992), 449–473.

[38] I. Mundet i Riera, A Hitchin-Kobayashi correspondence for Kähler fibrations, *J. Reine Angew. Math.* **528** (2000), 41–80.

[39] A. Kapustin and E. Witten, Electric-magnetic duality and the geometric Langlands program, *Commun. Number Theor. Phys.* **1** (2007), 1–236.

[40] F. Kirwan, *Cohomology of Quotients in Symplectic and Algebraic Geometry*, Mathematical Notes, vol. 31. Princeton, NJ: Princeton University Press (1984).

[41] F. Labourie, Existence d'applications harmoniques tordues à valeurs dans les variétés à courbure négative, *Proc. Am. Math. Soc.* **111** (1991), no. 3, 877–882.

[42] J. W. Milnor, On the existence of a connection with curvature zero, *Comment. Math. Helv.* **32** (1958), 216–223.

[43] M. S. Narasimhan and C. S. Seshadri, Stable and unitary vector bundles on a compact Riemann surface, *Ann. Math.* **82** (1965), 540–567.

[44] Roberto Rubio Núñez, Higgs bundles and Hermitian symmetric spaces, unpublished Ph.D. thesis, ICMAT Universidad Autónoma de Madrid (2012).

[45] A. Ramanathan, Stable principal bundles on a compact Riemann surface, *Math. Ann.* **213** (1975), 129–152.

[46] A. Schmitt, *Geometric Invariant Theory and Decorated Principal Bundles*, Zürich Lectures in Advanced Mathematics. Zurich: European Mathematical Society (2008).

[47] C. T. Simpson, Constructing variations of Hodge structure using Yang-Mills theory and applications to uniformization, *J. Am. Math. Soc.* **1** (1988), 867–918.

[48] C. T. Simpson, Higgs bundles and local systems, *Inst. Hautes Études Sci. Publ. Math.* **75** (1992), 5–95.

[49] D. Toledo, Harmonic maps from surfaces to certain Kaehler manifolds, *Math. Scand.* **45** (1979), 13–26.

[50] D. Toledo, Representations of surface groups in complex hyperbolic space, *J. Differential Geom.* **29** (1989), 125–133.

[51] V. G. Turaev, A cocycle of the symplectic first Chern class and the Maslov index, *Funct. Anal. Appl.* **18** (1984), 35–39.

4

Introduction to stability conditions

D. Huybrechts

Abstract

These are notes of a course given at the 'School on Moduli Spaces' at the Newton Institute in January 2011. The abstract theory of stability conditions (due to Bridgeland and Douglas) on abelian and triangulated categories is developed via tilting and t-structures. Special emphasis is put on the bounded derived category of coherent sheaves on smooth projective varieties (in particular for curves and K3 surfaces). The lectures were targeted at an audience with little prior knowledge of triangulated categories and stability conditions but with a keen interest in vector bundles on curves.

Contents

The title of the actual lecture course also mentioned derived categories prominently. And indeed, the original idea was to give an introduction to the basic techniques used to study $D^b(X)$, the bounded derived category of

Moduli Spaces, eds. L. Brambila-Paz, O. García-Prada, P. Newstead and R. Thomas. Published by Cambridge University Press. © Cambridge University Press 2014.

coherent sheaves on a (smooth projective) variety X, and, at the same time, to acquaint the audience with the slightly technical notion of stability conditions on $D^b(X)$ as invented by Bridgeland following work of Douglas. The lectures were delivered in this spirit and this writeup tries to reflect the actual lectures, but the emphasis has been shifted towards stability conditions considerably. It seemed worthwhile to spend most of the lectures just on stability conditions and to present some of the arguments used to study this new notion in detail.

These notes are meant to be a gentle introduction to stability conditions and not as a survey of the area, although we collect a few pointers to the literature in the last section. We start out by recalling stability for vector bundles on curves and slowly move to the more abstract version provided by stability conditions on abelian and triangulated categories. Roughly, a stability condition on a triangulated category can be thought of as a refinement of a bounded t-structure and we shall explain this relation carefully. Bridgeland endows the space of all stability conditions with a natural topology. This gives rise to a completely new kind of moduli space which has been much studied over the last years. The ultimate hope is that a good understanding of the space of stability conditions leads to a better grip on the category itself. The best example for this is an intriguing conjecture of Bridgeland describing the group of autoequivalences of the derived category of a K3 surface as a fundamental group of an explicit 'period domain' for the space of stability conditions.

Stability conditions on $D^b(X)$ will be studied for X smooth and projective of dimension one or two, but we will not touch upon the many results for X only quasi-projective or for more algebraic categories coming from quiver representations and there are many more results and aspects that are not covered by these lectures, e.g. wall crossing phenomena. Originally, stability conditions were invented in order to study $D^b(X)$ for projective Calabi–Yau threefolds, but up to this date and in spite of many attempts, not a single stability condition has been constructed in this situation. A glance at the discussion in the case of surfaces quickly shows why this is so complicated, but see [4, 5] for attempts in this direction.

The material covered in these lectures is based almost entirely on the two articles [9] and [10] by Bridgeland. Frequently, we add examples, hint at related results and highlight certain aspects, but we also take the liberty to leave out unpleasant technical points of the discussion. Only Section 4.5 contains material that is not completely covered by the existing literature. Here, the conjecture of Bridgeland is rephrased in terms of classical moduli stacks and their fundamental groups.

Acknowledgements: I wish to thank the organizers of the 'School on Moduli Spaces' at the Newton Institute for inviting me and the audience for a stimulating and demanding atmosphere. Parts of the material were also used for lectures on stability conditions at Peking University in the fall of 2006 and at Ann Arbor in the spring of 2011. I would like to thank both institutions for their hospitality. I am grateful to Pawel Sosna and the referee for a careful reading of the first version and the many comments.

4.1 Torsion theories and t-structures

This first lecture begins in Section 4.1.1 with a review of stability for vector bundles on algebraic curves which we will rephrase in terms of phases in order to motivate the notion of a stability condition. Similarly, we first give examples for decomposing the abelian category $\mathrm{Coh}(C)$ of coherent sheaves on a curve C and turn this later, in Section 4.1.2, into the abstract concept of a torsion theory for abelian categories. Its triangulated counterpart, t-structures, will be recalled as well, cf. Section 4.1.3. The final part of the first lecture is devoted to the interplay between torsion theories and t-structures via tilting.

4.1.1 μ-stability on curves (and surfaces): recollections

Consider a smooth projective curve C over an algebraically closed field $k = \bar{k}$. Let $\mathrm{Coh}(C)$ denote the abelian category of coherent sheaves on C, which we will consider with its natural k-linear structure.

Recall that a coherent sheaf $E \in \mathrm{Coh}(C)$ is called μ-*stable* (resp. μ-*semistable*) if E is torsion free (i.e. locally free) and for all proper subsheaves $0 \neq F \subset E$ one has $\mu(F) < \mu(E)$ (resp. $\mu(F) \leq \mu(E)$). Here, $\mu(\ ) = \frac{\deg(\ )}{\mathrm{rk}(\ )}$ is the slope. We shall rewrite this in terms of a stability function which is better suited for the more general notion of stability conditions on abelian or even triangulated categories.

Define,

$$Z(E) := -\deg(E) + i \cdot \mathrm{rk}(E).$$

Then, the *phase* $\phi(E) \in (0, 1]$ of a sheaf $0 \neq E$ is defined uniquely by the condition

$$Z(E) \in \exp(i\pi\phi(E)) \cdot \mathbb{R}_{>0}.$$

The *stability function* Z defines a map

$$Z : \mathrm{Coh}(C) \setminus \{0\} \to \overline{\mathbb{H}} := \mathbb{H} \cup \mathbb{R}_{<0}.$$

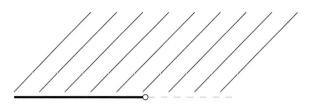

Warning: Later, the phase of an object in a triangulated category is only well defined if the object is semistable or at least contained in the heart of the associated t-structure.

Remark 4.1. i) If $\text{rk}(E) > 0$, e.g. when E is locally free, then $Z(E) \in \mathbb{H}$. More precisely, $Z(E) \in \mathbb{R}_{<0}$ if and only if E is torsion. In particular, $Z(k(x)) = -1$.

ii) Note that Z is additive, i.e. $Z(E_2) = Z(E_1) + Z(E_3)$ for any short exact sequence $0 \to E_1 \to E_2 \to E_3 \to 0$. Thus, it factorizes over the *Grothendieck group* $K(C) = K(\text{Coh}(C))$ of the abelian category $\text{Coh}(C)$, i.e. $Z : \text{Coh}(C) \to K(C) \to \mathbb{C}$. The map $K(C) \to \mathbb{C}$ is an additive group homomorphism.

The following easy observation is important for motivating the notion of a stability condition later on.

Lemma 4.2. *Suppose $E \in \text{Coh}(C)$ is locally free. Then E is μ-stable if and only if for all proper subsheaves $0 \neq F \subset E$ the following inequality of phases holds true:*

$$\phi(F) < \phi(E). \tag{4.1}$$

Proof. Since E is locally free (and thus all non-trivial subsheaves $F \subset E$ are), we can divide by the rank. Thus, $\frac{Z(E)}{\text{rk}(E)} = -\mu(E) + i$ and then $\mu(F) < \mu(E)$ if and only if $-\mu(E) < -\mu(F)$ if and only if $\phi(E) > \phi(F)$.

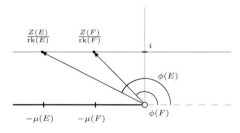

We leave it to the reader to rephrase μ-semistability as a weak inequality for phases.

Exercise 4.3. What happens if we forget about the assumption that E is locally free? Clearly, (4.1) for arbitrary $E \in \text{Coh}(C)$ is equivalent to E being either a μ-stable sheaf (and by definition in particular locally free) or $E \simeq k(x)$ for some closed point $x \in C$. The weak form of (4.1) is equivalent to E being either a μ-semistable sheaf (and in particular locally free) or a torsion sheaf.

Thus, it seems natural to define (semi)stability (instead of μ-(semi)stability) in the abelian category $\text{Coh}(C)$ in terms of the (weak) inequality (4.1). It allows us to treat vector bundles and torsion sheaves on the same footing. So from now on: ***Use phases rather than slopes***.

Exercise 4.4. Observe the useful formulae: $\mu(E) = -\cot(\pi \phi(E))$ and $\pi \phi(E) = \text{arcot}(-\mu(E))$.

Let us continue with the review of the classical theory of stable vector bundles on curves. The next step consists of establishing the existence of *Harder–Narasimhan* and *Jordan–Hölder* filtrations which filter any given sheaf such that the quotients are semistable resp. stable.
Every $E \in \text{Coh}(C)$ admits a unique (Harder–Narasimhan) filtration $0 = E_0 \subset E_1 \subsetneq E_2 \subsetneq \ldots \subsetneq E_n = E$ such that the quotients $A_1 := E_1/E_0, \ldots A_n = E_n/E_{n-1}$ are semistable sheaves of phase $\phi_1 > \ldots > \phi_n$.

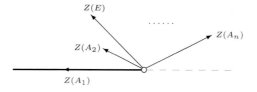

Thus, E_1 is the torsion of E, which might be trivial. To simplify notations, we implicitly allow $E_1 = 0$, although ϕ_1 would not be well defined in this case. But the other inclusions are strict. The A_i are called the *semistable factors* of E. They are unique.
As a refinement of the Harder–Narasimhan filtration, one can also construct a (Jordan–Hölder) filtration $0 = E_0 \subsetneq E_1 \subsetneq E_2 \subsetneq \ldots \subsetneq E_n = E$ of any $E \in \text{Coh}(C)$. This time, the quotients $A_1 := E_1/E_0, \ldots, A_n = E_n/E_{n-1}$ are stable sheaves of phase $\phi_1 \geq \ldots \geq \phi_n$.
Note that for E not locally free, the first few A_i are of the form $k(x_i)$, $x_i \in C$. The A_i are called the *stable factors* of E. The Jordan–Hölder filtration is in general not unique and the stable factors are unique only up to permutation.

Exercise 4.5. Here are a few principles that hold true in the general context of stability conditions on abelian or triangulated categories with identical proofs.

i) If $E, F \in \mathrm{Coh}(C)$ are semistable (resp. stable) such that $\phi(E) > \phi(F)$ (resp. $\phi(E) \geq \phi(F)$), then

$$\mathrm{Hom}(E, F) = 0.$$

ii) If $E, F \in \mathrm{Coh}(C)$ are stable such that $\phi(E) \geq \phi(F)$, then

$$\text{either } E \simeq F \text{ or } \mathrm{Hom}(E, F) = 0.$$

iii) If E is stable, then $\mathrm{End}(E) \simeq k$. (Recall $k = \bar{k}$.)

Exercise 4.6. If E is semistable, then all stable factors A_i of E have the same phase $\phi(E)$. Suppose $\mathrm{Hom}(A_{i_0}, E) \neq 0$ for some stable factor A_{i_0} of E, then there exists a short exact sequence

$$0 \to E' \to E \to E'' \to 0$$

with E', E'' semistable of phase $\phi(E)$, such that all stable factors of E' are isomorphic to A_{i_0} and $\mathrm{Hom}(E', E'') = 0$ (or, equivalently, $\mathrm{Hom}(A_{i_0}, E'') = 0$). Note that $E'' = 0$ if and only all A_i are isomorphic to A_{i_0}.

Definition 4.7. For $\phi \in (0, 1]$ one defines

$$\mathcal{P}(\phi) := \{E \in \mathrm{Coh}(C) \mid \text{semistable of phase } \phi\},$$

which we will consider as a full linear subcategory of $\mathrm{Coh}(C)$.

Note that $\mathcal{P}(1) \subset \mathrm{Coh}(C)$ is the full subcategory of torsion sheaves. More generally, $\mathcal{P}(\phi) \subset \mathrm{Coh}(C)$ are full abelian subcategories of *finite length*, i.e. ascending and descending chains of subobjects stabilize. In fact, the *minimal*[1] objects in $\mathcal{P}(\phi)$, i.e. those that do not contain any proper subobjects in $\mathcal{P}(\phi)$, are exactly the stable sheaves of phase ϕ. The finite length of $\mathcal{P}(\phi)$ is thus a consequence of the existence of the finite(!) Jordan–Hölder filtration. In the general context, the finite length condition is tricky. Here, it follows easily as the stability function $Z = -\deg + i \cdot \mathrm{rk}$ is rational (cf. Proposition 4.45).

For any interval $I \subset (0, 1]$ one defines $\mathcal{P}(I)$ as the full subcategory of sheaves $E \in \mathrm{Coh}(C)$ with (semi)stable factors A_i having phase $\phi(A_i)$ in I. They are only additive subcategories of $\mathrm{Coh}(C)$ in general.

---

[1] In representation theory, they would be called *simple*, but this has another meaning for sheaves. Also note that the minimal objects in the category $\mathrm{Coh}(C)$ are the point sheaves $k(x)$, $x \in C$. In particular, $\mathrm{Coh}(C)$ is not of finite length.

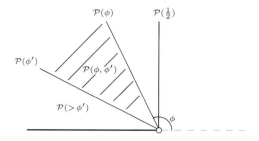

The next remark will lead us naturally to the notion of a torsion theory.

Remark 4.8. i) Let us look at the following special case: For $\varphi \in (0, 1]$ consider $\mathcal{T}_\varphi := \mathcal{P}(\varphi, 1]$ and $\mathcal{F}_\varphi := \mathcal{P}(0, \varphi]$. Then for all $E \in \text{Coh}(C)$ there exists a unique short exact sequence

$$0 \to E' \to E \to E'' \to 0$$

with $E' \in \mathcal{T}_\varphi$ and $E'' \in \mathcal{F}_\varphi$. The uniqueness follows from the observation that $\text{Hom}(\mathcal{T}_\varphi, \mathcal{F}_\varphi) = 0$, i.e. there are no non-trivial homomorphisms from any object in $\mathcal{T}_\varphi$ to any object in $\mathcal{F}_\varphi$.

ii) Similarly, if we let $\mathcal{T} := \mathcal{P}(1)$ and $\mathcal{F} := \mathcal{P}(0, 1)$, one obtains as above a short exact sequence with E' being the torsion of E.

Remark 4.9. What happens if we pass from curves to surfaces? So let S be a smooth projective surface with an ample (or big and nef) divisor H viewed as an element of the real vector space $\text{NS}(S) \otimes \mathbb{R}$. Then $\deg_H(E) = (c_1(E).H)$ is well-defined and $\mu_H(E) = \deg_H(E)/\text{rk}(E)$ can be used to define μ-stability for torsion free sheaves. A priori, one could now try to play the same game with the function $Z_H = -\deg_H + i \cdot \text{rk}$, but there are immediate problems: i) $Z_H(E) = 0$ for sheaves supported in dimension zero, so Z_H takes values in $\overline{\mathbb{H}} \cup \{0\}$. ii) There is no Jordan–Hölder filtration for sheaves of dimension one, i.e. those supported on the curve. The rank (on S) of such a sheaf is zero and its degree only reflects its rank as a sheaf on its support. Thus, for a curve $C \subset S$ all bundles on C would be semistable of phase 1 on S. (But see [35], where it is shown that the quotient of $\text{Coh}(S)$ by the subcategory of 0-dimensional sheaves together with Z_H behaves almost like $\text{Coh}(C)$.)

Although the situation seems more complicated than for curves, it has one interesting feature that is not present in the case of curves: The function Z_H itself depends on a parameter, namely $H \in \text{NS}(S) \otimes \mathbb{R}$. We will come back to this later.

Since it is known that on higher dimensional varieties one should rather work with Gieseker stability than μ-stability, i.e. taking the full Hilbert polynomial into account for defining stability, one could try to adapt the approach here accordingly (sce [16]), which taken literally leads to a theory in which the stability function will take values in a higher dimensional space and not simply in $\mathbb{C} = \mathbb{R}^2$.

4.1.2 Torsion theories in abelian categories

The abstract notion of a torsion theory has been first introduced by Dickson in [14], but was implicitly already present in earlier work of Gabriel. The two standard references are [17, 3]. In the following, we denote by $\mathcal{A}$ an abelian category. Usually, $\mathcal{A}$ is linear over some field k, but this will not be important for now.

The following notion is the abstract version of the two examples in Remark 4.8.

Definition 4.10. A *torsion theory* (or *torsion pair*) for $\mathcal{A}$ consists of a pair of full subcategories $\mathcal{T}, \mathcal{F} \subset \mathcal{A}$ such that:
  i) $\mathrm{Hom}(\mathcal{T}, \mathcal{F}) = 0$ and
  ii) For any $E \in \mathcal{A}$ there exists a short exact sequence

$$0 \to E' \to E \to E'' \to 0 \qquad\qquad (4.2)$$

with $E' \in \mathcal{T}$ and $E'' \in \mathcal{F}$.

For obvious reasons, one calls the (in general only additive) subcategories $\mathcal{T} \subset \mathcal{A}$ and $\mathcal{F} \subset \mathcal{A}$ the torsion part resp. torsion free part of the torsion theory.

Exercise 4.11. i) The short exact sequence (4.2) is unique. For this, one uses $\mathrm{Hom}(\mathcal{T}, \mathcal{F}) = 0$.
  ii) The inclusion $\mathcal{T} \subset \mathcal{A}$ admits a right adjoint $T_{\mathcal{T}} : \mathcal{A} \to \mathcal{T}$ given by mapping $E \in \mathcal{A}$ to its torsion part $T_{\mathcal{T}}(E) := E'$ in (4.2), i.e. for all $T \in \mathcal{T}$ one has $\mathrm{Hom}_{\mathcal{A}}(T, E) = \mathrm{Hom}_{\mathcal{T}}(T, T_{\mathcal{T}}(E))$.
  iii) Similarly, $\mathcal{F} \subset \mathcal{A}$ admits the left adjoint $\mathcal{A} \to \mathcal{F}$, $E \mapsto E/T_{\mathcal{T}}(E)$, i.e. $\mathrm{Hom}_{\mathcal{A}}(E, F) = \mathrm{Hom}_{\mathcal{F}}(E/T_{\mathcal{T}}(E), F)$ for all $F \in \mathcal{F}$.
  iv) The subcategories $\mathcal{T}, \mathcal{F} \subset \mathcal{A}$ are closed under extensions. Moreover, $\mathcal{T}$ is closed under quotients and $\mathcal{F}$ is closed under subobjects.
  v) If $\mathcal{T}^\perp$ is defined as the full subcategory of all objects $E \in \mathcal{A}$ such that $\mathrm{Hom}(T, E) = 0$ for all $T \in \mathcal{T}$. Then $\mathcal{F} = \mathcal{T}^\perp$. Similarly, $\mathcal{T} = {}^\perp\mathcal{F}$.

Remark 4.12. In the same manner, one can prove the following useful fact (see e.g. [3, Prop. 1.2]): For a full additive subcategory $\mathcal{T} \subset \mathcal{A}$, which is closed under isomorphisms, $(\mathcal{T}, \mathcal{T}^\perp)$ defines a torsion pair if and only if $\mathcal{T} \subset \mathcal{A}$ admits a right adjoint $T_{\mathcal{T}} : \mathcal{A} \to \mathcal{T}$ and $\mathcal{T}$ is closed under right exact sequences (i.e. if $T_1 \to E \to T_2 \to 0$ is exact with $T_1, T_2 \in \mathcal{T}$, then also $E \in \mathcal{T}$).

In Section 4.1.4 we shall explain how to tilt an abelian category with respect to a given torsion theory. This is a construction that takes place naturally within the bounded derived category. For a more direct construction not using derived categories, see Noohi's article [38]. He associates to a torsion theory $(\mathcal{T}, \mathcal{F})$ for an abelian category directly a new abelian category $\mathcal{B}$ the objects of which are of the form $[\varphi : E^{-1} \to E^0]$ with $\ker(\varphi) \in \mathcal{F}$ and $\mathrm{coker}(\varphi) \in \mathcal{T}$. The definition of morphisms in $\mathcal{B}$ is a little more involved but still very explicit. It is also shown in [38] that $\mathcal{B}$ describes a category that is equivalent to the Happel–Reiten–Smalø tilt to be discussed below.

4.1.3 t-structures on triangulated categories

In the following we shall denote by $\mathcal{D}$ a triangulated category, e.g. $\mathcal{D}$ could be the bounded derived category $\mathrm{D}^b(\mathcal{A})$ of an abelian category $\mathcal{A}$. A triangulated category $\mathcal{D}$ is an additive (but not abelian!) category endowed with two additional structures: The *shift functor*, an equivalence $\mathcal{D} \xrightarrow{\sim} \mathcal{D}$, $E \mapsto E[1]$, and a collection of *exact triangles* $E \to F \to G \to E[1]$ replacing short exact sequences in abelian categories. They are subject to axioms TR 1-4, see e.g. [15, 19, 26, 37, 45]. E.g. one requires that with $E \to F \to G \to E[1]$ also $F \to G \to E[1] \to F[1]$ is an exact triangle. Another consequence of the axioms is that for any such exact triangle one obtains long exact sequences

$$\mathrm{Hom}^{i-1}(A, G) \to \mathrm{Hom}^i(A, E) \to \mathrm{Hom}^i(A, F)$$

and

$$\mathrm{Hom}^i(F, A) \to \mathrm{Hom}^i(E, A) \to \mathrm{Hom}^{i+1}(G, A).$$

Here, $\mathrm{Hom}^i(A, B) := \mathrm{Hom}(A, B[i]) =: \mathrm{Ext}^i(A, B)$.

In the following, we shall often simply write $E \to F \to G$ for an exact triangle $E \to F \to G \to E[1]$.

Definition 4.13. A *t-structure* on a triangulated category $\mathcal{D}$ consists of a pair of full additive subcategories $(\mathcal{T}, \mathcal{F})$ such that:
   i) $\mathcal{F} = \mathcal{T}^\perp$.

ii) For all $E \in \mathcal{D}$ there exists an exact triangle

$$E' \to E \to E'' \qquad (4.3)$$

with $E' \in \mathcal{T}$ and $E'' \in \mathcal{F}$.

iii) $\mathcal{T}[1] \subset \mathcal{T}$.

The first two conditions are reminiscent of the definition of a torsion theory for abelian categories and we will comment on this relation below. The last condition takes the triangulated structure into account. Note that we do not require $\mathcal{T}[1] = \mathcal{T}$.[2]

To stress the analogy to torsion theories, we used the notation $\mathcal{T}$ and $\mathcal{F}$. More commonly however, one writes $\mathcal{D}^{\leq 0} := \mathcal{T}$ and $\mathcal{D}^{\geq 1} = \mathcal{D}^{>0} := \mathcal{F}$. With $\mathcal{D}^{\leq -i} := \mathcal{D}^{\leq 0}[i]$ one gets

$$\ldots \subset \mathcal{D}^{\leq -2} \subset \mathcal{D}^{\leq -1} \subset \mathcal{D}^{\leq 0} \subset \mathcal{D}^{\leq 1} \subset \ldots.$$

As for torsion theories, the inclusion $\mathcal{D}^{\leq 0} \subset \mathcal{D}$ has a right adjoint $\tau^{\leq 0} : \mathcal{D} \to \mathcal{D}^{\leq 0}$, $E \mapsto \tau^{\leq 0} E := E'$ (as in (4.3)). Similarly, $\tau^{\geq 1} : \mathcal{D} \to \mathcal{D}^{\geq 1}$ defines a left adjoint to the inclusion $\mathcal{D}^{\geq 1} \subset \mathcal{D}$. Thus, (4.3) can be written as an exact triangle

$$\tau^{\leq 0} E \to E \to \tau^{\geq 1} E.$$

The *heart* of a t-structure is defined as

$$\mathcal{A} := \mathcal{D}^{\leq 0} \cap \mathcal{D}^{\geq 0},$$

which in the notation of Definition 4.13 is $\mathcal{A} = \mathcal{T} \cap \mathcal{F}[1]$. It is an abelian category and short exact sequences in $\mathcal{A}$ are precisely the exact triangles in $\mathcal{D}$ with objects in $\mathcal{A}$. Moreover, one has cohomology functors $H^i : \mathcal{D} \to \mathcal{A}$, $E \mapsto (\tau^{\geq i} \tau^{\leq i} E)[i]$.

A t-structure on a triangulated category $\mathcal{D}$ is *bounded* if every $E \in \mathcal{D}$ is contained in $\mathcal{D}^{\leq n} \cap \mathcal{D}^{\geq -n}$ for $n \gg 0$.

Example 4.14. For $\mathcal{D} = D^b(\mathcal{A})$ the standard bounded t-structure is given by $\mathcal{D}^{\leq 0} := \{E \mid H^i(E) = 0 \; i > 0\}$. Then, $\mathcal{D}^{\geq 0} = \{E \mid H^i(E) = 0 \; i < 0\}$ and its heart is the original abelian category $\mathcal{A} \subset \mathcal{D}$ in degree zero.

Remark 4.15. i) Not every triangulated category $\mathcal{D}$ admits a bounded t-structure. In fact, the existence of a bounded t-structure on $\mathcal{D}$ implies that $\mathcal{D}$ is *idempotent closed* (or, *Karoubian*). This is a folklore result (see [29]).

---

[2] Note that $\mathcal{T} = \mathcal{D}$ and $\mathcal{F} = 0$ defines a t-structure, which of course satisfies $\mathcal{T}[1] = \mathcal{T}$. But it is neither bounded nor very useful.

Recall that a morphism $e : E \to E$ in $\mathcal{D}$ is idempotent if $e^2 = e$. An idempotent morphism $e : E \to E$ is split if $e = b \circ a : E \to F \to E$ with $a \circ b = \mathrm{id}$ or, equivalently, if $E \simeq F \oplus F'$ and e is the composition of the natural projection and inclusion $E \twoheadrightarrow F \hookrightarrow E$. (For the equivalence of the two descriptions the triangulated structure is needed.) The category $\mathcal{D}$ is called idempotent closed if every idempotent is split. For a given t-structure on $\mathcal{D}$ any idempotent $e : E \to E$ yields idempotent morphisms $\tau^{\leq 0} e : \tau^{\leq 0} E \to \tau^{\leq 0} E$ and $\tau^{>0} e : \tau^{>0} E \to \tau^{>0} E$. One shows that e is split if $\tau^{\leq 0} e$ and $\tau^{>0} e$ are split. For a bounded t-structure this allows one to reduce to objects in the (shifted) heart $\mathcal{A}$. But clearly, due to the existence of kernels and cokernels in abelian categories, any idempotent in $\mathcal{A}$ splits.

ii) Suppose $\mathcal{D}' \subset \mathcal{D}$ is a triangulated subcategory. It is called *dense* if every object $F \in \mathcal{D}$ is a direct summand of an object in $\mathcal{D}'$, i.e. there exists $F' \in \mathcal{D}$ with $E \simeq F \oplus F' \in \mathcal{D}'$. If $F \notin \mathcal{D}'$, then the induced natural idempotent $E \to E$ splits only in $\mathcal{D}$, but not in $\mathcal{D}'$. For a concrete example, take the smallest full triangulated subcategory $\mathcal{D}' \subset D^{\mathrm{b}}(\mathrm{Coh}(C))$ that contains all line bundles $\mathcal{O}(np)$, where C is a curve of genus > 0 and $p \in C$ is a fixed closed point. In particular, $\mathcal{D}'$ is dense in $\mathcal{D}$ but itself not idempotent closed.

Remark 4.16. For the following see [9, Lemma 3.2]. If $\mathcal{D}$ is a triangulated category and $\mathcal{A} \subset \mathcal{D}$ is a full additive subcategory. Then $\mathcal{A}$ is the heart of a bounded t-structure if and only if

i) $\mathrm{Hom}(\mathcal{A}[k_1], \mathcal{A}[k_2]) = 0$ for $k_1 > k_2$ and

ii) For any $E \in \mathcal{D}$ there exists a diagram

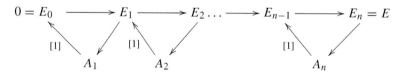

where $E_{i-1} \to E_i \to A_i$ are exact triangles with $A_i \in \mathcal{A}[k_i]$ and $k_1 > \ldots > k_n$.

Note that due to i) the diagram is unique. If i) and ii) hold, then the corresponding bounded t-structure is defined by $\mathcal{D}^{\leq 0} := \{E \mid A_i = 0 \text{ for } k_i > 0\}$. Conversely, if the bounded t-structure is given, then $E_1 := \tau^{\leq -k_1}(E) = A_1 \in \mathcal{A}[k_i]$, where $-k_1 = \min\{i \mid H^i(E) \neq 0\}$.

In general, t-structures are not preserved by equivalences. Positively speaking, if $\Phi : \mathcal{D} \xrightarrow{\sim} \mathcal{D}$ is an exact autoequivalence of a triangulated category and

$\mathcal{D}^{\leq 0} \subset \mathcal{D}$ defines a t-structure, then $\Phi(\mathcal{D}^{\leq 0}) \subset \mathcal{D}$ defines a new t-structure which is usually different from the original one.

4.1.4 Torsion theories versus t-structures

So far, we have seen t-structures as an analogue of torsion theories for abelian categories adapted to the triangulated setting. Another aspect of torsion theories for abelian categories is that they lead to new t-structures on triangulated categories.

Let us start with an abelian category $\mathcal{A}$ and a torsion theory for it given by $\mathcal{T}, \mathcal{F} \subset \mathcal{A}$. Then consider the *tilt* of $\mathcal{A}$ with respect to $(\mathcal{T}, \mathcal{F})$ defined as the full additive subcategory $\mathcal{A}^{\sharp} := Tilt_{(\mathcal{T},\mathcal{F})}(\mathcal{A}) \subset \mathrm{D}^b(\mathcal{A})$ of all objects $E \in \mathrm{D}^b(\mathcal{A})$ with

$$H^0(E) \in \mathcal{T}, \ H^{-1}(E) \in \mathcal{F}, \text{ and } H^i(E) = 0 \text{ for } i \neq 0, -1.$$

Note that the definition makes sense in the more general context when $\mathcal{A}$ is the heart of a t-structure on a triangulated category $\mathcal{D}$. In this case, the H^i are the cohomology functors that come with the t-structure.

For any $E \in \mathcal{A}$ there exists a short exact sequence in $\mathcal{A}$ or, equivalently, an exact triangle in $\mathcal{D}$ of the form $T \to E \to F$ with $T \in \mathcal{T}$ and $F \in \mathcal{F}$. The latter corresponds to a class in $\mathrm{Ext}^1(F, T)$. On the other hand, for $E \in \mathcal{A}^{\sharp}$ there exists an exact triangle $F[1] \to E \to T$, which corresponds to a class in $\mathrm{Ext}^2(T, F)$.

The torsion theory $(\mathcal{T}, \mathcal{F})$ for $\mathcal{A}$ gives rise to the torsion theory $(\mathcal{F}[1], \mathcal{T})$ for the tilt $\mathcal{A}^{\sharp}$. If $\mathcal{T}$ and $\mathcal{F}$ are both non-trivial, then $\mathcal{A} \neq \mathcal{A}^{\sharp}$ (even after shift). If $\mathcal{F} = 0$, then $\mathcal{A} = \mathcal{A}^{\sharp}$ and if $\mathcal{T} = 0$, then $\mathcal{A}^{\sharp} = \mathcal{A}[1]$.

The important fact is the following result proved in [17], which we leave as an exercise (use Remark 4.16).

Proposition 4.17. *The category* $\mathcal{A}^{\sharp}$ *is the heart of a bounded t-structure on* $\mathrm{D}^b(\mathcal{A})$. $\qquad\qquad\qquad\qquad\qquad\qquad\qquad\qquad\qquad\qquad\qquad\qquad\square$

The analogous statement for the heart $\mathcal{A}$ of a bounded t-structure on a triangulated category $\mathcal{D}$ also holds.

Remark 4.18. At this point it is natural to wonder what the relation between $\mathrm{D}^b(\mathcal{A})$ and $\mathrm{D}^b(\mathcal{A}^{\sharp})$ is. This question is addressed in [17] and [7]. E.g. it is proved that the bounded derived categories of $\mathcal{A}$ and its tilt $\mathcal{A}^{\sharp}$ are equivalent, if the torsion theory $(\mathcal{T}, \mathcal{F})$ for $\mathcal{A}$ is *cotilting*. This means that for all $E \in \mathcal{A}$ there exists a 'torsion free' object $F \in \mathcal{F}$ and an epimorphism

$F \twoheadrightarrow E$. Similarly, a torsion theory is *tilting* if for all $E \in \mathcal{A}$ there exists a 'torsion' object $T \in \mathcal{T}$ and a monomorphism $E \hookrightarrow T$. One verifies that $(\mathcal{T}, \mathcal{F})$ is tilting for $\mathcal{A}$ if and only if $(\mathcal{F}[1], \mathcal{T})$ is cotilting for $\mathcal{A}^\sharp$ (see [17]).

Our standard example (see Remark 4.8) starts with the torsion theory $(\mathcal{T}_\varphi, \mathcal{F}_\varphi)$, $\varphi \in (0, 1]$, for Coh(C), where as before C is a smooth projective curve. We call its tilt $\mathcal{A}_\varphi \subset D^b(C)$. This construction describes a *family of bounded t-structures* that depends on the parameter $\varphi \in (0, 1]$. Later however, we will see that this family is induced by a natural $\widetilde{GL}^+(2, \mathbb{R})$-action on the space of stability conditions. But there are more interesting families of t-structures that are not of this form.

The following picture of the image of $\mathcal{A} = \text{Coh}(C)$ and $\mathcal{A}_\varphi$ under the stability function $Z = -\deg + i \cdot \text{rk}$ might be helpful to visualize the tilt. In particular, for a μ-stable vector bundle E of slope $\mu(E)$ one has $E \in \mathcal{T}_\varphi \subset \mathcal{A}_\varphi$ if $\mu(E) < -\cot(\pi\varphi)$, but $E[1] \in \mathcal{F}_\varphi \subset \mathcal{A}_\varphi$ if $\mu(E) \geq -\cot(\pi\varphi)$.

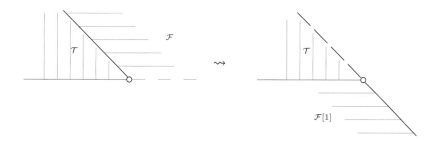

Remark 4.19. In [41] it is shown that for the various φ the tilts $\mathcal{A}_\varphi \subset D^b(C)$ for an elliptic curve C describe the categories of holomorphic vector bundles on non-commutative deformations of C. Roughly, for $C = \mathbb{C}/(\mathbb{Z} + \tau\mathbb{Z})$ and $\theta \in \mathbb{R} \setminus \mathbb{Q}$ one considers the algebra A_θ of series $\sum a_{m,n} U_1^m \cdot U_2^n$ with rapidly decreasing coefficients and the rule $U_1 \cdot U_2 = \exp(2\pi i \theta) U_2 \cdot U_1$. A holomorphic vector bundle on the non-commutative 2-torus $T_{\tau,\theta}$ is by definition an A_θ-module E with a 'connection' $\nabla : E \to E$ satisfying the Leibniz rule with respect to the derivative δ given by $\delta(U_1^m \cdot U_2^n) = (2\pi i)(m\tau + n)U_1^m \cdot U_2^n$. Then for $\varphi = \text{arcot}(-\theta)$ the category of holomorphic vector bundles on $T_{\tau,\theta}$ is equivalent to $\mathcal{A}_\varphi$.

The relation between tilts of the heart $\mathcal{A}$ of a t-structure on a triangulated category $\mathcal{D}$ and new t-structures on $\mathcal{D}$ is clarified by the following result, see [42, Lem. 1.1.2], [3, Thm. 3.1] or [46, Prop. 2.3].

Proposition 4.20. *If $\mathcal{A} = \mathcal{D}^{\leq 0} \cap \mathcal{D}^{\geq 0}$ denotes the heart of a t-structure on a triangulated category $\mathcal{D}$ given by $\mathcal{D}^{\leq 0} \subset \mathcal{D}$. Then there exists a natural bijection between:*

i) Torsion theories for $\mathcal{A}$ and

ii) t-structures on $\mathcal{D}$ given by $\mathcal{D}'^{\leq 0} \subset \mathcal{D}$ satisfying $\mathcal{D}^{\leq 0}[1] \subset \mathcal{D}'^{\leq 0} \subset \mathcal{D}^{\leq 0}$.

Proof. We only define the maps: A torsion theory $(\mathcal{T}, \mathcal{F})$ for $\mathcal{A}$ yields a t-structure $\mathcal{D}'^{\leq 0} := \{E \mid H^i_{\mathcal{A}}(E) = 0 \; i > 0 \text{ and } H^0_{\mathcal{A}}(E) \in \mathcal{T}\}$. Here, $H^i_{\mathcal{A}}$ denotes the cohomology with respect to the original t-structure which, therefore, takes values in $\mathcal{A}$. Conversely, if $\mathcal{D}'^{\leq 0}$ is given, then one defines a torsion theory by $\mathcal{T} := \mathcal{A} \cap \mathcal{D}'^{\leq 0}$ which can also be written as $\{E \in \mathcal{A} \mid E \in H^0_{\mathcal{A}'}(\mathcal{D}'^{\leq 0})\}$. $\qquad\square$

So roughly, one can tilt until $\mathcal{D}'^{\leq 0}$ leaves the slice $\mathcal{D}^{\leq -1} \subset \mathcal{D}^{\leq 0}$. The explicit description of torsion theories versus t-structures is later used often in the analysis of the space of stability conditions.

4.2 Stability conditions: definition and examples

In this second lecture, slicings and stability conditions are defined on arbitrary triangulated categories. Section 4.2.3 discusses the natural action of $\widetilde{\mathrm{GL}}^+(2, \mathbb{R})$ and $\mathrm{Aut}(\mathcal{D})$ on the space $\mathrm{Stab}(\mathcal{D})$ of stability conditions and in Section 4.2.4 it is shown that the $\widetilde{\mathrm{GL}}^+(2, \mathbb{R})$-action is transitive when $\mathcal{D}$ is the bounded derived category $\mathrm{D}^b(C)$ of a curve of genus $g(C) > 0$.

4.2.1 Slicings

We shall introduce this notion simultaneously for abelian categories $\mathcal{A}$ and triangulated categories $\mathcal{D}$. It is maybe easier to grasp for abelian categories and a slicing on a triangulated category is in fact nothing but a bounded t-structure with a slicing of the abelian category given by its heart (cf. Proposition 4.24). The analogous notions of a torsion theory and a t-structure are both refined by the notion of a slicing which formalizes the example of the categories $\mathcal{P}(\phi) \subset \mathrm{Coh}(C)$ of semistable sheaves of phase ϕ on a curve (see Definition 4.7).

Definition 4.21. A *slicing* $\mathcal{P} = \{\mathcal{P}(\phi)\}$ of $\mathcal{A}$ (resp. $\mathcal{D}$) consists of full additive (not necessarily abelian) subcategories $\mathcal{P}(\phi) \subset \mathcal{A}$, $\phi \in (0, 1]$ (resp. $\mathcal{P}(\phi) \subset \mathcal{D}, \phi \in \mathbb{R}$) such that:

i) $\mathrm{Hom}(\mathcal{P}(\phi_1), \mathcal{P}(\phi_2)) = 0$ if $\phi_1 > \phi_2$.

ii) For all $0 \neq E \in \mathcal{A}$ there exists a filtration $0 = E_0 \subset E_1 \subset \ldots \subset E_n = E$ with $A_i := E_i/E_{i-1} \in \mathcal{P}(\phi_i)$ and $\phi_1 > \ldots > \phi_n$. (For $E \in \mathcal{D}$ the inclusions $E_{i-1} \subset E_i$ are replaced by morphisms $E_{i-1} \to E_i$ and the quotients A_i are defined as their cones.)

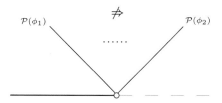

iii) This condition is for the triangulated case only: $\mathcal{P}(\phi)[1] = \mathcal{P}(\phi + 1)$ for all $\phi \in \mathbb{R}$.

As before, we will call $\mathcal{P}(\phi)$ the subcategory of *semistable* objects of phase ϕ. The minimal objects in $\mathcal{P}(\phi)$, i.e. those without any proper subobject in $\mathcal{P}(\phi)$, are called *stable* of phase ϕ. The filtration in ii) is the Harder–Narasimhan filtration, which is again unique due to i), and the A_i are the semistable factors of E. Also, for any interval I we let $\mathcal{P}(I)$ be the full subcategory of objects E, for which the semistable factors have phases in I. We repeat the warning that the phase $\phi(E)$ of an object is only well defined for semistable E.

Remark 4.22. Here are a few useful observations:

i) If $E \to B$ is a non-trivial morphism, then for at least one semistable factor A_i of E there exists a non-trivial morphism $A_i \to B$.

ii) If $A \in \mathcal{P}(\phi)$, $B \in \mathcal{P}(I)$, and $\phi > t$ for all $t \in I$, then $\mathrm{Hom}(A, B) = 0$.

iii) For the Harder–Narasimhan filtration of $E \in \mathcal{D}$ all the morphisms $E_i \to E$ are non-trivial and similarly the projection $E \to A_n$ is non-trivial.

iv) If E is not semistable, i.e. not contained in any of the $\mathcal{P}(\phi)$, then there exists a short exact sequence $0 \to A \to E \to B \to 0$ (resp. an exact triangle $A \to E \to B \to A[1]$) with $A, B \neq 0$ and $\mathrm{Hom}(A, B) = 0$.

Example 4.23. Consider a torsion theory $(\mathcal{T}, \mathcal{F})$ for an abelian category $\mathcal{A}$. Pick $1 \geq \phi_1 > \phi_2 > 0$ and let $\mathcal{P}(\phi_1) := \mathcal{T}$, $\mathcal{P}(\phi_2) := \mathcal{F}$, and $\mathcal{P}(\phi) = 0$ for $\phi \neq \phi_1, \phi_2$. This then defines a slicing of $\mathcal{A}$. In particular, it provides examples of slicings with slices $\mathcal{P}(\phi)$ which are not(!) abelian.

Similarly, if a t-structure on a triangulated category $\mathcal{D}$ has heart $\mathcal{A}$, then for any choice of $\phi_0 \in \mathbb{R}$ one can define a slicing of $\mathcal{D}$ by $\mathcal{P}(\phi) := \mathcal{A}[\phi - \phi_0]$ if $\phi - \phi_0 \in \mathbb{Z}$ and $\mathcal{P}(\phi) = 0$ otherwise.

Suppose $\mathcal{P} = \{\mathcal{P}(\phi)\}$ is a slicing of an abelian category $\mathcal{A}$. Pick $\phi_0 \in (0, 1]$. Then $\mathcal{T} := \mathcal{P}(> \phi_0)$ and $\mathcal{F} := \mathcal{P}(\le \phi_0)$ define a torsion theory for $\mathcal{A}$. In the triangulated setting this yields the following assertion where we choose $\phi_0 = 0$.

Proposition 4.24. *Let $\mathcal{D}$ be a triangulated category. There is a natural bijection between*

   i) Slicings of $\mathcal{D}$ and

   ii) Bounded t-structures on $\mathcal{D}$ together with a slicing of their heart.

Proof. Start with a slicing $\{\mathcal{P}(\phi)\}_{\phi \in \mathbb{R}}$ of $\mathcal{D}$. Then define a bounded t-structure on $\mathcal{D}$ by $\mathcal{D}^{\le 0} = \mathcal{P}(> 0)$.[3] Its heart is $\mathcal{A} = \mathcal{P}(0, 1]$ and $\{\mathcal{P}(\phi)\}_{\phi \in (0,1]}$ defines a slicing of it. Conversely, if a slicing $\{\mathcal{P}(\phi)\}_{\phi \in (0,1]}$ of the heart $\mathcal{A}$ of a bounded t-structure on $\mathcal{D}$ is given, define a slicing of $\mathcal{D}$ by $\mathcal{P}(\phi + k) := \mathcal{P}(\phi)[k]$, $k \in \mathbb{Z}$.   □

4.2.2 Stability conditions

To make slicings a really useful notion, they have to be linearized by a stability function, which is the abstract version of $Z = -\deg + i \cdot \mathrm{rk}$ on $\mathrm{D}^b(C)$. This will lead to the concept of stability conditions. The categories $\mathcal{P}(\phi)$ of semistable objects will then automatically be abelian.

There are two possible approaches to stability conditions. Either one starts with a slicing and searches for a compatible stability function or a natural stability function is given already and one verifies that it also defines a slicing, i.e. that Harder–Narasimhan filtrations exist.

Once more, we will first deal with the easier notion of stability conditions on abelian categories. So let $\mathcal{A}$ be an abelian category. As before, $K(\mathcal{A})$ will denote its Grothendieck group, i.e. the abelian group generated by objects of $\mathcal{A}$ subject to the relation $[E_2] = [E_1] + [E_3]$ for any short exact sequence $0 \to E_1 \to E_2 \to E_3 \to 0$.

Warning: Despite its easy definition, it is usually very difficult to control the Grothendieck group $K(\mathcal{A})$.

---

[3] This clash of notation really is unavoidable.

Definition 4.25. A *stability function* on an abelian category $\mathcal{A}$ is a linear map $Z : K(\mathcal{A}) \to \mathbb{C}$ such that $Z(E) = Z([E]) \in \overline{\mathbb{H}} = \mathbb{H} \cup \mathbb{R}_{<0}$ for any $0 \neq E \in \mathcal{A}$.

In particular, with respect to a given stability function Z any non-trivial object $E \in \mathcal{A}$ has a phase $\phi(E) \in (0, 1]$ which is determined by $Z(E) \in \exp(i\pi\phi(E)) \cdot \mathbb{R}_{>0}$.

Moreover, for a short exact sequence $0 \to E_1 \to E_2 \to E_3 \to 0$ the phases ϕ_i of E_i, $i = 1, 2, 3$ are related by $Z(E_2) = Z(E_1) + Z(E_3)$ as in:

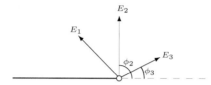

As in the case of $\mathrm{Coh}(C)$ discussed in Section 4.1.1, an object $E \in \mathcal{A}$ is called *semistable* with respect to a given stability function Z if it does not contain any subobject of phase $> \phi(E)$. Then, one defines the subcategory $\mathcal{P}(\phi)$ of semistable objects (with respect to Z) of phase ϕ as in Definition 4.7. A stability function is said to satisfy the *Harder–Narasimhan property* if any object has a finite filtration with semistable quotients A_i, i.e. if $\{\mathcal{P}(\phi)\}_{\phi \in (0,1]}$ is a slicing of $\mathcal{A}$.

Conversely, if a slicing $\{\mathcal{P}(\phi)\}_{\phi \in (0,1]}$ is given, then a stability function Z is *compatible* with it if $Z(E) \in \exp(i\pi\phi) \cdot \mathbb{R}_{>0}$ for all $0 \neq E \in \mathcal{P}(\phi)$.

Definition 4.26. A *stability condition* on an abelian category $\mathcal{A}$ is a pair $\sigma = (\mathcal{P}, Z)$ consisting of a slicing $\mathcal{P}$ and a compatible stability function Z.[4]

Remark 4.27. The datum of a stability condition is thus equivalent to the datum of a stability function $Z : K(\mathcal{A}) \to \mathbb{C}$ satisfying the Harder–Narasimhan property.

Corollary 4.28. *If* $\sigma = (\mathcal{P}, Z)$ *is a stability condition, then all categories* $\mathcal{P}(\phi)$ *are abelian.*

Proof. See [9, Lem. 5.2]. In order to make $\mathcal{P}(\phi)$ abelian, one has to show that kernels, images, etc., exist. Since $\mathcal{P}(\phi)$ is a subcategory of the given abelian category $\mathcal{A}$, one can consider a morphism $f : E \to F$ in $\mathcal{P}(\phi)$ as a morphism in $\mathcal{A}$ and take its kernel there. In order to conclude, one has to show that $\mathrm{Ker}(f)$

---

[4] Later, a finiteness condition will be added, see Definition 4.52, which in particular implies that any semistable object has a finite Jordan–Hölder filtration.

is in fact an object in $\mathcal{P}(\phi)$ (and similarly for image, cokernels, etc.). Suppose first that $\mathrm{Ker}(f)$ and $\mathrm{Im}(f)$ are semistable, i.e. $\mathrm{Ker}(f) \in \mathcal{P}(\psi)$ and $\mathrm{Im}(f) \in \mathcal{P}(\psi')$ for certain ψ, ψ'. Since $\mathrm{Ker}(f) \subset E$ and $\mathrm{Im}(f) \subset F$, semistability of E resp. F yields $\psi \leq \phi$ and $\psi' \leq \phi$. Thus, the images of $\mathrm{Ker}(f)$ and $\mathrm{Im}(f)$ under Z are to the right of the ray $Z(E) \cdot \mathbb{R}_{>0} = Z(F) \cdot \mathbb{R}_{>0}$ as pictured below.

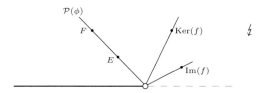

But this contradicts the linearity of Z which says $Z(E) = Z(\mathrm{Ker}(f)) + Z(\mathrm{Im}(f))$. For the general case, one uses the Harder–Narasimhan filtration of $\mathrm{Ker}(f)$ and $\mathrm{Im}(f)$ and again the linearity of Z to reduce to the case above. $\qquad\square$

Let us now pass to the case of a triangulated category $\mathcal{D}$. Its Grothendieck group $K(\mathcal{D})$ is defined to be the abelian group generated by objects E in $\mathcal{D}$ subject to the obvious relation defined by exact triangles. Note that with this definition one has $K(\mathcal{D}) \simeq K(\mathcal{A})$, whenever $\mathcal{A}$ is the heart of bounded t-structure on $\mathcal{D}$.

Warning: For triangulated subcategories $\mathcal{D}' \subset \mathcal{D}$ the induced map $K(\mathcal{D}') \to K(\mathcal{D})$ is in general not injective.[5]

Definition 4.29. A *stability function* on the triangulated category $\mathcal{D}$ is a linear map $Z : K(\mathcal{D}) \to \mathbb{C}$.

Note that we do not make any assumption on the image of Z. Indeed, the linearity of Z in particular implies that $Z(E[1]) = -Z(E)$, as $[E[1]] = -[E]$ in $K(\mathcal{D})$. Also note that there are always non-trivial objects in the kernel of Z. Indeed $Z(E \oplus E[1]) = 0$ for all E.

As before, we say that Z is compatible with a slicing $\{\mathcal{P}(\phi)\}_{\phi \in \mathbb{R}}$ of $\mathcal{D}$ if $Z(E) \in \exp(i\pi\phi) \cdot \mathbb{R}_{>0}$ for all $0 \neq E \in \mathcal{P}(\phi)$.

Warning: In contrast to the abelian case, the phase of an arbitrary object $E \in \mathcal{D}$ is not well defined (not even modulo $\mathbb{Z}$).

Definition 4.30. A *stability condition* on a triangulated category $\mathcal{D}$ is a pair $\sigma = (\mathcal{P}, Z)$ consisting of a slicing $\mathcal{P} = \{\mathcal{P}(\phi)\}_{\phi \in \mathbb{R}}$ of $\mathcal{D}$ and a compatible stability function $Z : K(\mathcal{D}) \to \mathbb{C}$.[6]

---

[5] For dense subcategories it is, see [44].
[6] Later, a finiteness condition will again be added, see Definition 4.52.

In particular, the *space of all stability conditions*

$$\mathrm{Stab}(\mathcal{D}) := \{\sigma = (\mathcal{P}, Z)\}$$

comes with a natural map

$$\pi : \mathrm{Stab}(\mathcal{D}) \to K(\mathcal{D})^* := \mathrm{Hom}(K(\mathcal{D}), \mathbb{C}), \quad \sigma = (\mathcal{P}, Z) \mapsto Z.$$

Example 4.31. Here is a pathological example (cf. [9, Ex. 5.6]) that we will want to avoid by adding local finiteness (see Definition 4.52) later on. Consider a curve C, $\varphi \in \mathbb{R} \setminus \mathbb{Q}$ and the induced torsion theory as in Section 4.1.4 with its tilt $\mathcal{A}_\varphi \subset \mathrm{D}^b(C)$. Define $Z : K(\mathcal{D}) = K(\mathcal{A}_\varphi) \to \mathbb{C}$ as $Z(E) = i \cdot (\deg(E) - \varphi \cdot \mathrm{rk}(E))$. Then $Z(E) \in i \cdot \mathbb{R}_{>0}$ for all $0 \neq E \in \mathcal{A}_\varphi$. In particular, all objects of $\mathcal{A}_\varphi$ are semistable of phase $\phi = \frac{1}{2}$, i.e. $\mathcal{P}(\frac{1}{2}) = \mathcal{A}_\varphi$ and $\mathcal{P}(\phi) = \{0\}$ for $\phi \neq \frac{1}{2}$. In this example, semistable objects do not necessarily admit finite Jordan–Hölder filtrations. In fact, for any stable vector bundle E of slope $< \varphi$ one obtains a strictly descending filtration $\ldots E_i[1] \subset E_{i-1}[1] \subset \ldots E_0[1] = E[1]$ in $\mathcal{P}(\frac{1}{2})$, where E_i is defined as the kernel of some arbitrarily chosen surjection $E_{i-1} \twoheadrightarrow k(x_i)$.

Building upon Proposition 4.24, one obtains (cf. [9, Prop. 5.3]):

Proposition 4.32. *Let $\mathcal{D}$ be a triangulated category. There is a natural bijection between*
  i) Stability conditions on $\mathcal{D}$ and
  ii) Bounded t-structures on $\mathcal{D}$ together with a stability condition on the heart. $\square$

By $\mathcal{A}_\sigma$ we shall denote the heart of a stability condition σ, so $\mathcal{A}_\sigma = \mathcal{P}(0, 1]$.

It may be instructive to picture the image of the hearts of the various t-structures associated with the slicing underlying a stability function on $\mathcal{D}$.

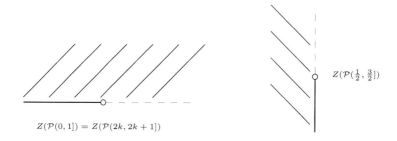

$$Z(\mathcal{P}(\frac{1}{2}, \frac{3}{2}])$$

$$Z(\mathcal{P}(0, 1]) = Z(\mathcal{P}(2k, 2k + 1])$$

Remark 4.33. To avoid the often mysterious $K(\mathcal{D})$, one usually fixes a finite rank quotient $K(\mathcal{D}) \twoheadrightarrow \Gamma \simeq \mathbb{Z}^d$ and requires that Z factors via Γ, i.e. $Z : K(\mathcal{D}) \twoheadrightarrow \Gamma \to \mathbb{C}$. A typical example is that of a k-linear triangulated category $\mathcal{D}$ with finite-dimensional $\bigoplus_i \operatorname{Hom}(E, F[i])$ for all $E, F \in \mathcal{D}$. Then one can consider the *numerical Grothendieck group*

$$N(\mathcal{D}) = \Gamma = K(\mathcal{D})/_\sim,$$

where $E \sim 0$ if $\chi(E, F) := \sum (-1)^i \dim \operatorname{Hom}(E, F[i]) = 0$ for all F. Stability conditions of this type will be called *numerical*.

In the geometric situation, i.e. when $\mathcal{D} = \mathrm{D}^b(X)$ with X a smooth complex projective variety, the numerical Grothendieck group $N(\mathrm{D}^b(X)) \otimes \mathbb{Q}$ should be, according to one of the standard conjectures, isomorphic to the algebraic cohomology $H^*_{\mathrm{alg}}(X, \mathbb{Q})$ (and according to the Hodge conjecture in fact to $\bigoplus H^{p,p}(X) \cap H^*(X, \mathbb{Q})$).

4.2.3 Aut($\mathcal{D}$)-action and $\widetilde{\mathrm{GL}}^+(2, \mathbb{R})$-action

We next come to a feature that is not present in the case of an abelian category. Let us start with the natural right action

$$\mathbb{C} \times \mathrm{GL}(2, \mathbb{R}) \to \mathbb{C}$$

given by $(z, M) \longmapsto M^{-1} \cdot z$ via the natural identification $\mathbb{C} \simeq \mathbb{R}^2$. This yields

$$K(\mathcal{D})^* \times \mathrm{GL}(2, \mathbb{R}) \to K(\mathcal{D})^*, \quad (Z, M) \longmapsto M^{-1} \circ Z.$$

Since the image of the heart $\mathcal{P}(0, 1]$ of a stability condition on $\mathcal{D}$ is contained in $\overline{\mathbb{H}}$, it might be helpful to picture the image of $\overline{\mathbb{H}}$ under some standard elements of $\mathrm{GL}(2, \mathbb{R})$. E.g. $-\mathrm{id} = \begin{pmatrix} -1 & 0 \\ 0 & -1 \end{pmatrix}$ acts as rotation by π (anti-clockwise) and $-i = \begin{pmatrix} 0 & -1 \\ 1 & 0 \end{pmatrix}$ as rotation by $\frac{\pi}{2}$ clockwise.

We would like to lift the action of $\mathrm{GL}(2, \mathbb{R})$ on $K(\mathcal{D})^*$ to an action on $\mathrm{Stab}(\mathcal{D})$ under the natural projection $\pi : \mathrm{Stab}(\mathcal{D}) \to K(\mathcal{D})^*$. Due to the phases of semistable objects in $\mathcal{D}$ being contained in $\mathbb{R}$ and not only in $(0, 1]$, one has to pass to the universal cover of $\mathrm{GL}(2, \mathbb{R})$ first. In fact, the same problem occurs when one wants to lift the action $\mathbb{C} \times \mathrm{GL}(2, \mathbb{R}) \to \mathbb{C}$ with respect to the exponential map $\exp : \mathbb{C} \to \mathbb{C}$. For stability conditions there is one more aspect to be taken into account. As one wants the property $\operatorname{Hom}(\mathcal{P}(\phi_1), \mathcal{P}(\phi_2)) = 0$ for $\phi_1 > \phi_2$, to be preserved under the action, one needs to restrict to the connected component $\mathrm{GL}^+(2, \mathbb{R})$ of invertible matrices with positive determinant.

Lemma 4.34. *The universal cover of* $\mathrm{GL}^+(2, \mathbb{R})$ *can be described explicitly as the group*

$$\widetilde{\mathrm{GL}}^+(2, \mathbb{R}) = \{(M, f) \mid M \in \mathrm{GL}^+(2, \mathbb{R}), \ f : \mathbb{R} \to \mathbb{R}, \ i), \ ii)\}$$

with:

i) f is increasing with $f(\phi + 1) = f(\phi) + 1$ for all $\phi \in \mathbb{R}$ and

ii) $M \cdot \exp(i\pi\phi) \in \exp(i\pi f(\phi)) \cdot \mathbb{R}_{>0}$.

Condition ii) simply says that the induced maps $\bar{M}$ and $\bar{f}$ on $(\mathbb{R}^2 \setminus \{0\})/\mathbb{R}_{>0} = S^1 = \mathbb{R}/2\mathbb{Z}$ coincide.

Then the action of $\mathrm{GL}^+(2, \mathbb{R})$ on $\mathbb{C}$ is lifted naturally to

$$\mathrm{Stab}(\mathcal{D}) \times \widetilde{\mathrm{GL}}^+(2, \mathbb{R}) \longrightarrow \mathrm{Stab}(\mathcal{D})$$

$$(\sigma = (\mathcal{P}, Z), (M, f)) \longmapsto \sigma' = (\mathcal{P}', Z')$$

with $Z' = M^{-1} \circ Z$ and $\mathcal{P}'(\phi) = \mathcal{P}(f(\phi))$.

For example, id can be lifted to $(\mathrm{id}, f : \longmapsto \phi + 2k) \in \widetilde{\mathrm{GL}}^+(2, \mathbb{R})$ with arbitrary $k \in \mathbb{Z}$. It acts on $\mathrm{Stab}(\mathcal{D})$ by fixing the stability function and by changing a given slicing $\{\mathcal{P}(\phi)\}_{\phi \in \mathbb{R}}$ to $\{\mathcal{P}'(\phi) = \mathcal{P}(\phi + 2k)\}_{\phi \in \mathbb{R}}$. Thus, in terms of its action on $\mathrm{Stab}(\mathcal{D})$ the universal cover of $\mathrm{GL}^+(2, \mathbb{R})$ can be seen as an extension

$$0 \longrightarrow \mathbb{Z}[2] \longrightarrow \widetilde{\mathrm{GL}}^+(2, \mathbb{R}) \longrightarrow \mathrm{GL}^+(2, \mathbb{R}) \longrightarrow 0.$$

Similarly, $-\mathrm{id}$ can be lifted to $(-\mathrm{id}, f : \phi \longmapsto \phi + 2k + 1) \in \widetilde{\mathrm{GL}}^+(2, \mathbb{R})$ with arbitrary $k \in \mathbb{Z}$. It sends a slicing $\mathcal{P}$ to $\{\mathcal{P}'(\phi) = \mathcal{P}(\phi + 2k + 1)\}$.

If one thinks of a stability condition as a way of decomposing the triangulated category in small slices, then stability conditions in the same $\widetilde{\mathrm{GL}}^+(2, \mathbb{R})$-orbit decompose the category essentially in the same way. In particular, the action rotates the slices around without changing the stability of an object. Compare the comments in Section 4.1.4. Thus, if one wants to really understand the different possibilities for stability conditions on $\mathcal{D}$ it is rather the quotient

$$\mathrm{Stab}(\mathcal{D}) \big/ \widetilde{\mathrm{GL}}^+(2, \mathbb{R})$$

one needs to describe.

Any linear exact autoequivalence Φ of $\mathcal{D}$, i.e. $\Phi \in \mathrm{Aut}(\mathcal{D})$, induces an action on the Grothendieck group $K(\mathcal{D})$ by $[E] \longmapsto [\Phi(E)]$ which we shall also denote Φ. Then $\mathrm{Aut}(\mathcal{D})$ acts on $\mathrm{Stab}(\mathcal{D})$ via

$$\mathrm{Aut}(\mathcal{D}) \times \mathrm{Stab}(\mathcal{D}) \longrightarrow \mathrm{Stab}(\mathcal{D})$$

$$(\Phi, \sigma = (\mathcal{P}, Z)) \longmapsto (\mathcal{P}', Z \circ \Phi^{-1}),$$

with $\mathcal{P}'(\phi) = \Phi(\mathcal{P}(\phi))$. The left $\mathrm{Aut}(\mathcal{D})$-action and the right $\widetilde{\mathrm{GL}}^+(2, \mathbb{R})$-action obviously commute.

4.2.4 Stability conditions on curves

We come back to the derived category $\mathrm{D}^b(C)$ of complexes of coherent sheaves on a smooth projective curve over an algebraically closed field k. It is possible to describe the space of numerical stability conditions completely, at least for $g(C) > 0$.[7]

We have tried to motivate the abstract notion of a stability condition by the example on $\mathrm{D}^b(C)$ with $Z = -\deg + i \cdot \mathrm{rk}$ and the standard bounded t-structure on $\mathrm{D}^b(C)$ with heart $\mathrm{Coh}(C)$. The Harder–Narasimhan property holds for Z and the additional finiteness condition we have alluded to is essentially the existence of finite Jordan–Hölder filtrations. Both are ultimately explained by the rationality of $-\deg + i \cdot \mathrm{rk}$ (see Remark 4.53).

The following result, due to Bridgeland [9] and Macrì [31], shows that up to the action of $\widetilde{\mathrm{GL}}^+(2, \mathbb{R})$ there is only one stability condition on $\mathrm{D}^b(X)$, namely the one with the classical choice $Z = -\deg + i \cdot \mathrm{rk}$ and the standard t-structure on $\mathrm{D}^b(C)$.

Theorem 4.35. *The space of numerical stability conditions* $\mathrm{Stab}(C) := \mathrm{Stab}(\mathrm{D}^b(C))$ *on* $\mathrm{D}^b(C) = \mathrm{D}^b(\mathrm{Coh}(C))$ *of a smooth projective curve* C *of genus* $g(C) > 0$ *over a field* $k = \bar{k}$ *consists of exactly one* $\widetilde{\mathrm{GL}}^+(2, \mathbb{R})$-*orbit:*

$$\mathrm{Stab}(C)/\widetilde{\mathrm{GL}}^+(2, \mathbb{R}) = \{\mathrm{pt}\}.$$

Proof. First, recall the following fact that works for arbitrary abelian categories of homological dimension ≤ 1: Any object $E \in \mathrm{D}^b(C)$ can be written as $E \simeq \bigoplus E_i[-i]$ with $E_i \in \mathrm{Coh}(C)$ (see e.g. [20, Cor. 3.15]). This is proved by induction on the length of a complex. If E is concentrated in degree $\geq i$, then there exists an exact triangle $H^i(E)[-i] \to E \to F$ with $H^j(F) = H^j(E)$ for $j > i$ and $= 0$ otherwise. So, we can assume $F \simeq \bigoplus F_j[-j]$. But then the boundary map is in $\mathrm{Ext}^1(F, H^i(E)[-i]) = \bigoplus_{j>i} \mathrm{Ext}^{1+j-i}(F_j, H^i(E))$ which is trivial, as $1 + j - i > 1$.

---

[7] There are, of course, easier examples one could look at. E.g. stability conditions on the abelian category of finite dimensional vector spaces and on its bounded derived category, the abelian category of Hodge structures, etc.

The key Lemma 4.36 below says that all point sheaves $k(x)$ and all line bundles $L \in \mathrm{Pic}(C)$ are stable with respect to any $\sigma \in \mathrm{Stab}(C)$. Using this, one concludes as follows. Let ϕ_x be the phase of the stable object $k(x)$ and let ϕ_L be the phase of the stable object $L \in \mathrm{Pic}(C)$. Since $\mathrm{Hom}(L, k(x)) \neq 0$ and by Serre duality also $\mathrm{Ext}^1(k(x), L) \simeq \mathrm{Hom}(L, k(x))^* \neq 0$, stability yields (see Exercise 4.5) $\phi_L < \phi_x$ and $\phi_x - 1 < \phi_L$.

The numerical Grothendieck group of C is $N(C) \simeq \mathbb{Z} \oplus \mathbb{Z}$ generated by rank and degree. E.g. $[k(x)] = (0, 1)$ and $[L] = (1, \deg(L))$. Thus, $Z : N(C) \otimes \mathbb{R} \simeq \mathbb{R}^2 \to \mathbb{C}$ can be pictured as

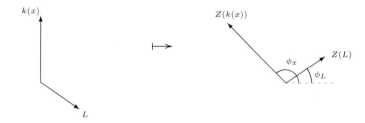

which shows that Z can be seen as an orientation preserving automorphism of $\mathbb{R}^2$ (under the identification of $N(C)$ with $\mathbb{Z}^2$ as chosen above). But then, by composing with a matrix in $\mathrm{GL}^+(2, \mathbb{R})$, it can be turned into $-\deg + i \cdot \mathrm{rk}$, which is rotation by $\frac{\pi}{2}$ anti-clockwise. In other words, in the $\widetilde{\mathrm{GL}}^+(2, \mathbb{R})$-orbit of any numerical stability condition σ one finds one stability condition, say σ', with stability function $Z = -\deg + i \cdot \mathrm{rk}$. Moreover, we may assume that for all points $x \in C$ the phase of $k(x)$ is $\phi_x = 1$ and hence L will have phase $\phi_L \in (0, 1)$. (A priori the phases of $k(x_1)$ and $k(x_2)$ for two distinct points could be different. But since the two sheaves are numerically equivalent, one has $\phi_{x_1} - \phi_{x_2} \in 2\mathbb{Z}$. Using $\phi_{x_i} - 1 < \phi_L < \phi_{x_i}$, $i = 1, 2$, for an arbitrary line bundle L, one finds that in fact $\phi_{x_1} = \phi_{x_2}$.) Thus, the heart $\mathcal{A}' = \mathcal{P}'(0, 1]$ of σ' contains all $k(x)$ and all line bundles $L \in \mathrm{Pic}(C)$. Since any coherent sheaf on C admits a filtration with quotients either isomorphic to point sheaves or line bundles, this shows $\mathrm{Coh}(C) \subset \mathcal{A}'$ and hence $\mathrm{Coh}(C) = \mathcal{A}'$, as both are hearts of bounded t-structures on $\mathrm{D}^b(C)$. $\qquad\square$

Lemma 4.36. *Suppose $\sigma = (\mathcal{P}, Z)$ is a numerical stability condition on $\mathrm{D}^b(C)$. Then all point sheaves $k(x)$, $x \in C$, and all line bundles $L \in \mathrm{Pic}(C)$ are stable with respect to σ.*

Proof. The following is a simplified version of an argument in [16].

i) Consider $E \in \mathrm{Coh}(C)$ and an exact triangle $A \to E \to B$ in $\mathrm{D}^b(C)$ with $\mathrm{Hom}^{<0}(A, B) = 0$. *Claim*: Then $A \simeq A_0 \oplus A_1[-1]$ and $B \simeq B_0 \oplus B_{-1}[1]$ with $A_i, B_i \in \mathrm{Coh}(C)$.

Write $A = \bigoplus A_i[-i]$ and $B = \bigoplus B_i[-i]$. The long cohomology sequence of the exact triangle yields an exact sequence

$$0 \longrightarrow B_{-1} \overset{\varphi}{\longrightarrow} A_0 \longrightarrow E \longrightarrow B_0 \overset{\psi}{\longrightarrow} A_1 \longrightarrow 0 \qquad (4.4)$$

and $B_{i-1} \simeq A_i$ for $i \neq 0, 1$. If $A_i \neq 0$ for some $i \neq 0, 1$, then

$$0 \neq \mathrm{Hom}(A_i[-i], B_{i-1}[-i]) = \mathrm{Hom}^{-1}(A_i[-i], B_{i-1}[-(i-1)])$$

and hence $\mathrm{Hom}^{-1}(A, B) \neq 0$, which contradicts the assumption.

ii) Consider $E \in \mathrm{Coh}(C)$ and an exact triangle $A \to E \to B$ in $\mathrm{D}^b(C)$ with $\mathrm{Hom}^{\leq 0}(A, B) = 0$. *Claim*: Then $A, B \in \mathrm{Coh}(C)$. (Here one uses the assumption $g(C) > 0$.)

Use the long exact sequence (4.4). If $\varphi \neq 0$, then twisting it with sections of ω_C yields non-trivial $B_{-1} \to A_0 \otimes \omega_C$. Hence, by Serre duality, $\mathrm{Hom}^1(A_0, B_{-1}) \neq 0$ and, therefore, $\mathrm{Hom}(A, B) \neq 0$, which contradicts the assumption. Thus, $\varphi = 0$ and hence $B_{-1} = 0$. The argument for proving $\psi = 0$ is similar.

iii) *Claim*: All point sheaves $k(x)$ and all line bundles L are semistable.

Apply ii) to $E = k(x)$ and $E = L$ by letting A be the first semistable factor. Then, $\mathrm{Hom}^{\leq 0}(A, B) = 0$ and by ii) $A, B \in \mathrm{Coh}(C)$. For $E = k(x)$ this immediately yields $B = 0$, as $k(x)$ has no proper non-trivial subsheaves. For $E = L$, the subsheaf A must also be a line bundle and hence B a torsion sheaf (or trivial). But if $B \neq 0$, then $\mathrm{Hom}(A, B) \neq 0$. Contradiction.

iv) *Claim*: All point sheaves $k(x)$ and all line bundles L are stable.

We apply again ii) to $E = k(x)$ and $E = L$. Let A_0 be a stable factor of E with $\mathrm{Hom}(A_0, E) \neq 0$. Then there exists an exact triangle $A \to E \to B$ with A, B semistable and such that all stable factors of A are isomorphic to A_0 and $\mathrm{Hom}(A, B) = 0$, cf. Exercise 4.6. (The vanishing of $\mathrm{Hom}^{<0}(A, B)$ follows directly from semistability.) Thus, $A, B \in \mathrm{Coh}(C)$ and as in iii) this shows $B = 0$, i.e. all stable factors of E are isomorphic to A_0. Hence, $[E] = n[A_0]$, where n is the number of stable factors. Since $[k(x)] = (0, 1)$ and $[L] = (1, \deg(L))$, one must have $n = 1$, i.e. $k(x)$ and L are stable. $\qquad\square$

Remark 4.37. i) The minimal objects in $\mathrm{Coh}(C)$ are the point sheaves $k(x)$. In the tilts of $\mathrm{Coh}(C)$ occuring as hearts of the other stability conditions on $\mathrm{Coh}(C)$ the description is more complicated, see e.g. [46, Lem. 2.4].

ii) For $g = 1$ the quotient of $\mathrm{Stab}(C)$ by the natural left action of the group $\mathrm{Aut}(\mathrm{D}^b(C))$ of all exact k-linear autoequivalences can be described as the quotient of $\mathrm{GL}^+(2, \mathbb{R})$ by $\mathrm{SL}(2, \mathbb{Z})$ which can also be interpreted as a $\mathbb{C}^*$-bundle over the moduli space of elliptic curves (see [9]). Indeed, $\mathrm{Aut}(\mathrm{D}^b(C))$ compares to $\mathrm{GL}^+(2, \mathbb{R})$ via the diagram

$$
\begin{array}{ccccccccc}
0 & \longrightarrow & \mathbb{Z}[2] \times (C \times \widehat{C}) & \longrightarrow & \mathrm{Aut}(\mathrm{D}^b(C)) & \longrightarrow & \mathrm{SL}(2, \mathbb{Z}) & \longrightarrow & 0 \\
 & & & & \big\downarrow & & \big\uparrow & & \\
0 & \longrightarrow & \mathbb{Z}[2] & \longrightarrow & \widetilde{\mathrm{GL}}^+(2, \mathbb{R}) & \longrightarrow & \mathrm{GL}^+(2, \mathbb{R}) & \longrightarrow & 0.
\end{array}
$$

Note that $C \times \widehat{C}$ acts, by translation resp. tensor product, trivially on $\mathrm{Stab}(C)$.

iii) Stability conditions on $\mathrm{D}^b(\mathbb{P}^1)$ can also be described. It turns out that $\mathrm{Stab}(\mathbb{P}^1)$ is connected and simply connected, see [31], and in fact $\mathrm{Stab}(\mathbb{P}^1) \simeq \mathbb{C}^2$, see [39].

4.3 Stability conditions on surfaces

We consider a smooth projective surface X over a field $k = \bar{k}$, but we shall also be interested in compact complex surfaces. For any ample class $\omega \in \mathrm{NS}(X) \otimes \mathbb{R}$ (or a Kähler class $\omega \in H^{1,1}(X, \mathbb{R})$) one defines the degree and the slope (if $\mathrm{rk}(E) \neq 0$) as

$$
\deg_\omega(E) := (c_1(E).\omega) \ \text{ resp. } \ \mu_\omega(E) := \frac{\deg_\omega(E)}{\mathrm{rk}(E)}.
$$

Fix $\beta \in \mathbb{R}$ and think of it as $\beta = (B.\omega)$ for some $B \in \mathrm{NS}(X) \otimes \mathbb{R}$. Then consider the full additive subcategories $\mathcal{T}_{(\omega,\beta)}, \mathcal{F}_{(\omega,\beta)} \subset \mathrm{Coh}(X)$:

$$
\mathcal{T}_{(\omega,\beta)} := \{E \in \mathrm{Coh}(X) \mid \forall E \twoheadrightarrow F \neq 0 \text{ torsion free} : \ \mu_\omega(F) > \beta\}
$$

and

$$
\mathcal{F}_{(\omega,\beta)} := \{E \in \mathrm{Coh}(X) \mid E \text{ torsion free and } \forall 0 \neq F \subset E : \ \mu_\omega(F) \leq \beta\}.
$$

We leave it as an exercise to verify that this defines a torsion theory for $\mathrm{Coh}(X)$. One needs to use the existence of the Harder–Narasimhan filtration and the fact that the saturation $F \subset F' \subset E$ of a subsheaf $F \subset E$, i.e. the minimal F' such that E/F' is torsion free, satisfies $\mu_\omega(F) \leq \mu_\omega(F')$. Note that for the existence of the Harder–Narasimhan filtration it would be enough to assume that ω is nef. (But Jordan–Hölder filtrations need a stronger positivity.)

The tilt of $\text{Coh}(X)$ with respect to the torsion theory $(\mathcal{T}_{(\omega,\beta)}, \mathcal{F}_{(\omega,\beta)})$ is denoted

$$\mathcal{A}(\omega, \beta) \subset D^b(X) := D^b(\text{Coh}(X))$$

or $\mathcal{A}(\exp(B + i\omega))$ if $\beta = (B.\omega)$.

Example 4.38. For a closed point $x \in X$ the skyscraper sheaf $k(x)$ is contained in $\mathcal{A}(\omega, \beta)$. If E is μ_ω-stable sheaf with $\mu_\omega(E) = \beta$, then $E[1] \in \mathcal{A}(\omega, \beta)$.

4.3.1 Classification of hearts

For the following see [10, Prop. 10.3].

Theorem 4.39. *If σ is a numerical stability condition on $D^b(X)$ such that all point sheaves $k(x)$, $x \in X$, are σ-stable of phase 1, then the heart $\mathcal{A}_\sigma$ of σ is of the form $\mathcal{A}_\sigma = \mathcal{A}(\omega, \beta)$ for some nef class $\omega \in \text{NS}(X)$ and $\beta \in \mathbb{R}$.*[8]

We will give the main technical arguments that go into the proof. They illustrate standard techniques in the study of $\text{Coh}(X)$ and $D^b(X)$.

Claim 1: i) If $E \in \mathcal{P}(0, 1]$, then $H^i(E) = 0$ for $i \neq 0, -1$ and $H^{-1}(E)$ is torsion free.

ii) If $E \in \mathcal{P}(1)$ is stable, then $E \simeq k(x)$ for some $x \in X$ or $E[-1] \in \text{Coh}(X)$, which in addition is locally free.

Proof. i) It is enough to show the assertion for E stable and not isomorphic to any point sheaf $k(x)$. Then stability of E and $k(x)$ implies

$$\text{Hom}^{<0}(E, k(x)) = \text{Hom}^{\leq 0}(k(x), E) = 0.$$

Thus, by Serre duality $\text{Ext}^i(E, k(x)) = 0$ for $i \neq 0, 1$.

Write $E \simeq E^\bullet$ with E^k locally free. Then $E^\bullet$ is not exact in degree k in a point x if and only if $H^k(E^\bullet \otimes k(x)) \neq 0$. But this cohomology also computes $\text{Ext}^k(E, k(x))^*$ and thus only $k = 0, 1$ are possible. This proves the first part of i).

Now consider the spectral sequence

$$E_2^{p,q} = \text{Ext}^p(H^{-q}(E), k(x)) \Rightarrow \text{Ext}^{p+q}(E, k(x)) \qquad (4.5)$$

---

[8] In fact, one proves $(\omega.C) > 0$ for all curves C. This is not quite enough to conclude that ω is ample. But if σ is 'good', i.e. contained in a maximal component, then ω will in addition be in the interior of the nef cone and hence ample.

for which $E_2^{p,q} = 0$ if $p \neq 0, 1, 2$. Let i and j be maximal resp. minimal with $H^i(E) \neq 0 \neq H^j(E)$. Then the marked boxes in the picture of (4.5) survive the passage to E_∞:

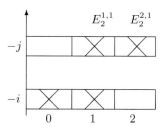

Note that for $x \in \mathrm{supp}(H^i(E))$, clearly $E_2^{0,-i} \neq 0$ and hence $E^{-i} \neq 0$ which proves $i \in \{0, -1\}$ confirming what we have seen already. But the spectral sequence also proves the second part of i). Indeed, suppose $H^{-1}(E)$ is not torsion free. If it has zero-dimensional torsion, then there exists a point $x \in X$ with $\mathrm{Hom}(k(x), H^{-1}(E)) \neq 0$ and thus by Serre duality $E_2^{2,1} \neq 0$ contradicting $E^3 = 0$. If the torsion of $H^{-1}(E)$ is purely one-dimensional, then there exists a point $x \in X$ with $\mathrm{Ext}^1(k(x), H^{-1}(E)) \neq 0$ (use that $\mathrm{Ext}^1(k(x), \mathcal{O}_C) \neq 0$ for a curve C and a smooth point $x \in C$). Thus, by Serre duality, $E_2^{1,1} \neq 0$ contradicting $E^2 = 0$. This concludes the proof of i).

For $k(x) \not\simeq E \in \mathcal{P}(1)$ stable one has $\mathrm{Hom}(E, k(x)) = 0$ (cf. Remark 4.22, ii)) and hence $i = j = -1$ in (4.5), i.e. $E \simeq F[1]$ for some torsion free $F \in \mathrm{Coh}(X)$. Moreover, F is locally free if and only if $\mathrm{Ext}^1(k(x), F) = 0$ for all $x \in X$. But $\mathrm{Ext}^1(k(x), F) \simeq \mathrm{Hom}(k(x), E) = 0$, again because E and $k(x)$ are non-isomorphic stable objects of the same phase. $\square$

Claim 2: i) If $E \in \mathrm{Coh}(X)$, then $E \in \mathcal{P}(-1, 1]$.

ii) If $E \in \mathrm{Coh}(X)$ is torsion, then $E \in \mathcal{P}(0, 1]$. (This part will be proved only later, but logically the assertion belongs here.)

Proof. For any $A \in \mathcal{P}(\phi)$ Claim 1 shows that then $H^i(A) = 0$ if $i \geq 0, \phi > 1$ or if $i \leq 0, \phi \leq -1$. Let $A \to E$ be the maximal semistable factor of E. Then, $\mathrm{Hom}(H^i(A)[-i], E) \neq 0$ for some i. But on the one hand, if $\phi(A) > 1$, then $H^i(A) = 0$ for $i \geq 0$, and on the other hand $\mathrm{Ext}^{<0} = 0$ on the abelian category $\mathrm{Coh}(X)$. Thus, $E \in \mathcal{P}(\leq 1)$.[9] A similar argument for the minimal semistable factor $E \to B$ proves $E \in \mathcal{P}(>-1)$. $\square$

Claim 3: Let $\mathcal{T} := \mathrm{Coh}(X) \cap \mathcal{P}(0, 1]$ and $\mathcal{F} := \mathrm{Coh}(X) \cap \mathcal{P}(-1, 0]$. Then $(\mathcal{T}, \mathcal{F})$ defines a torsion theory for $\mathrm{Coh}(X)$ with tilt $\mathcal{P}(0, 1] = \mathcal{A}_\sigma$.

---

[9] An alternative proof can be given by using the spectral sequence
$$E_2^{p,q} = \mathrm{Ext}^p(H^{-q}(A), E) \Rightarrow \mathrm{Ext}^{p+q}(A, E).$$

Proof. Apply Proposition 4.20 to compare the standard t-structure $\mathcal{D}^{\leq 0} \subset$ $D^b(X)$ with $\mathcal{D}^{\leq 0}_{\sigma} \subset D^b(X)$ given by σ. The assumptions $\mathcal{D}^{\leq 0}_{\sigma} \subset \mathcal{D}$ and $\mathcal{D}^{\leq -1} \subset \mathcal{D}^{\leq 0}_{\sigma}$ are easily verified. Here, $\mathcal{D}^{\leq 0}_{\sigma} = \mathcal{P}(>0)$. $\qquad\square$

Note that so far we have not described the torsion theory $(\mathcal{T}, \mathcal{F})$ explicitly, but ii) in Claim 2 can now be proven easily. Indeed, decompose a torsion sheaf $E \in \text{Coh}(X)$ with respect to the torsion theory $(\mathcal{T}, \mathcal{F})$ as $A \to E \to B$ with $A \in \mathcal{P}(0, 1]$ and $B \in \mathcal{P}(-1, 0]$. Then $H^0(B)$ is torsion free by Claim 1, i). But the long cohomology sequence yields a surjection $E = H^0(E) \twoheadrightarrow H^0(B)$ and an exact sequence $H^1(E) \to H^1(B) \to H^2(A)$. Thus, if E is torsion, then $H^0(B) = 0$, and, as $H^1(E) = 0 = H^2(A)$, also $H^1(B) = 0$. Therefore, $B = 0$ and hence $E \in \mathcal{P}(0, 1]$.

For the following we need that the stability condition is numerical.

Claim 4: The imaginary part of Z is of the form $\langle (0, \omega, \beta), v(E) \rangle$ for some $\beta \in \mathbb{R}$ and $\omega \in \text{NS}(X) \otimes \mathbb{R}$ with $(\omega.C) > 0$ for all curves $C \subset X$.

Proof. Here, $v(E) = \text{ch}(E)\sqrt{\text{td}(X)}$ is the Mukai vector of E, which for K3 surfaces equals $(\text{rk}(E), c_1(E), \chi(E) - \text{rk}(E))$. The Mukai pairing $\langle \ , \ \rangle$ is the usual intersection pairing with a sign in the pairing of H^0 and H^4. Since it is non-degenerate and Z is numerical, there exists a vector $w = (w_0, w_1, w_2) \in \text{NS}(X) \otimes \mathbb{C}$ such that $Z(E) = \langle w, v(E) \rangle$ for all E. As $Z(k(x)) \in \mathbb{R}_{<0}$ and $v(k(x)) = (0, 0, 1)$, one has $w_0 \in \mathbb{R}_{>0}$ and thus $\text{Im}(w) = (0, \omega, \beta)$.

Let $C \subset X$ be a curve. Then by Claim 2, $\mathcal{O}_C \in \mathcal{P}(0, 1]$. Since $v(\mathcal{O}_C) = (0, [C], \ )$, this yields $(\omega.C) = \text{Im}(Z(\mathcal{O}_C)) \geq 0$. If $(\omega.C) = 0$, then $Z(\mathcal{O}_C) \in \mathbb{R}_{<0}$ and hence $\mathcal{O}_C \in \mathcal{P}(1)$. The stable factors of $\mathcal{O}_C$, which are all in $\mathcal{P}(1)$, are by Claim 1 of the form $k(x)$ or $E[1]$ with $E \in \text{Coh}(X)$ locally free. Since their ranks add up to $\text{rk}(\mathcal{O}_C) = 0$, only point sheaves $k(x)$ can in fact occur. But this would yield the contradiction $(0, [C], *) = v(\mathcal{O}_C) = \sum v(k(x_i)) = (0, 0, n)$. $\qquad\square$

Claim 5: Let ω and β be as before. Suppose $E \in \text{Coh}(X)$ is torsion free and μ_ω-stable. Then
i) $E \in \mathcal{T}$ if and only if $\mu_\omega(E) > \beta$.
ii) $E \in \mathcal{F}$ if and only if $\mu_\omega(E) \leq \beta$.

Proof. Since $(\mathcal{T}, \mathcal{F})$ is a torsion theory, there exists a short exact sequence $0 \to A \to E \to B \to 0$ in $\text{Coh}(X)$ with $A \in \mathcal{T}$ and $B \in \mathcal{F}$. By Claim 2 all torsion sheaves are in $\mathcal{T}$. Thus, since $\text{Hom}(\mathcal{T}, \mathcal{F}) = 0$, the sheaf B must be torsion free.

By definition of $\mathcal{T}$ one has $A \in \mathcal{P}(0, 1]$ and thus $\text{Im}(Z(A)) \geq 0$ which is equivalent to $\mu_\omega(A) \geq \beta$. Similarly, for $B \in \mathcal{F}$, one has $B \in \mathcal{P}(-1, 0]$ and hence $\text{Im}(Z(B)) \leq 0$ or, equivalently, $\mu_\omega(B) \leq \beta$. If both $A \neq 0 \neq B$, this would contradict μ_ω-stability of E. Thus, $E = A \in \mathcal{T}$ or $E = B \in \mathcal{F}$. Clearly, if $\mu_\omega(E) > \beta$, then $E \in \mathcal{T}$. Similarly, if $\mu_\omega(E) < \beta$, then $E \in \mathcal{F}$. Thus, only the case $\mu_\omega(E) = \beta$, which is equivalent to $\text{Im}(Z(E)) = 0$, remains to be settled. Suppose $E \in \mathcal{T}$, then $E \in \mathcal{P}(1)$ and by Claim 1 all stable factors would be of the form $k(x)$ or $F[1]$ with F locally free. Thus, the sum of the ranks would be ≤ 0 contradicting the assumption that E is torsion free. $\square$

This finishes the proof of Theorem 4.39. Indeed, from Claim 2, ii) and Claim 5 one deduces $\mathcal{T}_{(\omega,\beta)} \subset \mathcal{T}$ and $\mathcal{F}_{(\omega,\beta)} \subset \mathcal{F}$. Since both, $(\mathcal{T}, \mathcal{F})$ and $(\mathcal{T}_{(\omega,\beta)}, \mathcal{F}_{(\omega,\beta)})$, are torsion theories for $\text{Coh}(X)$, they actually coincide. But by Claim 3, the tilt of $\text{Coh}(X)$ with respect to $(\mathcal{T}, \mathcal{F})$ is $\mathcal{A}_\sigma$, which, therefore, coincides with the tilt of $\text{Coh}(X)$ with respect to $(\mathcal{T}_{(\omega,\beta)}, \mathcal{F}_{(\omega,\beta)})$. But the latter is by definition just $\mathcal{A}(\omega, \beta)$. $\square$

Corollary 4.40. *If X is a smooth projective surface, then $\text{Coh}(X)$ cannot be the heart of a numerical stability condition.*

Proof. Suppose $\text{Coh}(X)$ is the heart of a stability condition. Since point sheaves $k(x) \in \text{Coh}(X)$ do not contain proper subsheaves, they are automatically stable. Now use $Z(I_{x_1,\ldots,x_n}) = Z(\mathcal{O}_X) - \sum Z(k(x_i)) = Z(\mathcal{O}_X) - nZ(k(x))$ which would be contained in the lower half plane $-\mathbb{H}$ except if $Z(k(x)) \in \mathbb{R}_{<0}$. Thus, all point sheaves $k(x)$ are stable of phase one. By Theorem 4.39, this shows $\text{Coh}(X) = \mathcal{A}(\omega, \beta)$. But on a projective surface there always exists a μ-stable vector bundle $E \in \text{Coh}(X)$ with $\mu_\omega(E) \leq \beta$ and thus $E[1] \in \mathcal{A}(\omega, \beta)$.[10] Contradiction. $\square$

Remark 4.41. Compare the corollary with Claim 3 which in particular shows that $\text{Coh}(X)$ is the tilt of the heart of a stability condition. Indeed, $\text{Coh}(X)[1]$ is the tilt of $\mathcal{A}(\omega, \beta)$ with respect to the torsion theory $\mathcal{T}_A := \mathcal{F}[1]$ and $\mathcal{F}_A := \mathcal{T}$. So, not any tilt of the heart of a stability condition is again the heart of a stability condition. The reason in our specific example is that $Z(\mathcal{F}_A)$ 'spreads out' over $\mathbb{H}$, e.g. $Z(k(x)) \in \mathbb{R}_{<0}$ whereas $\text{Re}(Z(E_n))/\text{Im}(Z(E_n)) \to \infty$ for E_n a sequence of μ-stable bundles with fixed slope $\mu > \beta$ but $c_2(E_n) \to \infty$.

Remark 4.42. If X is a K3 surface with $\text{Pic}(X) = 0$, in particular X is not projective, then $\text{Coh}(X)$ is the heart of a stability condition. As a stability function in this case one can choose $Z(E) = -r(\alpha + i\beta) - s$ with $\beta < 0$, where $v(E) = (r, 0, s)$. See [21].

[10] The argument is still valid although ω only satisfies $(\omega.C) > 0$.

Remark 4.43. For the following observations see [22].

i) Recall that the only minimal objects in $\mathrm{Coh}(X)$ are the point sheaves $k(x)$. For K3 surfaces the minimal objects in the tilted category $\mathcal{A}(\omega, \beta)$ are of the form $k(x)$ and $E[1]$, where E is a μ stable vector bundle with $\mu_\omega(E) = \beta$.

ii) By a classical theorem of Gabriel, $\mathrm{Coh}(X) \simeq \mathrm{Coh}(X')$ (as k-linear categories) if and only if $X \simeq X'$ (as varieties over k). But there exist non isomorphic (K3) surfaces with equivalent derived categories $\mathrm{D}^{\mathrm{b}}(X) \simeq \mathrm{D}^{\mathrm{b}}(X')$ (as k-linear triangulated categories). For projective K3 surfaces derived equivalence is determined by the tilted abelian categories: $\mathrm{D}^{\mathrm{b}}(X) \simeq \mathrm{D}^{\mathrm{b}}(X')$ if and only if there exist (ω, β) and (ω', β') on X resp. X' with $\mathcal{A}(\omega, \beta) \simeq \mathcal{A}(\omega', \beta')$.

4.3.2 Construction of hearts

For fixed ω and $\beta = (B.\omega)$ as before, we consider the function

$$Z(E) := \langle \exp(B + i\omega), v(E) \rangle.$$

Here, $v(E) = \mathrm{ch}(E)\sqrt{\mathrm{td}(X)} = (r, \ell, s)$ and $\exp(B + i\omega) = (1, B + i\omega, \frac{B^2 - \omega^2}{2} + i(B.\omega))$.

For the following assume that X is a K3 surface.[11]

Proposition 4.44. *If $\omega \in \mathrm{NS}(X) \otimes \mathbb{R}$ is ample and such that $Z(E) \notin \mathbb{R}_{\leq 0}$ for all spherical $E \in \mathrm{Coh}(X)$, then Z is a stability function on the abelian category $\mathcal{A}(\omega, \beta)$.*

Proof. As we will see, the assumption $Z(E) \notin \mathbb{R}_{\leq 0}$ is satisfied whenever $\omega^2 > 2$. For the definition of 'spherical' see Section 4.5.2.

Since Z is linear and $\mathcal{A}(\omega, \beta)$ is the tilt of $\mathrm{Coh}(X)$ with respect to $(\mathcal{T}_{(\omega,\beta)}, \mathcal{F}_{(\omega,\beta)})$, it suffices to show that $Z(E) \in \overline{\mathbb{H}}$ for $E \in \mathcal{T}_{(\omega,\beta)}$ and $-Z(E) \in \overline{\mathbb{H}}$ for $E \in \mathcal{F}_{(\omega,\beta)}$.

By definition of the torsion theory, $\mathrm{Im}(Z(E)) = \langle (0, \omega, \beta), v(E) \rangle = (\omega.\ell) - r\beta$, which is ≥ 0 for $E \in \mathcal{T}_{(\omega,\beta)}$ and ≤ 0 for $E \in \mathcal{F}_{(\omega,\beta)}$. Moreover, if $E \in \mathcal{T}_{(\omega,\beta)}$ is not torsion, then the inequality is strict. If $0 \neq E \in \mathrm{Coh}(X)$ is supported in dimension zero, then $\mathrm{Im}(Z(E)) = 0$, but $\mathrm{Re}(Z(E)) = -s = -h^0(E) < 0$. Hence, $Z(E) \in \mathbb{R}_{<0} \subset \overline{\mathbb{H}}$. If $0 \neq E \in \mathcal{T}_{(\omega,\beta)}$ has support in dimension one, then $\mathrm{Im}(Z(E)) > 0$. So lets consider $E \in \mathcal{F}_{(\omega,\beta)}$. One easily reduces to the case that E is μ-stable with slope $\mu_\omega(E) = \beta$. In particular, $(\ell - rB) \in \omega^\perp$ and thus, by Hodge index theorem, $(\ell - rB)^2 \leq 0$. Now one rewrites the real part of $Z(E)$ as

---

[11] For arbitrary X one would need to assume $\omega^2 \gg 0$.

$$\mathrm{Re}(Z(E)) = (B.\ell) - s - \frac{B^2 - \omega^2}{2} r = \frac{1}{2r} \left((\ell^2 - 2rs) + r^2\omega^2 - (\ell - rB)^2 \right).$$

Clearly, $r^2\omega^2 - (\ell - rB)^2 > 0$. And for the remaining summand observe that $(\ell^2 - 2rs) = v(E)^2 = -\chi(E, E) = -2 + \mathrm{ext}^1(E, E) \geq -2$ for any stable E. Thus, $\mathrm{Re}(Z(E)) \in \mathbb{R}_{>0}$ whenever E is not rigid (or $r\omega^2 > 2$). More precisely, it suffices to assume $Z(E) \notin \mathbb{R}_{\leq 0}$ for all spherical $E \in \mathrm{Coh}(X)$ to ensure that $Z(E[1]) \in \overline{\mathbb{H}}$ for all $E \in \mathcal{F}_{(\omega, \beta)}$. $\square$

We let ω be ample and assume that Z defines a stability function on $\mathcal{A}(\omega, \beta)$.

Proposition 4.45. *For rational $\omega, B \in \mathrm{NS}(X) \otimes \mathbb{Q}$, the stability function $Z = \langle \exp(B + i\omega), \ \rangle$ on $\mathcal{A}(\omega, \beta)$ satisfies the Harder–Narasimhan property and the local finiteness condition (cf. Definition 4.52). In particular, finite Jordan–Hölder filtrations exist.*

Proof. In order to show the existence of Harder–Narasimhan filtrations for all objects in $\mathcal{A} := \mathcal{A}(\omega, \beta)$, it suffices to show that any descending filtration $\ldots \subset E_{i+1} \subset E_i \subset \ldots \subset E_0 = E$ with increasing phases $\phi(E_{i+1}) > \phi(E_i) > \ldots > \phi(E_0)$ stabilizes and similarly for chains of quotients (cf. [9, Prop. 2.4]).

Since B and ω are rational, the image $Z(\mathcal{A}) \subset \mathbb{C}$ is discrete. Indeed, if $B, \omega \in \frac{1}{m}\mathrm{NS}(X)$, then $m^2 Z(\mathcal{A}) \subset \mathbb{Z}[i]$. Using linearity of Z, one deduces that $\mathrm{Im}(Z(E_{i+1})) \leq \mathrm{Im}(Z(E_i))$ for all i and hence, by discreteness of $Z(\mathcal{A})$, $\mathrm{Im}(Z(E_i)) \equiv \mathrm{const}$ for $i \gg 0$. But then $\mathrm{Im}(Z(E_i/E_{i+1})) = 0$ which is a contradiction.

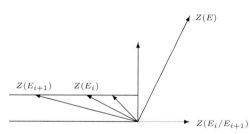

The argument to deal with chains of quotients is trickier, see [10, Prop. 7.1]. The local finiteness is again elementary (see [10, Lem. 4.4]): Fix any $\varepsilon < \frac{1}{2}$. Then for any ϕ the image of $Z : \mathcal{P}(\phi - \varepsilon, \phi + \varepsilon) \to \mathbb{C}$ will be contained in a region strictly smaller than a half plane. Thus, subobjects and quotients of a fixed $E \in \mathcal{P}(\phi - \varepsilon, \phi + \varepsilon)$ will have image in a bounded region, which for Z discrete implies the finiteness of ascending and descending chains.

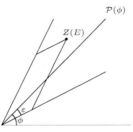

Corollary 4.46. *If ω, $B \in NS(X) \otimes \mathbb{Q}$ with ω ample and such that $Z(E) \notin \mathbb{R}_{\leq 0}$ for all spherical $E \in Coh(X)$ (e.g. $\omega^2 > 2$), then Z defines a stability condition on $\mathcal{A}(\omega, \beta)$ and, therefore, a stability condition on $D^b(X)$ with heart $\mathcal{A}(\omega, \beta)$.* □

In fact, the same result holds for real ω, $B \in NS(X) \otimes \mathbb{R}$, but this is proved a posteriori, after a detailed analysis of the component of $Stab(X)$ containing the stability conditions with rational Z constructed above.

4.4 The topological space of stability conditions

This lecture is devoted to fundamental results of Bridgeland concerning the topological structure of the space of stability conditions. Since a stability condition consists of a slicing and a stability function, the natural reflex is to construct topologies on the space of slicings and on the space of stability functions separately (the latter being a linear space) and use the product topology. This is roughly what happens, but details are intricate. We will only discuss the main aspects, in Section 4.4.1 for the space of slicings and in Section 4.4.2 for $Stab(\mathcal{D})$. The main result about the projection from stability conditions to stability functions being a local homeomorphism can be found in Section 4.4.3.

4.4.1 Topology of Slice($\mathcal{D}$)

As before, $\mathcal{D}$ will denote a triangulated category. The easier case of an abelian category is left to the reader. Recall that a slicing $\mathcal{P}$ of $\mathcal{D}$ consists of full additive subcategories $\mathcal{P}(\phi) \subset \mathcal{D}$, $\phi \in \mathbb{R}$ satisfying the conditions i)-iii) in Definition 4.21. In particular, any object $E \in \mathcal{D}$ has a filtration by exact triangles with factors $A_1 \in \mathcal{P}(\phi_1), \ldots, A_n \in \mathcal{P}(\phi_n)$ such that $\phi_1 > \ldots > \phi_n$. We will also write $\phi^+(E) := \phi_1$ and $\phi^-(E) := \phi_n$ (or, if the dependence on the given slicing needs to be stressed, $\phi_{\mathcal{P}}^{\pm}(E)$). With this notation, $\mathcal{P}(I) := \{E \in \mathcal{D} \mid \phi^{\pm}(E) \in I\}$ for any interval $I \subset \mathbb{R}$. The set of all slicings of $\mathcal{D}$ is denoted by $\mathrm{Slice}(\mathcal{D})$.

Definition 4.47. For $\mathcal{P}, \mathcal{Q} \in \mathrm{Slice}(\mathcal{D})$ one defines

$$d(\mathcal{P}, \mathcal{Q}) := \sup\{|\phi_{\mathcal{P}}^{\pm}(E) - \phi_{\mathcal{Q}}^{\pm}(E)| \mid 0 \neq E \in \mathcal{D}\}.$$

Obviously, $d(\mathcal{P}, \mathcal{Q}) \in [0, \infty]$.

Claim 1: With this definition, $d(\ ,\ )$ is a generalized metric, i.e. it satisfies all the axioms for a distance function except that the distance can be ∞.

Proof. The symmetry and the triangle inequality are straightforward to check. Suppose $d(\mathcal{P}, \mathcal{Q}) = 0$. Then $E \in \mathcal{P}(\phi)$, which is equivalent to $\phi_{\mathcal{P}}^{\pm}(E) = \phi$, implies $\phi_{\mathcal{Q}}^{\pm}(E) = \phi$ and thus $E \in \mathcal{Q}(\phi)$. Therefore, $\mathcal{P}(\phi) \subset \mathcal{Q}(\phi)$. Reversing the role of $\mathcal{P}$ and $\mathcal{Q}$ yields $\mathcal{P} = \mathcal{Q}$. □

Thus, $d(\ ,\ )$ defines a topology on $\mathrm{Slice}(\mathcal{D})$. Note that two slicings $\mathcal{P}, \mathcal{Q}$ with $d(\mathcal{P}, \mathcal{Q}) = \infty$ are contained in different connected components. Indeed, $U_{\mathcal{P}} := \{\mathcal{R} \mid d(\mathcal{P}, \mathcal{R}) < \infty\}$ is by definition an open neighbourhood of $\mathcal{P}$. An open neighbourhood $U_{\mathcal{Q}}$ of $\mathcal{Q}$ is defined similarly. Then use the triangle inequality to show that $U_{\mathcal{P}} \cap U_{\mathcal{Q}} = \emptyset$ and that $\mathrm{Slice}(\mathcal{D}) \setminus U_{\mathcal{P}}$ is open. Hence, $U_{\mathcal{P}}$ and $U_{\mathcal{Q}}$ are connected components of $\mathrm{Slice}(\mathcal{D})$.

Claim 2: $d(\mathcal{P}, \mathcal{Q}) = \inf\{\varepsilon \mid \mathcal{Q}(\phi) \subset \mathcal{P}[\phi - \varepsilon, \phi + \varepsilon]\ \forall \phi \in \mathbb{R}\}$.

This is best done as an exercise, but here is the argument anyway: Let's write $d := d(\mathcal{P}, \mathcal{Q})$ and $\delta := \inf\{\ \}$. The inequality $d \geq \delta$ is straightforward. For the other direction, one has to show that $|\phi_{\mathcal{P}}^{\pm}(E) - \phi_{\mathcal{Q}}^{\pm}(E)| \leq \delta$ for all $E \in \mathcal{D}$. Consider the Harder–Narasimhan filtration of E with respect to $\mathcal{Q}$ and denote its semistable factors by $A_1, \ldots, A_n$. Then $\phi_{\mathcal{Q}}^+(E) = \phi_{\mathcal{Q}}(A_1)$ and $\phi_{\mathcal{Q}}^-(E) = \phi_{\mathcal{Q}}(A_n)$. By definition of δ, one has $\phi_{\mathcal{P}}^+(A_i) \leq \phi_{\mathcal{Q}}(A_i) + \delta$ and $\phi_{\mathcal{P}}^-(A_i) \geq \phi_{\mathcal{Q}}(A_i) - \delta$. But $\phi_{\mathcal{P}}^+(E) \leq \max\{\phi_{\mathcal{P}}^+(A_i)\}$ and $\phi_{\mathcal{P}}^-(E) \geq \min\{\phi_{\mathcal{P}}^-(A_i)\}$. Hence, $\phi_{\mathcal{P}}^+(E) \leq \phi_{\mathcal{Q}}^+(E) + \delta$ and $\phi_{\mathcal{P}}^-(E) \geq \phi_{\mathcal{Q}}^-(E) - \delta$. Continuing along these lines eventually yields $|\phi_{\mathcal{P}}^{\pm}(E) - \phi_{\mathcal{Q}}^{\pm}(E)| \leq \delta$ and, therefore, $d \leq \delta$.

Proposition 4.48. *If for stability conditions $\sigma = (\mathcal{P}, Z)$ and $\tau = (\mathcal{Q}, W)$ one has $W = Z$ and $d(\mathcal{P}, \mathcal{Q}) < 1$, then $\sigma = \tau$. See* [9, Lem. 6.4].

Note that the strict inequality is needed, as e.g. $d(\mathcal{P}, \mathcal{P}[1]) = 1$.

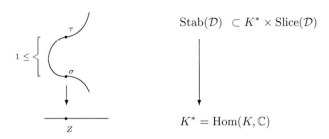

$$\text{Stab}(\mathcal{D}) \subset K^* \times \text{Slice}(\mathcal{D})$$

$$K^* = \text{Hom}(K, \mathbb{C})$$

Proof. Suppose there exists a $E \in \mathcal{P}(\phi) \setminus \mathcal{Q}(\phi)$.

i) If $E \in \mathcal{Q}(> \phi)$, then by assumption $E \in \mathcal{Q}(\phi, \phi + 1)$, i.e. $W(E)$ is contained in the half plane left to $Z(E)$ which contradicts $W = Z$.

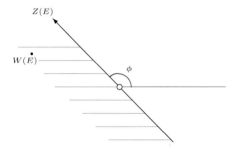

ii) For $E \in \mathcal{Q}(< \phi)$ the argument is similar.

iii) The remaining case is $E \in \mathcal{Q}(\phi - 1, \phi + 1)$. Then there exists an exact triangle

$$A \to E \to B$$

with $A \in \mathcal{Q}(\phi, \phi + 1)$ and $B \in \mathcal{Q}(\phi - 1, \phi]$. If $A \in \mathcal{P}(\leq \phi)$, then $A \in \mathcal{P}(\phi - 1, \phi]$, which leads to the same contradiction as above. Thus, there exists $0 \neq C \in \mathcal{P}(\psi)$ with $\psi > \phi$ and a non-trivial morphism $C \to A$ which factors via $B[-1]$ as $E \in \mathcal{P}(\phi)$:

$$
\begin{array}{ccccccc}
 & & C & & & & \\
 & {\scriptstyle g}\swarrow & {\scriptstyle \neq 0}\downarrow & {\scriptstyle =0}\searrow & & & \\
B[-1] & \longrightarrow & A & \longrightarrow & E & \longrightarrow & B
\end{array}
$$

But now $B[-1] \in \mathcal{Q}(\phi - 2, \phi - 1] \subset \mathcal{P}(\phi - 3, \phi]$ and, therefore, $g = 0$. Contradiction.   □

4.4.2 Topology of Stab($\mathcal{D}$)

Let Stab($\mathcal{D}$) be the space of stability conditions $\sigma = (\mathcal{P}, Z)$ for which Z factors via a fixed quotient $K(\mathcal{D}) \twoheadrightarrow \Gamma$. For simplicity, we will usually assume $\Gamma \simeq \mathbb{Z}^n$ and write $\Gamma^* = \mathrm{Hom}(\Gamma, \mathbb{C})$. Thus,

$$\mathrm{Stab}(\mathcal{D}) \subset \Gamma^* \times \mathrm{Slice}(\mathcal{D}).$$

It is then tempting, in particular when $\Gamma \simeq \mathbb{Z}^n$, to endow Stab($\mathcal{D}$) with the product topology. This is essentially what will happen, but not quite. Bridgeland defines a topology on Stab($\mathcal{D}$) such that for any connected component Stab($\mathcal{D}$)$^\circ$ there exists a linear topological subspace $V^\circ \subset \Gamma^*$ such that Stab($\mathcal{D}$)$^\circ \subset \mathrm{Slice}(\mathcal{D}) \times V^\circ$ with the induced topology. In particular, if Γ is of finite rank, then indeed the topology on each connected component of Stab($\mathcal{D}$) is induced by the product topology.

For a fixed stability condition $\sigma = (Z, \mathcal{P}) \in \mathrm{Stab}(\mathcal{D})$ a generalized norm $\| \ \|_\sigma : \Gamma^* \to [0, \infty]$ is defined by

$$\|U\|_\sigma := \sup \left\{ \frac{|U(E)|}{|Z(E)|} \ \Big| \ E \ \sigma\text{-semistable} \right\}.$$

Then let $V_\sigma := \{U \in \Gamma^* \mid \|U\|_\sigma < \infty\}$. It is easy to check that $\| \ \|_\sigma$ is indeed a generalized norm which becomes a finite norm on V_σ. For example, if $\|U\|_\sigma = 0$, then $U(E) = 0$ for all σ-semistable E and, using the existence of the Harder–Narasimhan filtration for all objects in $\mathcal{D}$, this implies $U = 0$.

Remark 4.49. The approach of Kontsevich and Soibelman (cf. [28]) to the topology on the space of stability conditions is slightly different. Roughly, the notion in [28] is stronger and leads to Bridgeland stability conditions with maximal $V_\sigma = \Gamma^*$. However, in the case of the stability conditions on K3 surfaces that have been introduced earlier and that will be studied further in the next section, both notions coincide a posteriori.

Here are a few details. In [28], a stability condition $\sigma = (\mathcal{P}, Z)$ is required to admit a quadratic form Q on $\Gamma \otimes \mathbb{R}$ such that:

i) Q is negative definite on $\mathrm{Ker}(Z)$ and

ii) If $|\phi^+(E) - \phi^-(E)| < 1$, then $Q([E]) \geq 0$. (In fact, this is only a weak version of what is really required in [28].)

In particular: iii) all semistable $0 \neq E$ satisfy $Q([E]) \geq 0$.

Note that i) and iii) together are equivalent to the existence of a constant C such that for all semistable $0 \neq E$ one has

$$\|[E]\| < C \cdot |Z(E)|,$$

which is called the *support property*. (Here, $\| \ \|$ is an arbitrary norm on $\Gamma \otimes \mathbb{C}$ which is assumed to be finite dimensional.)

The support property automatically implies $V_\sigma = \Gamma^*$. Indeed, since $U \in \Gamma^*$ is in particular continuous, there exists a constant D such that $|U(\alpha)| \leq D \cdot \|\alpha\|$ for all $\alpha \in \Gamma$. Combined with the support property, it shows that for any semistable $0 \neq E$ one has $|U(E)| < C \cdot D \cdot |Z(E)|$ and hence $\|U\|_\sigma < C \cdot D$.

The actual condition ii) in [28] implies the finiteness condition that we have avoided so far, see Definition 4.52. Indeed, the support property implies that $Z(\Gamma) \subset \mathbb{C}$ is discrete, see [26, Sect. 2.1], which in turn implies local finiteness (cf. Remark 4.53).

Now, in order to define a topology on $\mathrm{Stab}(\mathcal{D})$ Bridgeland considers for $\sigma = (\mathcal{P}, Z)$ the set

$$B_\varepsilon(\sigma) := \{\tau = (\mathcal{Q}, W) \mid \|W - Z\|_\sigma < \sin(\pi\varepsilon), \ d(\mathcal{P}, \mathcal{Q}) < \varepsilon\}.$$

Remark 4.50. The two inequalities are compatible in the following sense. If E is σ-semistable, then the condition $\|W - Z\|_\sigma < \sin(\pi\varepsilon)$ implies $|\phi_\tau(E) - \phi_\sigma(E)| < \varepsilon$, which confirms $d(\mathcal{P}, \mathcal{Q}) < \varepsilon$. Indeed, if φ is as below, then $\sin(\pi\varphi) \leq \frac{|(W-Z)(E)|}{|Z(E)|} < \sin(\pi\varepsilon)$.

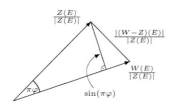

Claim: The sets $B_\varepsilon(\sigma)$ form the basis of a topology which on each connected component $\mathrm{Stab}(\mathcal{D})^\circ$ coincides with the topology induced by the product topology under the inclusion $\mathrm{Stab}(\mathcal{D}) \subset \Gamma^* \times \mathrm{Slice}(\mathcal{D})$.

The main technical point in the proof, is that for $\tau \in B_\varepsilon(\sigma)$ one always finds $C_1, C_2 > 0$ such that $C_1 \cdot \| \ \|_\sigma \leq \| \ \|_\tau \leq C_2 \cdot \| \ \|_\sigma$, i.e. $\| \ \|_\sigma \sim \| \ \|_\tau$. See [9, Lem. 6.2].

Note that $V_\sigma = V_\tau$ for σ, τ contained in the same connected component $\mathrm{Stab}(\mathcal{D})^\circ$.

4.4.3 Main result

We now come to the main result of [9]. Consider a connected component $\mathrm{Stab}(\mathcal{D})^\circ \subset \mathrm{Stab}(\mathcal{D})$ of the space of locally finite (see Definition 4.52) stability conditions on $\mathcal{D}$ with stability functions that factor via a fixed quotient $K(\mathcal{D}) \twoheadrightarrow \Gamma$. Let $V^{\circ} \subset \Gamma^*$ be the associated linear space (i.e. $V^{\circ} = V_{\sigma}$ for any $\sigma \in \text{Stab}(\mathcal{D})^{\circ}$).

Theorem 4.51. (Bridgeland) *The natural projection* $\text{Stab}(\mathcal{D})^{\circ} \to V^{\circ}$ *is a local homeomorphism.*

In fact, Bridgeland proves that for small ε the projection yields a homeomorphism $B_{\varepsilon}(\sigma) \xrightarrow{\sim} B_{\sin(\pi\varepsilon)}(Z)$.

Definition 4.52. A stability condition σ is *locally finite* if there exists a $0 < \eta$ such that $\mathcal{P}(\phi - \eta, \phi + \eta)$ is a category of finite type for all $\phi \in \mathbb{R}$.

A category is of finite type if it is artinian and noetherian, i.e. descending and ascending chains of subobjects stabilize. In particular, all abelian $\mathcal{P}(\phi)$ are of finite type which is equivalent to saying that any semistable object has a finite filtration with stable quotients.

The condition 'locally finite' is needed when it comes to proving the main lemma that says that for small ε the inequality $\|W - Z\|_{\sigma} < \sin(\pi\varepsilon)$ implies that there exists a stability condition $\tau = (\mathcal{Q}, W) \in B_{\varepsilon}(\sigma)$, i.e. such that $d(\mathcal{P}, \mathcal{Q}) < \varepsilon$. For the details of the proof we have to refer to the original [9].

Remark 4.53. The local finiteness of a stability condition holds automatically when the image of Z is discrete [10, Lem. 4.4], cf. proof of Proposition 4.45.

Warning: If a stability function Z on the heart of a bounded t-structure is given, then knowing that the image of Z is discrete does not automatically give that Z has the Harder–Narasimhan property.

Examples of stability conditions that are not locally finite are easily constructed, see e.g. Example 4.31.

Remark 4.54. A connected component $\text{Stab}(\mathcal{D})^{\circ}$ is called *full* if $V^{\circ} = \Gamma^*$. Then in [47] it is shown that any full component is complete with respect to the metric

$$d(\sigma, \tau) := \sup_{0 \neq E \in \mathcal{D}} \left\{ |\phi_{\mathcal{P}}^{\pm}(E) - \phi_{\mathcal{Q}}^{\pm}(E)|, \ |\log \frac{m_{\sigma}(E)}{m_{\tau}(E)}| \right\}$$

(which induces the topology on $\text{Stab}(\mathcal{D})$ as described before). Here, $\sigma = (\mathcal{P}, Z)$, $\tau = (\mathcal{Q}, W)$ and $m_{\sigma}(E)$ is the mass of E, i.e. $m_{\sigma}(E) = \sum |Z(\Lambda_i)|$ if A_i are the semistable factors of E.

4.5 Stability conditions on K3 surfaces

The last lecture discusses results of Bridgeland on stability conditions for K3 surfaces. In Section 4.3 explicit examples have been constructed and a characterization of those in terms of stability of point sheaves has been proved. Applying autoequivalences of $D^b(X)$ produces more examples and Bridgeland in [10] describes in this way a whole connected component of $\mathrm{Stab}(X)$. We will sketch the main steps of the argument and present an intriguing conjecture, due to Bridgeland, predicting that this distinguished connected component is simply connected. At the end, we will rephrase the conjecture in terms of the fundamental group of a certain moduli stack.

4.5.1 Main theorem and conjecture

For a triangulated category $\mathcal{D}$, we have introduced $\mathrm{Stab}(\mathcal{D})$, the space of locally finite stability conditions $\sigma = (\mathcal{P}, Z)$ on $\mathcal{D}$. Implicitly, the stability function $Z : K(\mathcal{D}) \to \mathbb{C}$ is assumed to factor through a fixed quotient $K(\mathcal{D}) \twoheadrightarrow \Gamma$, which in the application is obtained by quotienting $K(\mathcal{D})$ by numerical equivalence.

Following [9], we have equipped $\mathrm{Stab}(\mathcal{D})$ with a topology which on each connected component $\mathrm{Stab}(\mathcal{D})^\circ \subset \mathrm{Stab}(\mathcal{D})$ is induced by the product topology on $\Gamma^* \times \mathrm{Slice}(\mathcal{D})$. Moreover, according to Theorem 4.51, for each $\mathrm{Stab}(\mathcal{D})^\circ$ there exists a linear subspace $V^\circ \subset \Gamma^*$ such that the projection $\mathrm{Stab}(\mathcal{D})^\circ \to V^\circ$ is a local homeomorphism. Ideally, this would be a topological covering of an open subset of V°, but the covering property is known to be violated in examples, see e.g. [34]. However, in the case we are interested in here, namely $\mathcal{D} = D^b(X)$ with X a smooth projective K3 surface, the covering property has been verified in [10]. But we first need to describe the image.

In Section 4.3.2, stability conditions $\sigma = (\mathcal{P}, Z)$ on $D^b(X)$ have been constructed with $Z(E) = \langle \exp(B + i\omega), v(E) \rangle$ and heart $\mathcal{A}(\omega, \beta) = \mathcal{P}(0, 1]$ constructed as the tilt of $\mathrm{Coh}(X)$ with respect to a certain torsion theory $(\mathcal{T}_{(\omega,\beta)}, \mathcal{F}_{(\omega,\beta)})$. Here, $\omega, B \in \mathrm{NS}(X) \otimes \mathbb{Q}$, such that ω is ample with $\omega^2 > 2$, and $\beta = (B.\omega)$. The class

$$\exp(B + i\omega) = (1, \, B + i\omega, \, \frac{B^2 - \omega^2}{2} + i(B.\omega))$$

is contained in $N(X) \otimes \mathbb{Q}$, where $N(X)$ is the extended Néron–Severi lattice $N(X) = H^*_{alg}(X, \mathbb{Z}) = \mathbb{Z} \oplus \mathrm{NS}(X) \oplus \mathbb{Z}$. Using the non-degenerate Mukai pairing, we shall tacitly identify $N(X)^* = \mathrm{Hom}(N(X), \mathbb{C})$ with $N(X) \otimes \mathbb{C}$. In particular, any stability function Z can be written as $Z = \langle w, \ \rangle$ and thus the map $\sigma = (Z, \mathcal{P}) \longmapsto Z$ is viewed as a map

$$\text{Stab}(X) \to N(X) \otimes \mathbb{C}.$$

Elements in $N(X) \otimes \mathbb{C}$ of the form $\exp(B + i\omega)$ are contained in a distinguished subset $\mathcal{P}_0^+(X)$, which shall be introduced next.

Definition 4.55. Let

$$\mathcal{P}(X) \subset N(X) \otimes \mathbb{C}$$

be the open set of all $\Omega \in N(X) \otimes \mathbb{C}$ such that $\text{Re}(\Omega), \text{Im}(\Omega)$ span a positive plane in $N(X) \otimes \mathbb{R}$.

Here, $N(X) \otimes \mathbb{R}$ is endowed with the Mukai pairing, i.e. the standard intersection pairing with an extra sign on $H^0 \oplus H^4$. In particular, $N(X)$ is a lattice of signature $(2, \rho(X))$.

Example 4.56. For any $\omega \in \text{NS}(X) \otimes \mathbb{R}$ with $\omega^2 > 0$ and arbitrary $B \in \text{NS}(X) \otimes \mathbb{R}$, the class $\exp(B + i\omega)$ is contained in $\mathcal{P}(X)$. Indeed, $\text{Re}(\exp(i\omega)) = (1, 0, -\omega^2/2)$ and $\text{Im}(\exp(i\omega)) = \omega$ are orthogonal of square ω^2. Multiplication with $\exp(B)$ is an orthogonal transformation of $N(X) \otimes \mathbb{R}$, which proves the claim.

The orientations of two given positive planes in $N(X) \otimes \mathbb{R}$ can be compared via orthogonal projection. This leads to a decomposition

$$\mathcal{P}(X) = \mathcal{P}^+(X) \sqcup \mathcal{P}^-(X)$$

in two connected components. We can distinguish one, say $\mathcal{P}^+(X)$, by requiring that it contains $\exp(i\omega)$ with ω ample.

Instead of $\omega^2 > 2$ we assumed in Proposition 4.44 that $Z(E) \notin \mathbb{R}_{\leq 0}$ for all spherical $E \in \text{Coh}(X)$. The latter motivates a further shrinking of $\mathcal{P}^+(X)$.

Let $\Delta := \{\delta \in N(X) \mid \delta^2 = -2\}$ which in particular contains the classical (-2)-classes in $\text{NS}(X)$ such as $[C]$ with $\mathbb{P}^1 \simeq C \subset X$, but also $(1, 0, 1)$. Then one defines

$$\mathcal{P}_0^+(X) := \mathcal{P}^+(X) \setminus \bigcup_{\delta \in \Delta} \delta^\perp.$$

Since classes in Δ are real, the condition $\Omega \in \delta^\perp$ is equivalent to $\langle \text{Re}(\Omega), \delta \rangle = \langle \text{Im}(\Omega), \delta \rangle = 0$.

The following is the main result of [10]. Let $\mathrm{Stab}(X)^{\circ} \subset \mathrm{Stab}(X)$ be the connected component of numerical stability conditions containing those constructed in Corollary 4.46.

Theorem 4.57. *The natural projection* $\sigma = (\mathcal{P}, Z) \mapsto Z$ *defines a topological covering*

$$\mathrm{Stab}(X)^{\circ} \longrightarrow \mathcal{P}_0^+(X)$$

with $\mathrm{Aut}_0^{\circ}(\mathrm{D}^b(X))$ *as group of deck transformations.*

Here, $\mathrm{Aut}_0^{\circ}(\mathrm{D}^b(X))$ is the subgroup of $\mathrm{Aut}(\mathrm{D}^b(X))$ of all linear exact auto-equivalences $\Phi : \mathrm{D}^b(X) \xrightarrow{\sim} \mathrm{D}^b(X)$ that preserve the distinguished component $\mathrm{Stab}(X)^{\circ}$ and such that $\Phi = \mathrm{id}$ on $N(X)$.

Conjecture 4.58. (Bridgeland) *The component* $\mathrm{Stab}(X)^{\circ}$ *is simply connected and, therefore,* $\pi_1(\mathcal{P}_0^+(X)) \xrightarrow{\sim} \mathrm{Aut}_0^{\circ}(\mathrm{D}^b(X))$.

Remark 4.59. Stronger versions of this conjecture exist. E.g. one could conjecture that $\mathrm{Stab}(X)^{\circ}$ is the only connected component (of maximal dimension) of $\mathrm{Stab}(X)$ or, slightly weaker, that $\mathrm{Aut}(\mathrm{D}^b(X))$ preserves $\mathrm{Stab}(X)^{\circ}$.[12] Both would imply that

$$\mathrm{Aut}_0^{\circ}(\mathrm{D}^b(X)) = \mathrm{Aut}_0(\mathrm{D}^b(X)) = \mathrm{Ker}\left(\mathrm{Aut}(\mathrm{D}^b(X)) \to O(H^*(X, \mathbb{Z}))\right).$$

Since the image of the representation $\mathrm{Aut}(\mathrm{D}^b(X)) \to O(H^*(X, \mathbb{Z}))$ has been described (see [23]), this would eventually yield a complete description of $\mathrm{Aut}(\mathrm{D}^b(X))$. Note that before Bridgeland phrased the conjecture above, one had no idea how to describe $\mathrm{Aut}(\mathrm{D}^b(X))$, even conjecturally. Describing the highly complex group $\mathrm{Aut}(\mathrm{D}^b(X))$ in terms of a certain fundamental group (see also Section 4.5.4) is as explicit as a description can possibly get. It also fits well with Kontsevich's homological mirror symmetry which relates $\mathrm{Aut}(\mathrm{D}^b(X))$ of any Calabi–Yau variety X with the fundamental group of the moduli space of complex structures on its mirror [27].

Up to now, the conjecture has not been verified for a single projective K3 surface. The cases of generic non-projective K3 surfaces and generically twisted projective K3 surfaces (X, α) have been successfully treated in [21].

---

[12] In [23] it was shown that $\mathrm{Aut}(\mathrm{D}^b(X))$ at least preserves $\mathcal{P}_0^+(X)$.

4.5.2 Autoequivalences

Before discussing some aspects of the proof of Theorem 4.57, it is probably useful to give a few explicit examples of autoequivalences of $D^b(X)$ and to explain how loops around $\delta^\perp$, the generators of $\pi_1(\mathcal{P}_0^+(X))$, can possibly be responsible for elements in $\mathrm{Aut}(D^b(X))$.

As Mukai observed in [36], any $\Phi \in \mathrm{Aut}(D^b(X))$ induces an isometry Φ^H of $H^*(X, \mathbb{Z})$ (as before, endowed with the usual intersection pairing modified by a sign on $H^0 \oplus H^4$) that also respects the weight two Hodge structure given by $H^{2,0}$. By definition, $\mathrm{Aut}_0(D^b(X)) = \{\Phi \mid \Phi^H = \mathrm{id}\}$.

i) Any automorphism $f : X \xrightarrow{\sim} X$ induces naturally an autoequivalence: $\mathrm{Aut}(X) \hookrightarrow \mathrm{Aut}(D^b(X))$, $f \mapsto f_*$. The induced action on $H^*(X, \mathbb{Z})$ is just the standard one.

ii) If $L \in \mathrm{Pic}(X)$, then $E \mapsto E \otimes L$ defines an autoequivalence. Thus, $\mathrm{Pic}(X) \hookrightarrow \mathrm{Aut}(D^b(X))$. The induced action on $H^*(X, \mathbb{Z})$ is described by multiplication with $\exp(c_1(L))$.

iii) Recall that an object $E \in D^b(X)$ is called *spherical* if $\mathrm{Ext}^*(E, E)$ is two-dimensional, i.e.

$$\mathrm{Ext}^*(E, E) \simeq H^*(S^2, \mathbb{C}).$$

In particular, the Mukai vector $v(E) \in N(X)$ of a spherical object E is a (-2)-class, i.e. $v(E) \in \Delta$. Any spherical object $E \in D^b(X)$ induces an autoequivalence T_E which is called the *spherical twist* (or *Seidel–Thomas twist*) associated with E. It can be described as the Fourier–Mukai transform with Fourier–Mukai kernel given by the cone of the trace map $E^* \boxtimes E \to \mathcal{O}_\Delta$.

The spherical twist sends an object $F \in D^b(X)$ to the cone of the evaluation map $\mathrm{Ext}^*(E, F) \otimes E \to F$. The important features are: $T_E(E) \simeq E[1]$ and $T_E(F) \simeq F$ for any $F \in E^\perp$. For details see e.g. [20]. The induced action T_E^H on cohomology is given by reflection $s_{v(E)}$ in $v(E)^\perp$. But note that T_E itself is not of order two. The reflections $s_{v(E)}$ are well known classically for the case $E = \mathcal{O}_C(1)$ where $\mathbb{P}^1 \simeq C \subset X$. Then $v(\mathcal{O}_C(1)) = (0, [C], 0)$ and thus $T_{\mathcal{O}_C(1)}^H = s_{v(\mathcal{O}_C(1))}$ acts as identity on $H^0 \oplus H^4$ and as the reflection $s_{[C]}$ on H^2. The latter is important for the description of the ample cone of a K3 surface.

The mysterious part of $\mathrm{Aut}(D^b(X))$ is the subgroup that is generated by spherical twists T_E and, very roughly, Bridgeland's conjecture sets it in relation to $\pi_1(\mathcal{P}_0^+(X))$. But even the construction of a homomorphism

$$\pi_1(\mathcal{P}_0^+(X)) \to \mathrm{Aut}(D^b(X)), \tag{4.6}$$

one of the main achievements of [10], is highly non-trivial. The naive idea that a loop around $\delta^{\perp}$ is mapped to T_E^2, where $v(E) = \delta$, has among others the obvious flaw that such an E is not unique. Theorem 4.57 not only says that (4.6) exists but determines its image explicitly as $\mathrm{Aut}_0^0(\mathrm{D}^b(X))$.

4.5.3 Building up Stab$(X)^o$

Recall that in Section 4.3.2 stability conditions $\sigma(\omega, B)$ on $\mathrm{D}^b(X)$ were constructed with heart $\mathcal{A}(\omega, \beta)$ and stability function $Z = \langle \exp(B + i\omega), \ \rangle$. Here, $\beta = (B.\omega)$ with $B, \omega \in N(X) \otimes \mathbb{R}$, ω ample and such that $Z(\delta) \notin \mathbb{R}_{\leq 0}$ for all $\delta \in \Delta$ of positive rank.[13] Let

$$V(X) \subset \mathrm{Stab}(X)^o$$

be the open subset of all these stability conditions. As by construction a $\sigma(B, \omega) \in V(X)$ only depends on the stability function $Z = \langle \exp(B + i\omega), \ \rangle$, the projection $\mathrm{Stab}(X) \to N(X)^* = \mathrm{Hom}(N(X), \mathbb{C}) \simeq N(X) \otimes \mathbb{C}$ identifies $V(X)$ with an open subset of $N(X) \otimes \mathbb{C}$ and, in fact, under this identification $V(X) \subset \mathcal{P}_0^+(X)$ (cf. [10, Prop. 11.2]).

Applying the natural action of $\widetilde{\mathrm{GL}}^+(2, \mathbb{R})$ on $\mathrm{Stab}(X)$ (see Section 4.2.3) yields more stability conditions and one introduces

$$U(X) := V(X) \cdot \widetilde{\mathrm{GL}}^+(2, \mathbb{R}) \subset \mathrm{Stab}(X)^o,$$

which can in fact also be seen as a $\widetilde{\mathrm{GL}}^+(2, \mathbb{R})$-bundle over $V(X)$ (or rather its image in $\mathcal{P}_0^+(X)$). Note that $V(X)$ can also be considered naturally as a subset in the quotients $\mathrm{Stab}(X)^o/\widetilde{\mathrm{GL}}^+(2, \mathbb{R})$ and $\mathcal{P}_0^+(X)/\mathrm{GL}^+(2, \mathbb{R})$ which will be studied further in Section 4.5.4.

As a consequence of the discussion in Section 4.3.1 one obtains the following characterization of $U(X)$ (see [10, Prop. 10.3]).

Corollary 4.60. *A stability condition $\sigma \in \mathrm{Stab}(X)^o$ is contained in $U(X)$ if and only if all point sheaves $k(x)$ are σ-stable of the same phase.* $\square$

In fact, it would be enough to assume that σ is contained in a good component, see footnote page 204.

The next step in the program is to study the boundary $\partial U(X) \subset \mathrm{Stab}(X)^o$ and explain how to pass beyond it by applying spherical twists. Bridgeland studies two kinds of boundary components, of type $A^{\pm}$ and of type C.

---

[13] Strictly speaking, in Section 4.3.2 the result was only proved for rational such classes B, ω. It is true without the rationality, but in [10] the proof is given after the analysis of $\mathrm{Stab}(X)^o$.

Boundary components of type A**.** Consider a sequence $\sigma_t = (Z_t = \langle\exp(B_t + i\omega_t), \rangle, \mathcal{P}_t) \in V(X)$ converging to $\sigma = \sigma_0 = (Z = \langle\exp(B + i\omega), \rangle, \mathcal{P}) \in \partial U(X) \subset \mathrm{Stab}(X)^{\circ}$. One reason for σ not being in $V(X)$ is that there exists a $\delta = (r, \ell, s) \in \Delta$ of positive rank with $Z(\delta) \in \mathbb{R}_{\leq 0}$ and, since σ is in the boundary, in fact $Z(\delta) = 0$ (cf. Proposition 4.44). The imaginary part of the equation reads $\mu(\delta) := (\ell.\omega)/r = (B.\omega)$ and for fixed B this defines a decomposition of the positive cone into the two regions $\mu > (B.\omega)$ and $\mu < (B.\omega)$:

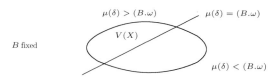

However, not all of $\mu(\delta) = (B.\omega)$ is part of the boundary. Indeed, the real part of $Z(\delta)$ can be written as $(B.\ell) - s - r\frac{B^2-\omega^2}{2}$. So again for fixed B, the boundary looks more like this (with two holes defined by $\delta^{\perp}$):

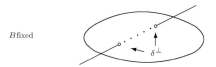

Remark 4.61. Here is what this means in terms of a spherical bundle A with $v(A) = \delta$. Suppose A is μ_{ω_t}-stable. Since $\mu_{\omega_t}(A) > (B.\omega)$, the bundle A is an object in $\mathcal{A}(\omega_t, \beta_t)$ for $t > 0$. As ω_t crosses the wall $\mu = (B.\omega)$, the bundle A can obviously not be contained in the heart of the stability condition any longer. However, if ω_t crosses the dotted part, then $A[1]$ will, but something more drastic happens when the solid part of $\mu = (B.\omega)$ is crossed.

The set $U(X)$ is determined by the stability of the point sheaves $k(x)$. Every bundle A as before admits a non-trivial homomorphism $A \to k(x)$ to any $k(x)$. For ω_t passing through the solid part of $\mu = (B.\omega)$, the object A becomes an object in $\mathcal{P}(1)$ and the existence of the non-trivial $A \to k(x)$ shows that $k(x)$ can no longer be stable and, therefore, $\sigma \notin U(X)$. On the other hand, passing through the dotted part the object A becomes semistable of phase 0 and, in particular, the non-trivial $A \to k(x)$ does not contradict stability of $k(x)$. Thus, σ is still in $U(X)$.[14] The image under the stability function looks likes this:

---

[14] The discussion is simplified by assuming that ω_t stays ample when passing through $\mu = (B.[C])$.

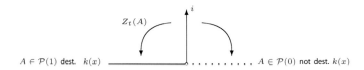

$A \in \mathcal{P}(1)$ dest. $k(x)$ .... $A \in \mathcal{P}(0)$ not dest. $k(x)$

Boundary components of type C. So far we have not taken into account the condition that ω has to be ample. It is well known that the ample cone inside the positive cone $\mathcal{C}_X \subset \mathrm{NS}(X) \otimes \mathbb{R}$ is cut out by the condition $(\omega.[C]) > 0$ for all (-2)-curves $\mathbb{P}^1 \simeq C \subset X$. Suppose now that in fact ω is still contained in the positive cone $\mathcal{C}_X$ but in the part of the boundary of the ample cone described by $(\omega.[C]) = 0$. Then $\exp(B+i\omega) \in \mathcal{P}_0^+(X)$ implies $\langle \exp(B+i\omega), (0, [C], k) \rangle \neq 0$ for all $k \in \mathbb{Z}$. Thus, on the line of all $(B.[C])$ the boundary of $\partial \mathcal{P}_0^+(X)$ looks like

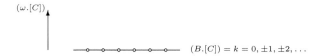

Remark 4.62. Recall, that the torsion sheaves $\mathcal{O}_C(k - 1)$ (with Mukai vector $(0, [C], k)$) are torsion sheaves on X and thus contained in $\mathcal{T}_{(\omega_t, \beta_t)} \subset \mathcal{A}(\omega_t, \beta_t)$. What happens with them when ω_t passes through the wall defined by $[C]$? This depends on the position of B. More precisely, when passing through the ray $(B.[C]) > k$ the real part of $Z(\mathcal{O}_C(k - 1))$ is positive and, therefore, $\mathcal{O}_C(k - 1)$ (and in fact all $\mathcal{O}_C(\ell)$ with $\ell \geq k - 1$) can no longer be in the heart of σ, they have to be shifted. However, passing through $(B.[C]) < k$ a priori does not affect the stability of $\mathcal{O}_C(k-1)$. But the stability of $\mathcal{O}_C(k-1)$ is not what matters for the description of the boundary of $U(X)$. Instead, one has to study the stability of all point sheaves $k(x)$ as before. For x contained in C, there exists a non-trivial $\mathcal{O}_C(k) \to k(x)$ and depending on whether one passes through $(B.[C]) < k$ or $(B.[C]) > k$ this affects the stability of $k(x)$ or not. However, if $x \notin C$, then the stability of $k(x)$ is not affected by the process at all. Thus, the boundary of $U(X)$ that is caused by non-trivial $\mathcal{O}_C(k - 1) \to k(x)$ looks like this

———————————○ · · · · ·

$(B.[C]) < k$

but for varying k they fill up a straight line with all $(0, [C], k)^\perp$ removed.

The two types of boundary above, type A and type C, are fundamentally different. Their images in $\mathcal{P}_0^+(X)$ look like a real boundary for type C and like a ray removed for type A:

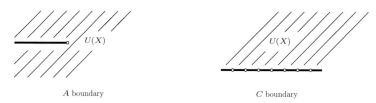

A boundary   C boundary

The above oversimplified discussion was meant to motivate the following result [10, Thm. 12.1]. The details of the proof are quite lengthy and the case ii) below did not appear above.

Proposition 4.63. *For a general point of the boundary* $\sigma = (Z, \mathcal{P}) \in \partial U(X)$ *one of the following possibilities occur:*

i) The stable factors of $k(x)$, $x \in X$, *are isomorphic to a spherical bundle* A *of rank* r *and to* $T_A(k(x))$. *More precisely, there exists an exact sequence in* $\mathcal{P}(\phi(k(x)))$ *of the form* $0 \to A^{\oplus r} \to k(x) \to T_A(k(x)) \to 0$.

ii) The stable factors of $k(x)$, $x \in X$, *are isomorphic to* $A[2]$ *with* A *a spherical bundle of rank* r *and to* $T_A^{-1}(k(x))$. *There exists an exact sequence in* $\mathcal{P}(\phi(k(x)))$ *of the form* $0 \to T_A^{-1}(k(x)) \to k(x) \to A^{\oplus r}[2] \to 0$.

iii) There exists a curve $\mathbb{P}^1 \simeq C \subset X$ *such that all* $k(x) \in \mathcal{P}(1)$ *are stable for* $x \notin C$ *and have stable factors* $\mathcal{O}_C(k+1)$ *and* $\mathcal{O}_C(k)[1]$ *for some* k, *i.e. there exists an exact sequence* $0 \to \mathcal{O}_C(k+1) \to k(x) \to \mathcal{O}_C(k)[1] \to 0$.

The picture above of the images in $\mathcal{P}_0^+(X)$ of the two kinds of boundary of $U(X)$ also suggests how to get back into $U(X)$ by autoequivalences. Suppose $\sigma_t, t \in (-\varepsilon, \varepsilon)$, with $\sigma_{t>0} \in U(X)$ and $\sigma_{t<0} \notin U(X)$. For a type C boundary, a reflection in $(0, [C], k)^{\perp}$ which lifts to the spherical twist $T_{\mathcal{O}_C(k)}$ interchanges the two sides of the boundary, i.e. $T_{\mathcal{O}_C(k)}(\sigma_{t<0}) \in U(X)$. This is clear from the picture in $\mathcal{P}_0^+(X)$ and is proved in [10, Sect. 13] using Proposition 4.63. For a type A boundary, one uses the square T_A^2, i.e. $T_A^2(\sigma_{t<0}) \in U(X)$.

Remark 4.64. Thus, more precisely Bridgeland shows

$$\mathrm{Aut}_0^\circ(\mathrm{D}^b(X)) = \langle T_{\mathcal{O}_C(k)}, T_A^2 \rangle \cap \mathrm{Aut}_0(\mathrm{D}^b(X)),$$

where all $\mathbb{P}^1 \simeq C \subset X$ and all $k \in \mathbb{Z}$ really occur. Which of the spherical bundles A really occur is not known. In particular, it is not clear if every spherical bundle on X is μ-stable with respect to at least one ample ω. For $\mathrm{Pic}(X) \simeq \mathbb{Z}$ every spherical bundle is μ-stable, see [36].

Building up $\mathrm{Stab}(X)^\circ$ from $U(X)$ eventually allows Bridgeland to deduce the existence of a group homomorphisms

$$\pi_1(\mathcal{P}_0^+(X)) \to \langle T_{\mathcal{O}_C(k)}, T_A^2 \rangle \subset \mathrm{Aut}(\mathrm{D}^b(X)).$$

Our discussion does not do justice to [10], it can at most serve as an illustration. Many of the intricacies have not been mentioned. E.g. why is $\pi(\mathrm{Stab}(X)^{\mathrm{o}})$ exactly $\mathcal{P}_0^+(X)$? Why is $\pi : \mathrm{Stab}(X)^{\mathrm{o}} \to \mathcal{P}_0^+(X)$ really a topological cover? We also have not explained how to use the analysis of $\mathrm{Stab}(X)^{\mathrm{o}}$ to finish the proof of Corollary 4.46 for non rational ω, B.

Related questions of this type are addressed in [19]. For example it is shown that T_A for any Gieseker stable spherical bundle A preserves the distinguished component. The special role of spherical objects in the study of $\mathrm{Stab}(X)^{\mathrm{o}}$ is further discussed in [24].

4.5.4 Moduli space rephrasing

Instead of $\mathcal{P}_0^+(X)$ one can consider the classical period domain

$$D \subset \mathbb{P}(N(X) \otimes \mathbb{C})$$

of all $x \in \mathbb{P}(N(X) \otimes \mathbb{C})$ with $(x.x) = 0$ and $(x.\bar{x}) > 0$, which is an open set of the smooth quadric defined by $(x.x) = 0$. Its two connected components $D^\pm \subset D$ are interchanged by complex conjugation. Similarly to the definition of $\mathcal{P}_0^+(X)$ one sets

$$D_0 := D \setminus \bigcup_{\delta \in \Delta} \delta^\perp$$

and $D_0^\pm := D^\pm \cap D_0$.

The natural projections $\mathcal{P}(X) \to D$ and $\mathcal{P}_0^+(X) \to D_0^+$ are $\mathrm{GL}^+(2, \mathbb{R})$-bundles and the Serre spectral sequence yields an exact sequence

$$\mathbb{Z} \simeq \pi_1(\mathrm{GL}^+(2, \mathbb{R})) \to \pi_1(\mathcal{P}_0^+(X)) \to \pi_1(D_0^+) \to 1.$$

Conjecture 4.58 is therefore equivalent to

$$\pi_1(D_0^+) \simeq \mathrm{Aut}_0^{\mathrm{o}}(\mathrm{D}^{\mathrm{b}}(X)) / _{\mathbb{Z}[2]}.$$

Note however that $\pi_1(\mathcal{P}_0^+(X))$ and $\pi_1(D_0^+)$ are usually not finitely generated which makes them slightly unpleasant to work with. But they can be related to the fundamental group of a quasi-projective variety, which is of course finitely generated, as follows.

Period domains of the type D_0 as above are well studied in moduli theory of K3 surfaces. Dividing out by the subgroup of the orthogonal group $\mathrm{O}(H^2(X, \mathbb{Z}))$ fixing the polarization would yield the moduli space of polarized K3 surfaces. In analogy, one considers here the subgroup

$$\Gamma := \{g \in \mathrm{O}(N(X)) \mid g_{N(X)^*/N(X)} = \mathrm{id}\}$$

which coincides with the group of orthogonal transformations of $N(X)$ that can be extended to an orthogonal transformation of $H^*(X, \mathbb{Z})$ acting trivially on the transcendental lattice. The lift of Γ to $\mathrm{Aut}(\mathrm{D}^b(X))$ is the group of symplectic autoequivalences

$$\mathrm{Aut}(\mathrm{D}^b(X))_s := \{\Phi \in \mathrm{Aut}(\mathrm{D}^b(X)) \mid \Phi^H|_{T(X)} = \mathrm{id}\}.$$

Restricting to $\mathrm{Aut}^o(\mathrm{D}^b(X))_s = \mathrm{Aut}^o(\mathrm{D}^b(X)) \cap \mathrm{Aut}(\mathrm{D}^b(X))_s$ yields a group that acts naturally on the distinguished component $\mathrm{Stab}(X)^o$. Then

$$\mathrm{Aut}^o(\mathrm{D}^b(X))_s \setminus {}^{\mathrm{Stab}(X)^o}\!/\widetilde{\mathrm{GL}}^+(2, \mathbb{R}) \simeq D_0/\Gamma.$$

Thus, the (distinguished component of the) space of stability conditions $\mathrm{Stab}(X)$ leads naturally, by identifying stability conditions which only differ by $\widetilde{\mathrm{GL}}^+(2, \mathbb{R})$ and $\mathrm{Aut}^o(\mathrm{D}^b(X))_s$, to the quasi-projective variety D_0/Γ. At this point one starts wondering whether there is a more functorial approach towards stability conditions that awaits to be unraveled.

In any case, this point of view allows one to rephrase Bridgeland's original conjecture as a conjecture that only involves finitely generated groups.

Conjecture 4.65. *There exists a natural isomorphism*

$$\mathrm{Aut}(\mathrm{D}^b(X))_s/\mathbb{Z}[2] \simeq \pi_1^{\mathrm{st}}([D_0/\Gamma]).$$

While the difference between the stacky fundamental group $\pi_1^{\mathrm{st}}([D/\Gamma])$ and the ordinary $\pi_1(D/\Gamma)$ is huge, for the stack $[D_0/\Gamma]$ the two only differ by the usually very small subgroup of Γ of elements with fixed points in D_0 which sometimes is even trivial. In any case, one always has an exact sequence $1 \to \pi_1(D_0) \to \pi_1^{\mathrm{st}}([D_0/\Gamma]) \to \Gamma \to \mathbb{Z}/2\mathbb{Z} \to 0$.

Remark 4.66. There is another way of looking at this picture. The period domains D_0 and their quotients $[D_0/\Gamma]$ are well studied spaces. Essentially, D_0 is the complement of the infinite hyperplane arrangement $\bigcup \delta^\perp$. The universal cover of such spaces can rarely be described explicitly and/or in a meaningful way (e.g. with a moduli theoretic interpretation). But the moduli space of stability conditions $\mathrm{Stab}(X)^o/\widetilde{\mathrm{GL}}^+(2, \mathbb{R})$ provides such a description provided it really is simply connected as predicted by Bridgeland's conjecture. Once this is settled, one would ask whether it is actually contractible which would make D_0 a $K(\pi, 1)$-space. Questions of this type have recently also been addressed by Allcock in [2].

Geodesics in D_0 converging to cusps, which are in bijection with Fourier–Mukai partners of X (see [30]), have been studied in detail in [18]. They are shown to be related to so-called linear degenerations in $\mathrm{Stab}(X)^{\circ}$.

4.6 Further results

There was not enough time in the lectures nor is there enough space here to enter the description of $\mathrm{Stab}(\mathcal{D})$ for other situations, only the case of curves and (K3) surfaces has been discussed in some detail. In fact, there are not many other examples of the type $\mathcal{D} = \mathrm{D}^{\mathrm{b}}(X)$ with X smooth and projective one really understands well. In particular, Calabi–Yau threefolds remain elusive. But there are examples of non-compact X and, closely related, more algebraic triangulated categories $\mathcal{D}$ for which the theory is quite well understood, see e.g. [11, 46] for more information.

4.6.1 Non-compact cases

i) The derived category $\mathrm{D}^{\mathrm{b}}(X)$ of a K3 surface X is an example of a *K3 category*, i.e. a triangulated category for which the double shift [2] is a Serre functor. As we have seen, a description of $\mathrm{Stab}(\mathrm{D}^{\mathrm{b}}(X))$ or $\mathrm{Aut}(\mathrm{D}^{\mathrm{b}}(X))$ is non-trivial (and still not completely understood). The case of local K3 surfaces is more accessible. More precisely, for the minimal resolution $\pi : X \to \mathbb{C}^2/G$ of a Kleinian singularity one can consider K3 categories $\mathcal{D} \subset \hat{\mathcal{D}} \subset \mathrm{D}^{\mathrm{b}}(X)$ of complexes supported on the exceptional divisor (resp. with vanishing $R\pi_*$). The spaces $\mathrm{Stab}(\mathcal{D})$ and $\mathrm{Stab}(\hat{\mathcal{D}})$ are studied in detail in [25, 43] for A_n-singularities and in [8, 13] in general. The analogue of Bridgeland's conjecture, originally formulated for projective K3 surfaces, has been proved in the local situation for the category $\mathcal{D}$.

ii) Triangulated categories with a Serre functor given by the triple shift [3] are called *CY3-categories*. The most prominent examples is $\mathrm{D}^{\mathrm{b}}(X)$ with X a smooth projective Calabi–Yau threefold. More accessible examples are provided by local Calabi–Yau manifolds. For example, if X is the total space of $\omega_{\mathbb{P}^2}$, then the bounded derived category $\mathcal{D} = \mathcal{D}_0(X)$ of coherent sheaves on X supported on the zero section of $X \to \mathbb{P}^2$ is a CY3-category. The study of $\mathrm{Stab}(\mathcal{D}_0(X))$ has been initiated in [12] and a complete connected component of $\mathrm{Stab}(\mathcal{D}_0(X))$ has been described in great detail in [6]. In particular, it is shown to be simply connected, which in the case of $\mathrm{D}^{\mathrm{b}}(X)$ of a compact K3 surface is still conjectural (see Conjecture 4.58). Categories of this type, i.e. derived categories of coherent sheaves supported on the zero section of the canonical bundle of a Fano surface, often allow for a more combinatorial approach via quiver representations, see [11] for a quick introduction.

4.6.2 Compact cases

i) Generic non-projective K3 surfaces X and generic twisted but projective K3 surfaces (X, α) have been studied in [21] (see [40] for related results). As in these cases there are no spherical objects, the space of stability conditions is much easier to describe. Stability conditions on $D^b(Y)$ of an Enriques surface are in [33] related to stability conditions on the covering K3 surface X. In fact, for generic Y it is shown that a connected component of $\mathrm{Stab}(Y)$ is isomorphic to the component $\mathrm{Stab}(X)^o$ discussed in the earlier sections.

ii) In [4] the authors come close to constructing stability conditions on $D^b(X)$ for projective Calabi–Yau threefolds or, in fact, arbitrary projective threefolds. The remaining problem is related to certain Chern class inequalities which have been further studied in [5].

iii) Stability conditions on $D^b(\mathbb{P}^n)$ are more accessible due to the existence of full exceptional collections. In particular the case of $\mathbb{P}^1$ is completely understood, see [31, 39]. For $n > 1$ see [32], where one also finds results on Del Pezzo surfaces.

References

[1] D. Abramovich, A. Polishchuk *Sheaves of t-structures and valuative criteria for stable complexes.* J. Reine Angew. Math. 590 (2006), 89–130.

[2] D. Allcock *Completions, branched covers, Artin groups and singularity theory.* arXiv:1106.3459v1.

[3] A. Beligiannis, I. Reiten *Homological and homotopical aspects of torsion theories.* Mem. AMS 188 (2007).

[4] A. Bayer, E. Macrì, Y. Toda *Bridgeland stability conditions on threefolds I: Bogomolov-Gieseker type inequalities.* arXiv:1103.5010v1.

[5] A. Bayer, A. Bertram, E. Macrì, Y. Toda *Bridgeland stability conditions on threefolds II: An application to Fujita's conjecture.* arXiv:1106.3430v1.

[6] A. Bayer, E. Macrì *The space of stability conditions on the local projective plane.* Duke Math. J. 160 (2011), 263–322.

[7] A. Bondal, M. Van den Bergh *Generators and representability of functors in commutative and noncommutative geometry.* Moscow Math. J. 3 (2003), 1–36.

[8] C. Brav, H. Thomas *Braid groups and Kleinian singularities.* Math. Ann. 351 (2011), 1005–1017.

[9] T. Bridgeland *Stability conditions on triangulated categories.* Ann. of Math. 166 (2007), 317–345.

[10] T. Bridgeland *Stability conditions on K3 surfaces.* Duke Math. J. 141 (2008), 241–291.

[11] T. Bridgeland *Spaces of stability conditions.* Algebraic geometry-Seattle 2005. Proc. Sympos. Pure Math. 80, Part 1, (2009), 1–21.

[12] T. Bridgeland *Stability conditions on a non-compact Calabi–Yau threefold.* Comm. Math. Phys. 266 (2006), 715–733.

[13] T. Bridgeland *Stability conditions and Kleinian singularities.* Int. Math. Res. Not. 21 (2009), 4142–4157.

[14] S. Dickson *A torsion theory for abelian categories.* Trans. AMS 121 (1966) 223–235.

[15] S. Gelfand, Y. Manin *Methods of homological algebra.* Springer Monographs in Math. (2002).

[16] A. Gorodentsev, S. Kuleshov, A. Rudakov *t-stabilities and t-structures on triangulated categories.* Izv. Math. 68 (2004), 749–781.

[17] D. Happel, I. Reiten, S. Smalø *Tilting in abelian categories and quasitilted algebras.* Mem. AMS 120 (1996).

[18] H. Hartmann *Cusps of the Kähler moduli space and stability conditions on K3 surfaces.* Math. Ann. 354 (2012), 1–42.

[19] R. Hartshorne *Residues and duality.* Springer LNM 20 (1966).

[20] D. Huybrechts *Fourier–Mukai transforms in algebraic geometry.* Oxford Mathematical Monographs (2006).

[21] D. Huybrechts, E. Macrì, P. Stellari *Stability conditions for generic K3 categories.* Compos. Math. 144 (2008), 134–162.

[22] D. Huybrechts *Derived and abelian equivalence of K3 surface.* J. Alg. Geom. 17 (2008), 375–400.

[23] D. Huybrechts, E. Macrì, P. Stellari *Derived equivalences of K3 surfaces and orientation.* Duke Math. J. 149 (2009), 461–507.

[24] D. Huybrechts *Stability conditions via spherical objects.* Math. Z. 271 (2012), 1253–1270.

[25] A. Ishii, K. Ueda, H. Uehara *Stability conditions on A_n-singularities.* J. Diff. Geom. 84 (2010), 87–126.

[26] M. Kashiwara, P. Schapira *Sheaves on manifolds.* Grundlehren math. Wiss. 292 (1990).

[27] M. Kontsevich *Triangulated Categories and Geometry.* Course at the École Normale Superieure (1998).

[28] M. Kontsevich, Y. Soibelman *Stability structures, motivic Donaldson–Thomas invariants and cluster transformations.* arXiv:0811.2435.

[29] J. Le, X.-W. Chen *Karoubianness of a triangulated category.* J. Alg. 310 (2007), 452–457.

[30] S. Ma *Fourier–Mukai partners of a K3 surface and the cusps of its Kähler moduli.* Int. J. Math. 20 (2009), 727–750.

[31] E. Macrì *Stability conditions on curves.* Math. Res. Lett. 14 (2007), 657–672.

[32] E. Macrì *Some examples of spaces of stability conditions on derived categories.* math.AG/0411613.

[33] E. Macrì, S. Mehrotra, P. Stellari *Inducing stability conditions.* J. Alg. Geom. 18 (2009), 605–649.

[34] S. Meinhardt *Stability conditions on generic complex tori.* to appear in Int. J. Math.

[35] S. Meinhardt, H. Partsch *Quotient categories, stability conditions, and birational geometry.* arXiv:0805.0492. to appear in Geometriae Dedicata.

[36] S. Mukai *On the moduli space of bundles on K3 surfaces, I.* In: Vector Bundles on Algebraic Varieties, Bombay (1984), 341–413.

[37] A. Neeman *Triangulated categories.* Annals Math. Stud. 148 (2001).

[38] B. Noohi *Explicit HRS-tilting.* J. Noncommut. Geom. 3 (2009), 223–259.

[39] S. Okada *Stability manifold of* $\mathbb{P}^1$. J. Alg. Geom. 15 (2006), 487–505.

[40] S. Okada *On stability manifolds of Calabi–Yau surfaces.* Int. Math. Res. Not. Article ID 58743 (2006), 1–16.

[41] A. Polishchuk *Classification of holomorphic vector bundles on non-commutative two-tori.* Doc. Math. 9 (2004), 163–181.

[42] A. Polishchuk *Constant families of t-structures on derived categories of coherent sheaves.* Moscow Math. J. 7 (2007), 109–134.

[43] R. Thomas *Stability conditions and the braid group.* Comm. Anal. Geom. 14 (2006), 135–161.

[44] R. Thomason *The classification of triangulated subcategories.* Comp. Math. 105 (1997), 1–27.

[45] J.-L. Verdier *Des catégories dérivées des catégories abéliennes.* Astérisque 239 (1996).

[46] J. Woolf *Stability conditions, torsion theories and tilting.* J. LMS (2) 82 (2010), 663–682.

[47] J. Woolf *Some metric properties of spaces of stability conditions.* arXiv:1108.2668v1

5

An introduction to d-manifolds and derived differential geometry

Dominic Joyce

Abstract

This is a survey of [20]. We introduce a 2-category **dMan** of *d-manifolds*, new geometric objects which are 'derived' smooth manifolds, in the sense of the 'derived algebraic geometry' of Toën and Lurie. They are a 2-category truncation of the 'derived manifolds' of Spivak [30]. Manifolds **Man** embed in **dMan** as a full (2-)subcategory. There are also 2-categories **dMan**$^{\mathbf{b}}$, **dMan**$^{\mathbf{c}}$ of *d-manifolds with boundary* and *with corners*, and orbifold versions of these **dOrb**, **dOrb**$^{\mathbf{b}}$, **dOrb**$^{\mathbf{c}}$, *d-orbifolds*.

Much of differential geometry extends very nicely to d-manifolds – immersions, submersions, submanifolds, transverse fibre products, orientations, etc. Compact oriented d-manifolds have virtual classes.

Many areas of symplectic geometry involve 'counting' moduli spaces $\overline{\mathcal{M}}_{g,m}(J, \beta)$ of J-holomorphic curves to define invariants, Floer homology theories, etc. Such $\overline{\mathcal{M}}_{g,m}(J, \beta)$ are given the structure of *Kuranishi spaces* in the work of Fukaya, Oh, Ohta and Ono [9], but there are problems with the theory. The author believes the 'correct' definition of Kuranishi spaces is that they are d-orbifolds with corners. D-manifolds and d-orbifolds will have applications in symplectic geometry, and elsewhere.

For brevity, this survey focusses on d-manifolds without boundary. A longer and more detailed summary of the book is given in [21].

Moduli Spaces, eds. L. Brambila-Paz, O. García-Prada, P. Newstead and R. Thomas. Published by Cambridge University Press. © Cambridge University Press 2014.

Contents

5.1 Introduction

This is a survey of [20], and describes the author's new theory of 'derived differential geometry'. The objects in this theory are *d-manifolds*, 'derived' versions of smooth manifolds, which form a (strict) 2-category **dMan**. There are also 2-categories of *d-manifolds with boundary* **dMan**[b] and *d-manifolds with corners* **dMan**[c], and orbifold versions of all these, *d-orbifolds* **dOrb**, **dOrb**[b], **dOrb**[c].

Here 'derived' is intended in the sense of *derived algebraic geometry*. The original motivating idea for derived algebraic geometry, as in Kontsevich [22] for instance, was that certain moduli schemes $\mathcal{M}$ appearing in enumerative invariant problems may be highly singular as schemes. However, it may be natural to realize $\mathcal{M}$ as a 1-categorical truncation of some 'derived' moduli space $\mathcal{M}$, a new kind of geometric object living in a higher category. The geometric structure on $\mathcal{M}$ should encode the full deformation theory of the moduli problem, the obstructions as well as the deformations. It was hoped that $\mathcal{M}$ would be 'smooth', and so in some sense simpler than its truncation $\mathcal{M}$.

Early work in derived algebraic geometry focussed on *dg-schemes*, as in Ciocan-Fontanine and Kapranov [6]. These have largely been replaced by the *derived stacks* of Toën and Vezzosi [32, 33], and the *structured spaces* of Lurie [24, 25]. *Derived differential geometry* aims to generalize these ideas to differential geometry and smooth manifolds. A brief note about it can be found in Lurie [25, §4.5]; the ideas are worked out in detail by Lurie's student David Spivak [30], who defines an ∞-category (simplicial category) of *derived manifolds*.

The author came to these questions from a different direction, symplectic geometry. Many important areas in symplectic geometry involve forming moduli spaces $\overline{\mathcal{M}}_{g,m}(J, \beta)$ of J-holomorphic curves in some symplectic manifold (M, ω), possibly with boundary in a Lagrangian L, and then 'counting' these moduli spaces to get 'invariants' with interesting properties. Such areas include Gromov–Witten invariants (open and closed), Lagrangian Floer cohomology, Symplectic Field Theory, contact homology, and Fukaya categories.

To do this 'counting', one needs to put a suitable geometric structure on $\overline{\mathcal{M}}_{g,m}(J, \beta)$ – something like the 'derived' moduli spaces $\mathcal{M}$ above – and use this to define a 'virtual class' or 'virtual chain' in $\mathbb{Z}$, $\mathbb{Q}$ or some homology theory. There is no general agreement on what geometric structure to use – compared to the elegance of algebraic geometry, this area is something of a mess. Two rival theories for geometric structures to put on moduli spaces $\overline{\mathcal{M}}_{g,m}(J, \beta)$ are the *Kuranishi spaces* of Fukaya, Oh, Ohta and Ono [10, 9] and the *polyfolds* of Hofer, Wysocki and Zehnder [12, 13, 14].

The theory of Kuranishi spaces in [10, 9] does not go far – they define Kuranishi spaces, and construct virtual cycles upon them, but they do not define morphisms between Kuranishi spaces, for instance. The author tried to study Kuranishi spaces as geometric spaces in their own right, but ran into problems, and became convinced that a new definition of Kuranishi space was needed. Upon reading Spivak's theory of derived manifolds [30], it became clear that some form of 'derived differential geometry' was required: *Kuranishi spaces in the sense of* [9, §A] *ought to be defined to be 'derived orbifolds with corners'*.

The author tried to read Lurie [24, 25] and Spivak [30] with a view to applications to Kuranishi spaces and symplectic geometry, but ran into problems of a different kind: the framework of [24, 25, 30] is formidably long, complex and abstract, and proved too difficult for a humble trainee symplectic geometer to understand, or use as a tool. So the author looked for a way to simplify the theory, while retaining the information needed for applications to symplectic geometry. The theory of d-manifolds and d-orbifolds of [20] is the result.

The essence of our simplification is this. Consider a 'derived' moduli space $\mathcal{M}$ of some objects E, e.g. vector bundles on some $\mathbb{C}$-scheme X. One expects $\mathcal{M}$ to have a 'cotangent complex' $\mathbb{L}_{\mathcal{M}}$, a complex in some derived category with cohomology $h^i(\mathbb{L}_{\mathcal{M}})|_E \cong \mathrm{Ext}^{1-i}(E, E)^*$ for $i \in \mathbb{Z}$. In general, $\mathbb{L}_{\mathcal{M}}$ can have nontrivial cohomology in many negative degrees, and because of this such objects $\mathcal{M}$ must form an ∞-category to properly describe their geometry.

However, the moduli spaces relevant to enumerative invariant problems are of a restricted kind: one considers only $\mathcal{M}$ such that $\mathbb{L}_{\mathcal{M}}$ has nontrivial cohomology only in degrees $-1, 0$, where $h^0(\mathbb{L}_{\mathcal{M}})$ encodes the (dual of the) deformations $\mathrm{Ext}^1(E, E)^*$, and $h^{-1}(\mathbb{L}_{\mathcal{M}})$ the (dual of the) obstructions $\mathrm{Ext}^2(E, E)^*$. As in Toën [32, §4.4.3], such derived spaces are called *quasi-smooth*, and this is a necessary condition on $\mathcal{M}$ for the construction of a virtual fundamental class.

Our d-manifolds are a 2-*category truncation* of Spivak's derived manifolds. Roughly speaking, this truncation replaces complexes in a derived category $D^b \mathrm{coh}(\mathcal{M})$ with a 2-category of complexes in degrees $-1, 0$ only. For general $\mathcal{M}$ this loses a lot of information, but for quasi-smooth $\mathcal{M}$, since $\mathbb{L}_{\mathcal{M}}$ is concentrated in degrees $-1, 0$, the important information is retained.

In the language of Toën and Vezzosi [32, 33], this should correspond to working with a subclass of derived schemes [32, §4.2] whose dg-algebras are of a special kind: they are 2-step supercommutative dg-algebras $A^{-1} \xrightarrow{\mathrm{d}} A^0$ such that $\mathrm{d}(A^{-1}) \cdot A^{-1} = 0$. This implies that $\mathrm{d}(A^{-1})$ is a square zero ideal in A^0, and A^{-1} is a module over $H^0\big(A^{-1} \xrightarrow{\mathrm{d}} A^0\big)$.

The set up of [20] is also long and complicated. But mostly this complexity comes from other sources: working over C^∞-rings, and including manifolds with boundary, manifolds with corners, and orbifolds. The 2-category style 'derived geometry' of [20] really is far simpler than those of [24, 25, 30, 32, 33].

Following Spivak [30], in order to be able to use the tools of algebraic geometry – schemes, stacks, quasicoherent sheaves – in differential geometry, our d-manifolds are built on the notions of C^∞-*ring* and C^∞-*scheme* that were invented in synthetic differential geometry, and developed further by the author in [18, 19]. We survey the C^∞-algebraic geometry we need in §5.2. Section 5.3 discusses the 2-category of *d-spaces* **dSpa**, which are 'derived' C^∞-schemes, and §5.4 describes the 2-category of d-manifolds **dMan**, and our theory of 'derived differential geometry'. Appendix A explains the basics of 2-categories.

For brevity, and to get the main ideas across as simply as possible, this survey will concentrate on d-manifolds without boundary, apart from a short section on d-manifolds with corners and d-orbifolds in §5.4.9. A longer summary of the book, including much more detail on d-manifolds with corners, d-orbifolds, and d-orbifolds with corners, is given in [21].

Acknowledgements. I would like to thank Jacob Lurie for helpful conversations.

5.2 C^∞-rings and C^∞-schemes

If X is a manifold then the $\mathbb{R}$-algebra $C^\infty(X)$ of smooth functions c : $X \to \mathbb{R}$ is a C^∞-*ring*. That is, for each smooth function $f : \mathbb{R}^n \to \mathbb{R}$ there is an n-fold operation $\Phi_f : C^\infty(X)^n \to C^\infty(X)$ acting by $\Phi_f : c_1, \dots, c_n \mapsto f(c_1, \dots, c_n)$, and these operations Φ_f satisfy many natural identities. Thus, $C^\infty(X)$ actually has a far richer algebraic structure than the obvious $\mathbb{R}$-algebra structure.

In [18] (surveyed in [19]) the author set out a version of algebraic geometry in which rings or algebras are replaced by C^∞-rings, focussing on C^∞-*schemes*, a category of geometric objects which generalize manifolds, and whose morphisms generalize smooth maps, *quasicoherent* and *coherent* *sheaves* on C^∞-schemes, and C^∞-*stacks*, in particular *Deligne–Mumford* C^∞-*stacks*, a 2-category of geometric objects which generalize orbifolds. Much of the material on C^∞-schemes was already known in synthetic differential geometry, see for instance Dubuc [8] and Moerdijk and Reyes [28].

5.2.1 C^∞-rings

Definition 5.1. A C^∞-*ring* is a set $\mathfrak{C}$ together with operations $\Phi_f : \mathfrak{C}^n \to \mathfrak{C}$ for all $n \geqslant 0$ and smooth maps $f : \mathbb{R}^n \to \mathbb{R}$, where by convention when $n = 0$ we define $\mathfrak{C}^0$ to be the single point $\{\emptyset\}$. These operations must satisfy the following relations: suppose $m, n \geqslant 0$, and $f_i : \mathbb{R}^n \to \mathbb{R}$ for $i = 1, \dots, m$ and $g : \mathbb{R}^m \to \mathbb{R}$ are smooth functions. Define a smooth function $h : \mathbb{R}^n \to \mathbb{R}$ by

$$h(x_1, \dots, x_n) = g\big(f_1(x_1, \dots, x_n), \dots, f_m(x_1 \dots, x_n)\big),$$

for all $(x_1, \dots, x_n) \in \mathbb{R}^n$. Then for all $(c_1, \dots, c_n) \in \mathfrak{C}^n$ we have

$$\Phi_h(c_1, \dots, c_n) = \Phi_g\big(\Phi_{f_1}(c_1, \dots, c_n), \dots, \Phi_{f_m}(c_1, \dots, c_n)\big).$$

We also require that for all $1 \leqslant j \leqslant n$, defining $\pi_j : \mathbb{R}^n \to \mathbb{R}$ by $\pi_j : (x_1, \dots, x_n) \mapsto x_j$, we have $\Phi_{\pi_j}(c_1, \dots, c_n) = c_j$ for all $(c_1, \dots, c_n) \in \mathfrak{C}^n$.

Usually we refer to $\mathfrak{C}$ as the C^∞-ring, leaving the operations Φ_f implicit.

A *morphism* between C^∞-rings $\left(\mathfrak{C}, (\Phi_f)_{f:\mathbb{R}^n \to \mathbb{R} \, C^\infty}\right)$, $\left(\mathfrak{D}, (\Psi_f)_{f:\mathbb{R}^n \to \mathbb{R}} \right.$ $\left. C^\infty\right)$ is a map $\phi : \mathfrak{C} \to \mathfrak{D}$ such that $\Psi_f\big(\phi(c_1), \ldots, \phi(c_n)\big) = \phi \circ$ $\Phi_f(c_1, \ldots, c_n)$ for all smooth $f : \mathbb{R}^n \to \mathbb{R}$ and $c_1, \ldots, c_n \in \mathfrak{C}$. We will write **$C^\infty$Rings** for the category of C^∞-rings.

Here is the motivating example:

Example 5.2. Let X be a manifold. Write $C^\infty(X)$ for the set of smooth functions $c : X \to \mathbb{R}$. For $n \geqslant 0$ and $f : \mathbb{R}^n \to \mathbb{R}$ smooth, define $\Phi_f : C^\infty(X)^n \to C^\infty(X)$ by

$$\big(\Phi_f(c_1, \ldots, c_n)\big)(x) = f\big(c_1(x), \ldots, c_n(x)\big), \qquad (5.1)$$

for all $c_1, \ldots, c_n \in C^\infty(X)$ and $x \in X$. It is easy to see that $C^\infty(X)$ and the operations Φ_f form a C^∞-ring.

Now let $f : X \to Y$ be a smooth map of manifolds. Then pullback $f^* : C^\infty(Y) \to C^\infty(X)$ mapping $f^* : c \longmapsto c \circ f$ is a morphism of C^∞-rings. Furthermore (at least for Y without boundary), every C^∞-ring morphism $\phi : C^\infty(Y) \to C^\infty(X)$ is of the form $\phi = f^*$ for a unique smooth map $f : X \to Y$.

Write **C^∞Rings^{op}** for the opposite category of **C^∞Rings**, with directions of morphisms reversed, and **Man** for the category of manifolds without boundary. Then we have a full and faithful functor $F_{\mathbf{Man}}^{\mathbf{C^\infty Rings}} : \mathbf{Man} \to \mathbf{C^\infty Rings}^{\mathrm{op}}$ acting by $F_{\mathbf{Man}}^{\mathbf{C^\infty Rings}}(X) = C^\infty(X)$ on objects and $F_{\mathbf{Man}}^{\mathbf{C^\infty Rings}}(f) = f^*$ on morphisms. This embeds **Man** as a full subcategory of **C^∞Rings^{op}**.

Note that C^∞-rings are far more general than those coming from manifolds. For example, if X is any topological space we could define a C^∞-ring $C^0(X)$ to be the set of *continuous* $c : X \to \mathbb{R}$, with operations Φ_f defined as in (5.1). For X a manifold with $\dim X > 0$, the C^∞-rings $C^\infty(X)$ and $C^0(X)$ are different.

Definition 5.3. Let $\mathfrak{C}$ be a C^∞-ring. Then we may give $\mathfrak{C}$ the structure of a *commutative $\mathbb{R}$-algebra*. Define addition '$+$' on $\mathfrak{C}$ by $c + c' = \Phi_f(c, c')$ for $c, c' \in \mathfrak{C}$, where $f : \mathbb{R}^2 \to \mathbb{R}$ is $f(x, y) = x + y$. Define multiplication '$\cdot$' on $\mathfrak{C}$ by $c \cdot c' = \Phi_g(c, c')$, where $g : \mathbb{R}^2 \to \mathbb{R}$ is $f(x, y) = xy$. Define scalar multiplication by $\lambda \in \mathbb{R}$ by $\lambda c = \Phi_{\lambda'}(c)$, where $\lambda' : \mathbb{R} \to \mathbb{R}$ is $\lambda'(x) = \lambda x$. Define elements $0, 1 \in \mathfrak{C}$ by $0 = \Phi_{0'}(\emptyset)$ and $1 = \Phi_{1'}(\emptyset)$, where $0' : \mathbb{R}^0 \to \mathbb{R}$ and $1' : \mathbb{R}^0 \to \mathbb{R}$ are the maps $0' : \emptyset \longmapsto 0$ and $1' : \emptyset \longmapsto 1$. One can show using the relations on the Φ_f that the axioms of a commutative $\mathbb{R}$-algebra are satisfied. In Example 5.2, this yields the obvious $\mathbb{R}$-algebra structure on the smooth functions $c : X \to \mathbb{R}$.

An *ideal* I in $\mathfrak{C}$ is an ideal $I \subset \mathfrak{C}$ in $\mathfrak{C}$ regarded as a commutative $\mathbb{R}$-algebra. Then we make the quotient $\mathfrak{C}/I$ into a C^∞-ring as follows. If $f : \mathbb{R}^n \to \mathbb{R}$ is smooth, define $\Phi^I_f : (\mathfrak{C}/I)^n \to \mathfrak{C}/I$ by

$$\big(\Phi^I_f(c_1 + I, \ldots, c_n + I)\big)(x) = f\big(c_1(x), \ldots, c_n(x)\big) + I.$$

Using Hadamard's Lemma, one can show that this is independent of the choice of representatives $c_1, \ldots, c_n$. Then $\big(\mathfrak{C}/I, (\Phi^I_f)_{f:\mathbb{R}^n \to \mathbb{R}} C^\infty\big)$ is a C^∞-ring.

A C^∞-ring $\mathfrak{C}$ is called *finitely generated* if there exist $c_1, \ldots, c_n$ in $\mathfrak{C}$ which generate $\mathfrak{C}$ over all C^∞-operations. That is, for each $c \in \mathfrak{C}$ there exists smooth $f : \mathbb{R}^n \to \mathbb{R}$ with $c = \Phi_f(c_1, \ldots, c_n)$. Given such $\mathfrak{C}, c_1, \ldots, c_n$, define $\phi : C^\infty(\mathbb{R}^n) \to \mathfrak{C}$ by $\phi(f) = \Phi_f(c_1, \ldots, c_n)$ for smooth $f : \mathbb{R}^n \to \mathbb{R}$, where $C^\infty(\mathbb{R}^n)$ is as in Example 5.2 with $X = \mathbb{R}^n$. Then ϕ is a surjective morphism of C^∞-rings, so $I = \mathrm{Ker}\,\phi$ is an ideal in $C^\infty(\mathbb{R}^n)$, and $\mathfrak{C} \cong C^\infty(\mathbb{R}^n)/I$ as a C^∞-ring. Thus, $\mathfrak{C}$ is finitely generated if and only if $\mathfrak{C} \cong C^\infty(\mathbb{R}^n)/I$ for some $n \geqslant 0$ and some ideal I in $C^\infty(\mathbb{R}^n)$.

5.2.2 C^∞-schemes

Next we summarize material in [18, §4] on C^∞-schemes.

Definition 5.4. A C^∞-*ringed space* $\underline{X} = (X, \mathcal{O}_X)$ is a topological space X with a sheaf $\mathcal{O}_X$ of C^∞-rings on X. A *morphism* $\underline{f} = (f, f^\sharp) : (X, \mathcal{O}_X) \to (Y, \mathcal{O}_Y)$ of C^∞ ringed spaces is a continuous map $f : X \to Y$ and a morphism $f^\sharp : f^{-1}(\mathcal{O}_Y) \to \mathcal{O}_X$ of sheaves of C^∞-rings on X, where $f^{-1}(\mathcal{O}_Y)$ is the inverse image sheaf. Write $\mathbf{C^\infty RS}$ for the category of C^∞-ringed spaces.

Here we follow [20, §B] in regarding $f^\sharp$ as a morphism $f^\sharp : f^{-1}(\mathcal{O}_Y) \to \mathcal{O}_X$ of sheaves on X, rather than [18, 19] in regarding $f^\sharp$ as a morphism $f^\sharp : \mathcal{O}_Y \to f_*(\mathcal{O}_X)$ of sheaves on Y. The two are equivalent as in [20, §B.4], since f^{-1}, f_* are adjoint functors. For the purposes of [20] it is convenient to write everything in terms of pullbacks (inverse images) of sheaves f^{-1}, f^*, rather than mixing pullbacks f^{-1}, f^* and pushforwards (direct images) f_*.

As in [8, Th. 8] there is a *spectrum functor* Spec : $\mathbf{C^\infty Rings}^{\mathrm{op}} \to \mathbf{C^\infty RS}$, defined explicitly in [18, Def. 4.5]. A C^∞-ringed space $\underline{X}$ is called an *affine* C^∞-*scheme* if it is isomorphic in $\mathbf{C^\infty RS}$ to Spec $\mathfrak{C}$ for some C^∞-ring $\mathfrak{C}$. A C^∞-ringed space $\underline{X} = (X, \mathcal{O}_X)$ is called a C^∞-*scheme* if X can be covered by open sets $U \subseteq X$ such that $(U, \mathcal{O}_X|_U)$ is an affine C^∞-scheme. Write $\mathbf{C^\infty Sch}$ for the full subcategory of C^∞-schemes in $\mathbf{C^\infty RS}$.

A C^∞-scheme $\underline{X} = (X, \mathcal{O}_X)$ is called *locally fair* if X can be covered by open $U \subseteq X$ with $(U, \mathcal{O}_X|_U) \cong$ Spec $\mathfrak{C}$ for some finitely generated C^∞-ring

$\mathfrak{C}$. Roughly speaking this means that $\underline{X}$ is locally finite-dimensional. Write $\mathbf{C^\infty Sch^{lf}}$ for the full subcategory of locally fair C^∞-schemes in $\mathbf{C^\infty Sch}$.

We call a C^∞-scheme $\underline{X}$ *separated, second countable, compact,* or *paracompact,* if the underlying topological space X is Hausdorff, second countable, compact, or paracompact, respectively.

We define a C^∞-scheme $\underline{X}$ for each manifold X.

Example 5.5. Let X be a manifold. Define a C^∞-ringed space $\underline{X} = (X, \mathcal{O}_X)$ to have topological space X and $\mathcal{O}_X(U) = C^\infty(U)$ for each open $U \subseteq X$, where $C^\infty(U)$ is the C^∞-ring of smooth maps $c : U \to \mathbb{R}$, and if $V \subseteq U \subseteq X$ are open define $\rho_{UV} : C^\infty(U) \to C^\infty(V)$ by $\rho_{UV} : c \mapsto c|_V$. Then $\underline{X} = (X, \mathcal{O}_X)$ is a local C^∞-ringed space. It is canonically isomorphic to $\mathrm{Spec}\, C^\infty(X)$, and so is an affine C^∞-scheme. It is locally fair.

Define a functor $F_{\mathbf{Man}}^{\mathbf{C^\infty Sch}} : \mathbf{Man} \to \mathbf{C^\infty Sch^{lf}} \subset \mathbf{C^\infty Sch}$ by $F_{\mathbf{Man}}^{\mathbf{C^\infty Sch}} = \mathrm{Spec} \circ F_{\mathbf{Man}}^{\mathbf{C^\infty Rings}}$. Then $F_{\mathbf{Man}}^{\mathbf{C^\infty Sch}}$ is full and faithful, and embeds $\mathbf{Man}$ as a full subcategory of $\mathbf{C^\infty Sch}$.

By [18, Cor. 4.14 & Th. 4.26] we have:

Theorem 5.6. *Fibre products and all finite limits exist in the category* $\mathbf{C^\infty Sch}$. *The subcategory* $\mathbf{C^\infty Sch^{lf}}$ *is closed under fibre products and all finite limits in* $\mathbf{C^\infty Sch}$. *The functor* $F_{\mathbf{Man}}^{\mathbf{C^\infty Sch}}$ *takes transverse fibre products in* $\mathbf{Man}$ *to fibre products in* $\mathbf{C^\infty Sch}$.

The proof of the existence of fibre products in $\mathbf{C^\infty Sch}$ follows that for fibre products of schemes in Hartshorne [11, Th. II.3.3], together with the existence of C^∞-scheme products $\underline{X} \times \underline{Y}$ of affine C^∞-schemes $\underline{X}, \underline{Y}$. The latter follows from the existence of coproducts $\mathfrak{C} \hat{\otimes} \mathfrak{D}$ in $\mathbf{C^\infty Rings}$ of C^∞-rings $\mathfrak{C}, \mathfrak{D}$. Here $\mathfrak{C} \hat{\otimes} \mathfrak{D}$ may be thought of as a 'completed tensor product' of $\mathfrak{C}, \mathfrak{D}$. The actual tensor product $\mathfrak{C} \otimes_\mathbb{R} \mathfrak{D}$ is naturally an $\mathbb{R}$-algebra but not a C^∞-ring, with an inclusion of $\mathbb{R}$-algebras $\mathfrak{C} \otimes_\mathbb{R} \mathfrak{D} \hookrightarrow \mathfrak{C} \hat{\otimes} \mathfrak{D}$, but $\mathfrak{C} \hat{\otimes} \mathfrak{D}$ is often much larger than $\mathfrak{C} \otimes_\mathbb{R} \mathfrak{D}$. For free C^∞-rings we have $C^\infty(\mathbb{R}^m) \hat{\otimes} C^\infty(\mathbb{R}^n) \cong C^\infty(\mathbb{R}^{m+n})$.

In [18, Def. 4.27 & Prop. 4.28] we discuss *partitions of unity* on C^∞-schemes.

Definition 5.7. Let $\underline{X} = (X, \mathcal{O}_X)$ be a C^∞-scheme. Consider a formal sum $\sum_{a \in A} c_a$, where A is an indexing set and $c_a \in \mathcal{O}_X(X)$ for $a \in A$. We say

$\sum_{a \in A} c_a$ is a *locally finite sum on $\underline{X}$* if X can be covered by open $U \subseteq X$ such that for all but finitely many $a \in A$ we have $\rho_{XU}(c_a) = 0$ in $\mathcal{O}_X(U)$.

By the sheaf axioms for $\mathcal{O}_X$, if $\sum_{a \in A} c_a$ is a locally finite sum there exists a unique $c \in \mathcal{O}_X(X)$ such that for all open $U \subseteq X$ with $\rho_{XU}(c_a) = 0$ in $\mathcal{O}_X(U)$ for all but finitely many $a \in A$, we have $\rho_{XU}(c) = \sum_{a \in A} \rho_{XU}(c_a)$ in $\mathcal{O}_X(U)$, where the sum makes sense as there are only finitely many nonzero terms. We call c the *limit* of $\sum_{a \in A} c_a$, written $\sum_{a \in A} c_a = c$.

Let $c \in \mathcal{O}_X(X)$. Suppose $V_i \subseteq X$ is open and $\rho_{XV_i}(c) = 0 \in \mathcal{O}_X(V_i)$ for $i \in I$, and let $V = \bigcup_{i \in I} V_i$. Then $V \subseteq X$ is open, and $\rho_{XV}(c) = 0 \in \mathcal{O}_X(V)$ as $\mathcal{O}_X$ is a sheaf. Thus taking the union of all open $V \subseteq X$ with $\rho_{XV}(c) = 0$ gives a unique maximal open set $V_c \subseteq X$ such that $\rho_{XV_c}(c) = 0 \in \mathcal{O}_X(V_c)$. Define the *support* $\operatorname{supp} c$ of c to be $X \setminus V_c$, so that $\operatorname{supp} c$ is closed in X. If $U \subseteq X$ is open, we say that c *is supported in U* if $\operatorname{supp} c \subseteq U$.

Let $\{U_a : a \in A\}$ be an open cover of X. A *partition of unity on $\underline{X}$ subordinate to $\{U_a : a \in A\}$* is $\{\eta_a : a \in A\}$ with $\eta_a \in \mathcal{O}_X(X)$ supported on U_a for $a \in A$, such that $\sum_{a \in A} \eta_a$ is a locally finite sum on $\underline{X}$ with $\sum_{a \in A} \eta_a = 1$.

Proposition 5.8. *Suppose $\underline{X}$ is a separated, paracompact, locally fair C^∞-scheme, and $\{\underline{U}_a : a \in A\}$ an open cover of $\underline{X}$. Then there exists a partition of unity $\{\eta_a : a \in A\}$ on $\underline{X}$ subordinate to $\{\underline{U}_a : a \in A\}$.*

Here are some differences between ordinary schemes and C^∞-schemes:

Remark 5.9. (i) If A is a ring or algebra, then points of the corresponding scheme $\operatorname{Spec} A$ are prime ideals in A. However, if $\mathfrak{C}$ is a C^∞-ring then (by definition) points of $\operatorname{Spec} \mathfrak{C}$ are maximal ideals in $\mathfrak{C}$ with residue field $\mathbb{R}$, or equivalently, $\mathbb{R}$-algebra morphisms $x : \mathfrak{C} \to \mathbb{R}$. This has the effect that if X is a manifold then points of $\operatorname{Spec} C^\infty(X)$ are just points of X.

(ii) In conventional algebraic geometry, affine schemes are a restrictive class. Central examples such as $\mathbb{CP}^n$ are not affine, and affine schemes are not closed under open subsets, so that $\mathbb{C}^2$ is affine but $\mathbb{C}^2 \setminus \{0\}$ is not. In contrast, affine C^∞-schemes are already general enough for many purposes. For example:

- All manifolds are fair affine C^∞-schemes.
- Open C^∞-subschemes of fair affine C^∞-schemes are fair and affine.
- If $\underline{X}$ is a separated, paracompact, locally fair C^∞-scheme then $\underline{X}$ is affine.

Affine C^∞-schemes are always separated (Hausdorff), so we need general C^∞-schemes to include non-Hausdorff behaviour.

(iii) In conventional algebraic geometry the Zariski topology is too coarse for many purposes, so one has to introduce the étale topology. In C^∞-algebraic geometry there is no need for this, as affine C^∞-schemes are Hausdorff.

(iv) Even very basic C^∞-rings such as $C^\infty(\mathbb{R}^n)$ for $n > 0$ are not noetherian as $\mathbb{R}$-algebras. So C^∞-schemes should be compared to non-noetherian schemes in conventional algebraic geometry.

(v) The existence of partitions of unity, as in Proposition 5.8, makes some things easier in C^∞-algebraic geometry than in conventional algebraic geometry. For example, geometric objects can often be 'glued together' over the subsets of an open cover using partitions of unity, and if $\mathcal{E}$ is a quasicoherent sheaf on a separated, paracompact, locally fair C^∞-scheme $\underline{X}$ then $H^i(\mathcal{E}) = 0$ for $i > 0$.

5.2.3 Modules over C^∞-rings, and cotangent modules

In [18, §5] we discuss modules over C^∞-rings.

Definition 5.10. Let $\mathfrak{C}$ be a C^∞-ring. A $\mathfrak{C}$-*module* M is a module over $\mathfrak{C}$ regarded as a commutative $\mathbb{R}$-algebra as in Definition 5.3. $\mathfrak{C}$-modules form an abelian category, which we write as $\mathfrak{C}$-mod. For example, $\mathfrak{C}$ is a $\mathfrak{C}$-module, and more generally $\mathfrak{C} \otimes_\mathbb{R} V$ is a $\mathfrak{C}$-module for any real vector space V. Let $\phi : \mathfrak{C} \to \mathfrak{D}$ be a morphism of C^∞-rings. If M is a $\mathfrak{C}$-module then $\phi_*(M) = M \otimes_\mathfrak{C} \mathfrak{D}$ is a $\mathfrak{D}$-module. This induces a functor $\phi_* : \mathfrak{C}$-mod $\to \mathfrak{D}$-mod.

Example 5.11. Let X be a manifold, and $E \to X$ a vector bundle. Write $C^\infty(E)$ for the vector space of smooth sections e of E. Then $C^\infty(X)$ acts on $C^\infty(E)$ by multiplication, so $C^\infty(E)$ is a $C^\infty(X)$-module.

In [18, §5.3] we define the *cotangent module* $\Omega_\mathfrak{C}$ of a C^∞-ring $\mathfrak{C}$.

Definition 5.12. Let $\mathfrak{C}$ be a C^∞-ring, and M a $\mathfrak{C}$-module. A C^∞-*derivation* is an $\mathbb{R}$-linear map d $: \mathfrak{C} \to M$ such that whenever $f : \mathbb{R}^n \to \mathbb{R}$ is a smooth map and $c_1, \ldots, c_n \in \mathfrak{C}$, we have

$$\mathrm{d}\Phi_f(c_1, \ldots, c_n) = \sum_{i=1}^n \Phi_{\frac{\partial f}{\partial x_i}}(c_1, \ldots, c_n) \cdot \mathrm{d}c_i.$$

We call such a pair M, d a *cotangent module* for $\mathfrak{C}$ if it has the universal property that for any $\mathfrak{C}$-module M' and C^∞-derivation d$' : \mathfrak{C} \to M'$, there exists a unique morphism of $\mathfrak{C}$-modules $\phi : M \to M'$ with d$' = \phi \circ$ d.

Define $\Omega_\mathfrak{C}$ to be the quotient of the free $\mathfrak{C}$-module with basis of symbols dc for $c \in \mathfrak{C}$ by the $\mathfrak{C}$-submodule spanned by all expressions of the form

$\mathrm{d}\Phi_f(c_1, \ldots, c_n) - \sum_{i=1}^{n} \Phi_{\frac{\partial f}{\partial x_i}}(c_1, \ldots, c_n) \cdot \mathrm{d}c_i$ for $f : \mathbb{R}^n \to \mathbb{R}$ smooth and $c_1, \ldots, c_n \in \mathfrak{C}$, and define $\mathrm{d}_{\mathfrak{C}} : \mathfrak{C} \to \Omega_{\mathfrak{C}}$ by $\mathrm{d}_{\mathfrak{C}} : c \mapsto \mathrm{d}c$. Then $\Omega_{\mathfrak{C}}, \mathrm{d}_{\mathfrak{C}}$ is a cotangent module for $\mathfrak{C}$. Thus cotangent modules always exist, and are unique up to unique isomorphism.

Let $\mathfrak{C}, \mathfrak{D}$ be C^∞-rings with cotangent modules $\Omega_{\mathfrak{C}}, \mathrm{d}_{\mathfrak{C}}, \Omega_{\mathfrak{D}}, \mathrm{d}_{\mathfrak{D}}$, and $\phi : \mathfrak{C} \to \mathfrak{D}$ be a morphism of C^∞-rings. Then ϕ makes $\Omega_{\mathfrak{D}}$ into a $\mathfrak{C}$-module, and there is a unique morphism $\Omega_\phi : \Omega_{\mathfrak{C}} \to \Omega_{\mathfrak{D}}$ in $\mathfrak{C}$-mod with $\mathrm{d}_{\mathfrak{D}} \circ \phi = \Omega_\phi \circ \mathrm{d}_{\mathfrak{C}}$. This induces a morphism $(\Omega_\phi)_* : \Omega_{\mathfrak{C}} \otimes_{\mathfrak{C}} \mathfrak{D} \to \Omega_{\mathfrak{D}}$ in $\mathfrak{D}$-mod with $(\Omega_\phi)_* \circ (\mathrm{d}_{\mathfrak{C}} \otimes \mathrm{id}_{\mathfrak{D}}) = \mathrm{d}_{\mathfrak{D}}$.

Example 5.13. Let X be a manifold. Then the cotangent bundle T^*X is a vector bundle over X, so as in Example 5.11 it yields a $C^\infty(X)$-module $C^\infty(T^*X)$. The exterior derivative $\mathrm{d} : C^\infty(X) \to C^\infty(T^*X)$ is a C^∞-derivation. These $C^\infty(T^*X), \mathrm{d}$ have the universal property in Definition 5.12, and so form a *cotangent module* for $C^\infty(X)$.

Now let X, Y be manifolds, and $f : X \to Y$ be smooth. Then $f^*(TY), TX$ are vector bundles over X, and the derivative of f is a vector bundle morphism $\mathrm{d}f : TX \to f^*(TY)$. The dual of this morphism is $\mathrm{d}f^* : f^*(T^*Y) \to T^*X$. This induces a morphism of $C^\infty(X)$-modules $(\mathrm{d}f^*)_* : C^\infty\big(f^*(T^*Y)\big) \to C^\infty(T^*X)$. This $(\mathrm{d}f^*)_*$ is identified with $(\Omega_{f^*})_*$ in Definition 5.12 under the natural isomorphism $C^\infty\big(f^*(T^*Y)\big) \cong C^\infty(T^*Y) \otimes_{C^\infty(Y)} C^\infty(X)$.

Definition 5.12 abstracts the notion of cotangent bundle of a manifold in a way that makes sense for any C^∞-ring.

5.2.4 Quasicoherent sheaves on C^∞-schemes

In [18, §6] we discuss sheaves of modules on C^∞-schemes.

Definition 5.14. Let $\underline{X} = (X, \mathcal{O}_X)$ be a C^∞-scheme. An $\mathcal{O}_X$-*module* $\mathcal{E}$ on $\underline{X}$ assigns a module $\mathcal{E}(U)$ over $\mathcal{O}_X(U)$ for each open set $U \subseteq X$, with $\mathcal{O}_X(U)$-action $\mu_U : \mathcal{O}_X(U) \times \mathcal{E}(U) \to \mathcal{E}(U)$, and a linear map $\mathcal{E}_{UV} : \mathcal{E}(U) \to \mathcal{E}(V)$ for each inclusion of open sets $V \subseteq U \subseteq X$, such that the following commutes:

$$
\begin{array}{ccc}
\mathcal{O}_X(U) \times \mathcal{E}(U) & \xrightarrow{\ \mu_U\ } & \mathcal{E}(U) \\
{\scriptstyle \rho_{UV} \times \mathcal{E}_{UV}} \big\downarrow & & \big\downarrow {\scriptstyle \mathcal{E}_{UV}} \\
\mathcal{O}_X(V) \times \mathcal{E}(V) & \xrightarrow{\ \mu_V\ } & \mathcal{E}(V),
\end{array}
$$

and all this data $\mathcal{E}(U), \mathcal{E}_{UV}$ satisfies the usual sheaf axioms [11, §II.1].

A *morphism of* $\mathcal{O}_X$-*modules* $\phi : \mathcal{E} \to \mathcal{F}$ assigns a morphism of $\mathcal{O}_X(U)$-modules $\phi(U) : \mathcal{E}(U) \to \mathcal{F}(U)$ for each open set $U \subseteq X$, such that $\phi(V) \circ \mathcal{E}_{UV} = \mathcal{F}_{UV} \circ \phi(U)$ for each inclusion of open sets $V \subseteq U \subseteq X$. Then $\mathcal{O}_X$-modules form an *abelian category*, which we write as $\mathcal{O}_X$-mod.

As in [18, §6.2], the spectrum functor Spec : $\mathbf{C}^\infty\mathbf{Rings}^{op} \to \mathbf{C}^\infty\mathbf{Sch}$ has a counterpart for modules: if $\mathfrak{C}$ is a C^∞-ring and $(X, \mathcal{O}_X) = \mathrm{Spec}\,\mathfrak{C}$ we can define a functor MSpec : $\mathfrak{C}$-mod $\to \mathcal{O}_X$-mod. If $\mathfrak{C}$ is a *fair* C^∞-ring, there is a full abelian subcategory $\mathfrak{C}$-mod^{co} of *complete* $\mathfrak{C}$-modules in $\mathfrak{C}$-mod, such that MSpec $|_{\mathfrak{C}\text{-mod}^{co}} : \mathfrak{C}$-mod$^{co} \to \mathcal{O}_X$-mod is an equivalence of categories, with quasi-inverse the global sections functor $\Gamma : \mathcal{O}_X$-mod $\to \mathfrak{C}$-mod^{co}. Let $\underline{X} = (X, \mathcal{O}_X)$ be a C^∞-scheme, and $\mathcal{E}$ an $\mathcal{O}_X$-module. We call $\mathcal{E}$ *quasicoherent* if $\underline{X}$ can be covered by open $\underline{U}$ with $\underline{U} \cong \mathrm{Spec}\,\mathfrak{C}$ for some C^∞-ring $\mathfrak{C}$, and under this identification $\mathcal{E}|_U \cong \mathrm{MSpec}\,M$ for some $\mathfrak{C}$-module M. We call $\mathcal{E}$ a *vector bundle of rank n* $\geqslant 0$ if $\underline{X}$ may be covered by open $\underline{U}$ such that $\mathcal{E}|_U \cong \mathcal{O}_U \otimes_{\mathbb{R}} \mathbb{R}^n$.

Write qcoh($\underline{X}$), vect($\underline{X}$) for the full subcategories of quasicoherent sheaves and vector bundles in $\mathcal{O}_X$-mod. Then qcoh($\underline{X}$) is an abelian category. Since MSpec : $\mathfrak{C}$-mod$^{co} \to \mathcal{O}_X$-mod is an equivalence for $\mathfrak{C}$ fair and $(X, \mathcal{O}_X) = \mathrm{Spec}\,\mathfrak{C}$, as in [18, Cor. 6.11] we see that if $\underline{X}$ is a locally fair C^∞-scheme then every $\mathcal{O}_X$-module $\mathcal{E}$ on $\underline{X}$ is quasicoherent, that is, qcoh($\underline{X}$) = $\mathcal{O}_X$-mod.

Remark 5.15. If $\underline{X}$ is a separated, paracompact, locally fair C^∞-scheme then vector bundles on $\underline{X}$ are projective objects in the abelian category qcoh($\underline{X}$).

Definition 5.16. Let $\underline{f} : \underline{X} \to \underline{Y}$ be a morphism of C^∞-schemes, and let $\mathcal{E}$ be an $\mathcal{O}_Y$-module. Define the *pullback* $\underline{f}^*(\mathcal{E})$, an $\mathcal{O}_X$-module, by $\underline{f}^*(\mathcal{E}) = f^{-1}(\mathcal{E}) \otimes_{f^{-1}(\mathcal{O}_Y)} \mathcal{O}_X$, where $f^{-1}(\mathcal{E}), f^{-1}(\mathcal{O}_Y)$ are inverse image sheaves. If $\phi : \mathcal{E} \to \mathcal{F}$ is a morphism in $\mathcal{O}_Y$-mod we have an induced morphism $\underline{f}^*(\phi) = f^{-1}(\phi) \otimes \mathrm{id}_{\mathcal{O}_X} : \underline{f}^*(\mathcal{E}) \to \underline{f}^*(\mathcal{F})$ in $\mathcal{O}_X$-mod. Then $\underline{f}^* : \mathcal{O}_Y$-mod $\to \mathcal{O}_X$-mod is a *right exact functor* between abelian categories, which restricts to a right exact functor $\underline{f}^* : \mathrm{qcoh}(\underline{Y}) \to \mathrm{qcoh}(\underline{X})$.

Remark 5.17. Pullbacks $\underline{f}^*(\mathcal{E})$ are characterized by a universal property, and so are unique up to canonical isomorphism, rather than unique. Our definition of $\underline{f}^*(\mathcal{E})$ is not functorial in $\underline{f}$. That is, if $\underline{f} : \underline{X} \to \underline{Y}, \underline{g} : \underline{Y} \to \underline{Z}$ are morphisms and $\mathcal{E} \in \mathcal{O}_Z$-mod then $(\underline{g} \circ \underline{f})^*(\mathcal{E})$ and $\underline{f}^*(\underline{g}^*(\mathcal{E}))$ are canonically isomorphic in $\mathcal{O}_X$-mod, but may not be equal. In [20] we keep track of these canonical isomorphisms, writing them as $I_{\underline{f},\underline{g}}(\mathcal{E}) : (\underline{g} \circ \underline{f})^*(\mathcal{E}) \to \underline{f}^*(\underline{g}^*(\mathcal{E}))$. However, in this survey, by an abuse of notation that is common in the literature, we will for simplicity omit the isomorphisms $I_{\underline{f},\underline{g}}(\mathcal{E})$, and identify $(\underline{g} \circ \underline{f})^*(\mathcal{E})$ with $\underline{f}^*(\underline{g}^*(\mathcal{E}))$.

Similarly, when f is the identity $\underline{\mathrm{id}}_{\underline{X}} : \underline{X} \to \underline{X}$ and $\mathcal{E} \in \mathcal{O}_X$-mod, we may not have $\underline{\mathrm{id}}_{\underline{X}}^*(\mathcal{E}) = \mathcal{E}$, but there is a canonical isomorphism $\delta_{\underline{X}}(\mathcal{E}) : \underline{\mathrm{id}}_{\underline{X}}^*(\mathcal{E}) \to \mathcal{E}$, which we keep track of in [20]. But here, for simplicity, by an abuse of notation we omit $\delta_{\underline{X}}(\mathcal{E})$, and identify $\underline{\mathrm{id}}_{\underline{X}}^*(\mathcal{E})$ with $\mathcal{E}$.

Example 5.18. Let X be a manifold, and $\underline{X}$ the associated C^∞-scheme from Example 5.5, so that $\mathcal{O}_X(U) = C^\infty(U)$ for all open $U \subseteq X$. Let $E \to X$ be a vector bundle. Define an $\mathcal{O}_X$-module $\mathcal{E}$ on $\underline{X}$ by $\mathcal{E}(U) = C^\infty(E|_U)$, the smooth sections of the vector bundle $E|_U \to U$, and for open $V \subseteq U \subseteq X$ define $\mathcal{E}_{UV} : \mathcal{E}(U) \to \mathcal{E}(V)$ by $\mathcal{E}_{UV} : e_U \mapsto e_U|_V$. Then $\mathcal{E} \in \mathrm{vect}(\underline{X})$ is a vector bundle on $\underline{X}$, which we think of as a lift of E from manifolds to C^∞-schemes.

Let $f : X \to Y$ be a smooth map of manifolds, and $\underline{f} : \underline{X} \to \underline{Y}$ the corresponding morphism of C^∞-schemes. Let $F \to Y$ be a vector bundle over Y, so that $f^*(F) \to X$ is a vector bundle over X. Let $\mathcal{F} \in \mathrm{vect}(\underline{Y})$ be the vector bundle over $\underline{Y}$ lifting F. Then $\underline{f}^*(\mathcal{F})$ is canonically isomorphic to the vector bundle over $\underline{X}$ lifting $f^*(F)$.

We define *cotangent sheaves*, the sheaf version of cotangent modules in §5.2.3.

Definition 5.19. Let $\underline{X}$ be a C^∞-scheme. Define $\mathcal{P}T^*\underline{X}$ to associate to each open $U \subseteq X$ the cotangent module $\Omega_{\mathcal{O}_X(U)}$, and to each inclusion of open sets $V \subseteq U \subseteq X$ the morphism of $\mathcal{O}_X(U)$-modules $\Omega_{\rho_{UV}} : \Omega_{\mathcal{O}_X(U)} \to \Omega_{\mathcal{O}_X(V)}$ associated to the morphism of C^∞-rings $\rho_{UV} : \mathcal{O}_X(U) \to \mathcal{O}_X(V)$. Then $\mathcal{P}T^*\underline{X}$ is a *presheaf of $\mathcal{O}_X$-modules on $\underline{X}$*. Define the *cotangent sheaf $T^*\underline{X}$ of $\underline{X}$* to be the sheafification of $\mathcal{P}T^*\underline{X}$, as an $\mathcal{O}_X$-module.

If $\underline{f} : \underline{X} \to \underline{Y}$ is a morphism of C^∞-schemes, then $\underline{f}^*(T^*\underline{Y})$ is the sheafification of the presheaf $\underline{f}^*(\mathcal{P}T^*\underline{Y})$ acting by

$$U \longmapsto \underline{f}^*(\mathcal{P}T^*\underline{Y})(U) = \lim_{\substack{\longrightarrow \\ V \supseteq f(U)}} \mathcal{P}T^*\underline{Y}(V) \otimes_{\mathcal{O}_Y(V)} \mathcal{O}_X(U)$$

$$= \lim_{\substack{\longrightarrow \\ V \supseteq f(U)}} \Omega_{\mathcal{O}_Y(V)} \otimes_{\mathcal{O}_Y(V)} \mathcal{O}_X(U).$$

Define a morphism of presheaves $\mathcal{P}\Omega_{\underline{f}} : \underline{f}^*(\mathcal{P}T^*\underline{Y}) \to \mathcal{P}T^*\underline{X}$ on X by

$$(\mathcal{P}\Omega_{\underline{f}})(U) = \lim_{\substack{\longrightarrow \\ V \supseteq f(U)}} (\Omega_{\rho_{f^{-1}(V)U} \circ f^\sharp(V)})_*,$$

where $(\Omega_{\rho_{f^{-1}(V)U} \circ f^\sharp(V)})_* : \Omega_{\mathcal{O}_Y(V)} \otimes_{\mathcal{O}_Y(V)} \mathcal{O}_X(U) \to \Omega_{\mathcal{O}_X(U)} = (\mathcal{P}T^*\underline{X})(U)$ is constructed as in Definition 5.12 from the C^∞-ring morphisms

$f^\sharp(V) : \mathcal{O}_Y(V) \to \mathcal{O}_X(f^{-1}(V))$ in $\underline{f}$ and $\rho_{f^{-1}(V)U} : \mathcal{O}_X(f^{-1}(V)) \to \mathcal{O}_X(U)$ in $\mathcal{O}_X$. Define $\Omega_{\underline{f}} : \underline{f}^*(T^*\underline{Y}) \to T^*\underline{X}$ to be the induced morphism of the associated sheaves.

Example 5.20. Let X be a manifold, and $\underline{X}$ the associated C^∞-scheme. Then $T^*\underline{X}$ is a vector bundle on $\underline{X}$, and is canonically isomorphic to the lift to C^∞-schemes from Example 5.18 of the cotangent vector bundle T^*X of X.

Here [18, Th. 6.16] are some properties of cotangent sheaves.

Theorem 5.21. (a) *Let* $\underline{f} : \underline{X} \to \underline{Y}$ *and* $\underline{g} : \underline{Y} \to \underline{Z}$ *be morphisms of* C^∞-*schemes. Then*

$$\Omega_{\underline{g} \circ \underline{f}} = \Omega_{\underline{f}} \circ \underline{f}^*(\Omega_{\underline{g}})$$

as morphisms $(\underline{g} \circ \underline{f})^*(T^*\underline{Z}) \to T^*\underline{X}$ *in* $\mathcal{O}_X$-*mod. Here* $\Omega_{\underline{g}} : \underline{g}^*(T^*\underline{Z}) \to T^*\underline{Y}$ *is a morphism in* $\mathcal{O}_Y$-*mod, so applying* $\underline{f}^*$ *gives* $\underline{f}^*(\Omega_{\underline{g}}) : (\underline{g} \circ \underline{f})^*(T^*\underline{Z}) = \underline{f}^*(\underline{g}^*(T^*\underline{Z})) \to \underline{f}^*(T^*\underline{Y})$ *in* $\mathcal{O}_X$-*mod.*
(b) *Suppose* $\underline{W}, \underline{X}, \underline{Y}, \underline{Z}$ *are locally fair* C^∞-*schemes with a Cartesian square*

$$\begin{array}{ccc} \underline{W} & \longrightarrow & \underline{Y} \\ {\scriptstyle\underline{e}}\downarrow & {\scriptstyle\substack{\underline{f}\\ \underline{g}}} & {\scriptstyle\underline{h}}\downarrow \\ \underline{X} & \longrightarrow & \underline{Z} \end{array}$$

in $\mathbf{C^\infty Sch^{lf}}$, *so that* $\underline{W} = \underline{X} \times_{\underline{Z}} \underline{Y}$. *Then the following is exact in* $\mathrm{qcoh}(\underline{W})$:

$$(\underline{g} \circ \underline{e})^*(T^*\underline{Z}) \xrightarrow{\underline{e}^*(\Omega_{\underline{g}}) \oplus -\underline{f}^*(\Omega_{\underline{h}})} \underline{e}^*(T^*\underline{X}) \oplus \underline{f}^*(T^*\underline{Y}) \xrightarrow{\Omega_{\underline{e}} \oplus \Omega_{\underline{f}}} T^*\underline{W} \longrightarrow 0.$$

5.3 The 2-category of d-spaces

We will now define the 2-category of *d-spaces* **dSpa**, following [20, §2]. D-spaces are 'derived' versions of C^∞-schemes. In §5.4 we will define the 2-category of d-manifolds **dMan** as a 2-subcategory of **dSpa**. For an introduction to 2-categories, see Appendix A.

5.3.1 The definition of d-spaces

Definition 5.22. A *d-space* $\mathbf{X}$ is a quintuple $\mathbf{X} = (\underline{X}, \mathcal{O}'_X, \mathcal{E}_X, \iota_X, \jmath_X)$ such that $\underline{X} = (X, \mathcal{O}_X)$ is a separated, second countable, locally fair C^∞-scheme, and $\mathcal{O}'_X, \mathcal{E}_X, \iota_X, \jmath_X$ fit into an exact sequence of sheaves on X

$$\mathcal{E}_X \xrightarrow{\jmath_X} \mathcal{O}'_X \xrightarrow{\iota_X} \mathcal{O}_X \longrightarrow 0,$$

satisfying the conditions:

(a) $\mathcal{O}'_X$ is a sheaf of C^∞-rings on X, with $\underline{X}' = (X, \mathcal{O}'_X)$ a C^∞-scheme.

(b) $\iota_X : \mathcal{O}'_X \to \mathcal{O}_X$ is a surjective morphism of sheaves of C^∞-rings on X. Its kernel $\mathcal{I}_X$ is a sheaf of ideals in $\mathcal{O}'_X$, which should be a sheaf of square zero ideals. Here a *square zero ideal* in a commutative $\mathbb{R}$-algebra A is an ideal I with $i \cdot j = 0$ for all $i, j \in I$. Then $\mathcal{I}_X$ is an $\mathcal{O}'_X$-module, but as $\mathcal{I}_X$ consists of square zero ideals and ι_X is surjective, the $\mathcal{O}'_X$-action factors through an $\mathcal{O}_X$-action. Hence $\mathcal{I}_X$ is an $\mathcal{O}_X$-module, and thus a quasicoherent sheaf on $\underline{X}$, as $\underline{X}$ is locally fair.

(c) $\mathcal{E}_X$ is a quasicoherent sheaf on $\underline{X}$, and $\jmath_X : \mathcal{E}_X \to \mathcal{I}_X$ is a surjective morphism in qcoh($\underline{X}$).

As $\underline{X}$ is locally fair, the underlying topological space X is locally homeomorphic to a closed subset of $\mathbb{R}^n$, so it is *locally compact*. But Hausdorff, second countable and locally compact imply paracompact, and thus $\underline{X}$ is *paracompact*.

The sheaf of C^∞-rings $\mathcal{O}'_X$ has a sheaf of cotangent modules $\Omega_{\mathcal{O}'_X}$, which is an $\mathcal{O}'_X$-module with exterior derivative d : $\mathcal{O}'_X \to \Omega_{\mathcal{O}'_X}$. Define $\mathcal{F}_X = \Omega_{\mathcal{O}'_X} \otimes_{\mathcal{O}'_X} \mathcal{O}_X$ to be the associated $\mathcal{O}_X$-module, a quasicoherent sheaf on $\underline{X}$, and set $\psi_X = \Omega_{\iota_X} \otimes \mathrm{id} : \mathcal{F}_X \to T^*\underline{X}$, a morphism in qcoh($\underline{X}$). Define $\phi_X : \mathcal{E}_X \to \mathcal{F}_X$ to be the composition of morphisms of sheaves of abelian groups on X:

$$\mathcal{E}_X \xrightarrow{\jmath_X} \mathcal{I}_X \xrightarrow{\mathrm{d}|_{\mathcal{I}_X}} \Omega_{\mathcal{O}'_X} \xrightarrow{\sim} \Omega_{\mathcal{O}'_X} \otimes_{\mathcal{O}'_X} \mathcal{O}'_X \xrightarrow{\mathrm{id}\otimes\iota_X} \Omega_{\mathcal{O}'_X} \otimes_{\mathcal{O}'_X} \mathcal{O}_X = \mathcal{F}_X.$$

It turns out that ϕ_X is actually a morphism of $\mathcal{O}_X$-modules, and the following sequence is exact in qcoh($\underline{X}$) :

$$\mathcal{E}_X \xrightarrow{\phi_X} \mathcal{F}_X \xrightarrow{\psi_X} T^*\underline{X} \longrightarrow 0.$$

The morphism $\phi_X : \mathcal{E}_X \to \mathcal{F}_X$ will be called the *virtual cotangent sheaf* of X, for reasons we explain in §5.4.3.

Let X, Y be d-spaces. A 1-*morphism* $f : X \to Y$ is a triple $f = (\underline{f}, f', f'')$, where $\underline{f} = (f, f^\sharp) : \underline{X} \to \underline{Y}$ is a morphism of C^∞-schemes, $f' : f^{-1}(\mathcal{O}'_Y) \to \mathcal{O}'_X$ a morphism of sheaves of C^∞-rings on X, and $f'' : \underline{f}^*(\mathcal{E}_Y) \to \mathcal{E}_X$ a morphism in qcoh($\underline{X}$), such that the following diagram of sheaves on X commutes:

$$f^{-1}(\mathcal{E}_Y) \otimes^{\mathrm{id}}_{f^{-1}(\mathcal{O}_Y)} f^{-1}(\mathcal{O}_Y) = f^{-1}(\mathcal{E}_Y) \longrightarrow f^{-1}(\mathcal{O}'_Y) \longrightarrow f^{-1}(\mathcal{O}_Y) \to 0$$

with vertical maps $\mathrm{id}\otimes f^\sharp$, $f^{-1}(\jmath_Y)$, $f^{-1}(\iota_Y)$, f', $f^\sharp$:

$$\underline{f}^*(\mathcal{E}_Y) = f^{-1}(\mathcal{E}_Y) \otimes^{f^\sharp}_{f^{-1}(\mathcal{O}_Y)} \mathcal{O}_X \xrightarrow{f''} \mathcal{E}_X \xrightarrow{\jmath_X} \mathcal{O}'_X \xrightarrow{\iota_X} \mathcal{O}_X \longrightarrow 0.$$

Define morphisms $f^2 = \Omega_{f'} \otimes \mathrm{id} : \underline{f}^*(\mathcal{F}_Y) \to \mathcal{F}_X$ and $f^3 = \Omega_f : \underline{f}^*(T^*\underline{Y}) \to T^*\underline{X}$ in qcoh($\underline{X}$). Then the following commutes in qcoh($\underline{X}$), with exact rows:

$$\begin{array}{ccccccc}
\underline{f}^*(\mathcal{E}_Y) & \xrightarrow{\underline{f}^*(\phi_Y)} & \underline{f}^*(\mathcal{F}_Y) & \xrightarrow{\underline{f}^*(\psi_Y)} & \underline{f}^*(T^*\underline{Y}) & \longrightarrow & 0 \\
\downarrow{f''} & & \downarrow{f^2} & & \downarrow{f^3} & & \\
\mathcal{E}_X & \xrightarrow{\phi_X} & \mathcal{F}_X & \xrightarrow{\psi_X} & T^*\underline{X} & \longrightarrow & 0.
\end{array} \qquad (5.2)$$

If X is a d-space, the *identity 1-morphism* $\mathbf{id}_X : X \to X$ is $\mathbf{id}_X = (\underline{\mathrm{id}}_X, \mathrm{id}_{\mathcal{O}'_X}, \mathrm{id}_{\mathcal{E}_X})$. Let X, Y, Z be d-spaces, and $f : X \to Y, g : Y \to Z$ be 1-morphisms. Define the *composition of 1-morphisms* $g \circ f : X \to Z$ to be

$$g \circ f = \big(\underline{g} \circ \underline{f}, f' \circ f^{-1}(g'), f'' \circ \underline{f}^*(g'')\big).$$

Let $f, g : X \to Y$ be 1-morphisms of d-spaces, where $f = (\underline{f}, f', f'')$ and $g = (\underline{g}, g', g'')$. Suppose $\underline{f} = \underline{g}$. A *2-morphism* $\eta : f \Rightarrow g$ is a morphism $\eta : \underline{f}^*(\mathcal{F}_Y) \to \mathcal{E}_X$ in qcoh($\underline{X}$), such that

$$g' = f' + \jmath_X \circ \eta \circ \big(\mathrm{id} \otimes (f^\sharp \circ f^{-1}(\iota_Y))\big) \circ \big(f^{-1}(d)\big)$$
$$\text{and} \qquad g'' = f'' + \eta \circ \underline{f}^*(\phi_Y).$$

Then $g^2 = f^2 + \phi_X \circ \eta$ and $g^3 = f^3$, so (5.2) for f, g combine to give a diagram

$$\begin{array}{ccccccc}
\underline{f}^*(\mathcal{E}_Y) & \xrightarrow{\underline{f}^*(\phi_Y)} & \underline{f}^*(\mathcal{F}_Y) & \xrightarrow{\underline{f}^*(\psi_Y)} & \underline{f}^*(T^*\underline{Y}) & \longrightarrow & 0 \\
\overline{f}''\downarrow\downarrow{\scriptstyle g''=f''+\eta\circ\underline{f}^*(\phi_Y)} & \overset{\eta}{\nwarrow} & f^2\downarrow\downarrow{\scriptstyle g^2=f^2+\phi_X\circ\eta} & & \downarrow{f^3=g^3} & & \\
\mathcal{E}_X & \xrightarrow{\phi_X} & \mathcal{F}_X & \xrightarrow{\psi_X} & T^*\underline{X} & \longrightarrow & 0.
\end{array} \qquad (5.3)$$

That is, η is a homotopy between the morphisms of complexes (5.2) from f, g.

If $f : X \to Y$ is a 1-morphism, the *identity 2-morphism* $\mathrm{id}_f : f \Rightarrow f$ is the zero morphism $0 : \underline{f}^*(\mathcal{F}_Y) \to \mathcal{E}_X$. Suppose X, Y are d-spaces, $f, g, h : X \to Y$ are 1-morphisms and $\eta : f \Rightarrow g, \zeta : g \Rightarrow h$ are 2-morphisms. The *vertical composition of* 2-*morphisms* $\zeta \odot \eta : f \Rightarrow h$ *as in* (5.14) *is* $\zeta \odot \eta = \zeta + \eta$.

Let X, Y, Z be d-spaces, $f, \tilde{f} : X \to Y$ and $g, \tilde{g} : Y \to Z$ be 1-morphisms, and $\eta : f \Rightarrow \tilde{f}, \zeta : g \Rightarrow \tilde{g}$ be 2-morphisms. The *horizontal composition of* 2-*morphisms* $\zeta * \eta : g \circ f \Rightarrow \tilde{g} \circ \tilde{f}$ as in (5.15) is

$$\zeta * \eta = \eta \circ \underline{f}^*(g^2) + f'' \circ \underline{f}^*(\zeta) + \eta \circ \underline{f}^*(\phi_Y) \circ \underline{f}^*(\zeta).$$

Regard the category $\mathbf{C}^\infty\mathbf{Sch}^{\mathrm{lf}}_{\mathrm{ssc}}$ of separated, second countable, locally fair C^∞-schemes as a 2-category with only identity 2-morphisms id_f for (1-)morphisms $\underline{f} : \underline{X} \to \underline{Y}$. Define a 2-functor $F^{\mathrm{dSpa}}_{\mathbf{C}^\infty\mathbf{Sch}} : \mathbf{C}^\infty\mathbf{Sch}^{\mathrm{lf}}_{\mathrm{ssc}} \to \mathbf{dSpa}$ to map $\underline{X}$ to $X = (\underline{X}, \mathcal{O}_X, 0, \mathrm{id}_{\mathcal{O}_X}, 0)$ on objects $\underline{X}$, to map $\underline{f}$ to $f = (\underline{f}, f^\sharp, 0)$ on (1-)morphisms $\underline{f} : \underline{X} \to \underline{Y}$, and to map identity 2-morphisms $\mathrm{id}_{\underline{f}} : \underline{f} \Rightarrow \underline{f}$ to identity 2-morphisms $\mathrm{id}_f : f \Rightarrow f$. Define a 2-functor $F^{\mathrm{dSpa}}_{\mathbf{Man}} : \mathbf{Man} \to \mathbf{dSpa}$ by $F^{\mathrm{dSpa}}_{\mathbf{Man}} = F^{\mathrm{dSpa}}_{\mathbf{C}^\infty\mathbf{Sch}} \circ F^{\mathbf{C}^\infty\mathbf{Sch}}_{\mathbf{Man}}$.

Write $\hat{\mathbf{C}}^\infty\mathbf{Sch}^{\mathrm{lf}}_{\mathrm{ssc}}$ for the full 2-subcategory of objects X in $\mathbf{dSpa}$ equivalent to $F^{\mathrm{dSpa}}_{\mathbf{C}^\infty\mathbf{Sch}}(\underline{X})$ for some $\underline{X}$ in $\mathbf{C}^\infty\mathbf{Sch}^{\mathrm{lf}}_{\mathrm{ssc}}$, and $\hat{\mathbf{Man}}$ for the full 2-subcategory of objects X in $\mathbf{dSpa}$ equivalent to $F^{\mathrm{dSpa}}_{\mathbf{Man}}(X)$ for some manifold X. When we say that a d-space X *is a* C^∞-*scheme*, or *is a manifold*, we mean that $X \in \hat{\mathbf{C}}^\infty\mathbf{Sch}^{\mathrm{lf}}_{\mathrm{ssc}}$, or $X \in \hat{\mathbf{Man}}$, respectively.

In [20, §2.2] we prove:

Theorem 5.23. **(a)** *Definition* 5.22 *defines a strict 2-category* $\mathbf{dSpa}$, *in which all 2-morphisms are 2-isomorphisms.*
(b) *For any 1-morphism* $f : X \to Y$ *in* $\mathbf{dSpa}$ *the 2-morphisms* $\eta : f \Rightarrow f$ *form an abelian group under vertical composition, and in fact a real vector space.*
(c) $F^{\mathrm{dSpa}}_{\mathbf{C}^\infty\mathbf{Sch}}$ *and* $F^{\mathrm{dSpa}}_{\mathbf{Man}}$ *in Definition* 5.22 *are full and faithful strict 2-functors. Hence* $\mathbf{C}^\infty\mathbf{Sch}^{\mathrm{lf}}_{\mathrm{ssc}}, \mathbf{Man}$ *and* $\hat{\mathbf{C}}^\infty\mathbf{Sch}^{\mathrm{lf}}_{\mathrm{ssc}}, \hat{\mathbf{Man}}$ *are equivalent 2-categories.*

Remark 5.24. One should think of a d-space $X = (\underline{X}, \mathcal{O}'_X, \mathcal{E}_X, \imath_X, \jmath_X)$ as being a C^∞-scheme $\underline{X}$, which is the 'classical' part of X and lives in a 1-category rather than a 2-category, together with some extra 'derived' information $\mathcal{O}'_X, \mathcal{E}_X, \imath_X, \jmath_X$. 2-morphisms in $\mathbf{dSpa}$ are wholly to do with this derived part. The sheaf $\mathcal{E}_X$ may be thought of as a (dual) 'obstruction sheaf' on $\underline{X}$.

5.3.2 Gluing d-spaces by equivalences

Next we discuss gluing of d-spaces and 1-morphisms on open d-subspaces.

Definition 5.25. Let $X = (\underline{X}, \mathcal{O}'_X, \mathcal{E}_X, \imath_X, \jmath_X)$ be a d-space. Suppose $\underline{U} \subseteq \underline{X}$ is an open C^∞-subscheme. Then $U = (\underline{U}, \mathcal{O}'_X|_{\underline{U}}, \mathcal{E}_X|_{\underline{U}}, \imath_X|_{\underline{U}}, \jmath_X|_{\underline{U}})$ is a d-space. We call U an *open d-subspace* of X. An *open cover* of a d-space X is a family $\{U_a : a \in A\}$ of open d-subspaces U_a of X with $\underline{X} = \bigcup_{a \in A} \underline{U}_a$.

As in [20, §2.4], we can glue 1-morphisms on open d-subspaces which are 2-isomorphic on the overlap. The proof uses partitions of unity, as in §5.2.2.

Proposition 5.26. *Suppose X, Y are d-spaces, $U, V \subseteq X$ are open d-subspaces with $X = U \cup V$, $f : U \to Y$ and $g : V \to Y$ are 1-morphisms, and $\eta : f|_{U \cap V} \Rightarrow g|_{U \cap V}$ is a 2-morphism. Then there exists a 1-morphism $h : X \to Y$ and 2-morphisms $\zeta : h|_U \Rightarrow f$, $\theta : h|_V \Rightarrow g$ such that $\theta|_{U \cap V} = \eta \odot \zeta|_{U \cap V} : h|_{U \cap V} \Rightarrow g|_{U \cap V}$. This h is unique up to 2-isomorphism, and independent up to 2-isomorphism of the choice of η.*

Equivalences $f : X \to Y$ in a 2-category are defined in Appendix A, and are the natural notion of when two objects X, Y are 'the same'. In [20, §2.4] we prove theorems on gluing d-spaces by equivalences. See Spivak [30, Lem. 6.8 & Prop. 6.9] for results similar to Theorem 5.27 for his 'local C^∞-ringed spaces', an ∞-categorical analogue of our d-spaces.

Theorem 5.27. *Suppose X, Y are d-spaces, $U \subseteq X$, $V \subseteq Y$ are open d-subspaces, and $f : U \to V$ is an equivalence in $\mathbf{dSpa}$. At the level of topological spaces, we have open $U \subseteq X$, $V \subseteq Y$ with a homeomorphism $f : U \to V$, so we can form the quotient topological space $Z := X \amalg_f Y = (X \amalg Y)/\sim$, where the equivalence relation $\sim$ on $X \amalg Y$ identifies $u \in U \subseteq X$ with $f(u) \in V \subseteq Y$.*

Suppose Z is Hausdorff. Then there exist a d-space Z with topological space Z, open d-subspaces $\hat{X}, \hat{Y}$ in Z with $Z = \hat{X} \cup \hat{Y}$, equivalences $g : X \to \hat{X}$ and $h : Y \to \hat{Y}$ in $\mathbf{dSpa}$ such that $g|_U$ and $h|_V$ are both equivalences with $\hat{X} \cap \hat{Y}$, and a 2-morphism $\eta : g|_U \Rightarrow h \circ f : U \to \hat{X} \cap \hat{Y}$. Furthermore, Z is independent of choices up to equivalence.

Theorem 5.28. *Suppose I is an indexing set, and $<$ is a total order on I, and X_i for $i \in I$ are d-spaces, and for all $i < j$ in I we are given open d-subspaces $U_{ij} \subseteq X_i$, $U_{ji} \subseteq X_j$ and an equivalence $e_{ij} : U_{ij} \to U_{ji}$, such that for all $i < j < k$ in I we have a 2-commutative diagram*

$$
\begin{array}{ccc}
& U_{ji} \cap U_{jk} & \\
{\scriptstyle e_{ij}|_{U_{ij} \cap U_{ik}}} \nearrow & \Downarrow {\scriptstyle \eta_{ijk}} & \searrow {\scriptstyle e_{jk}|_{U_{ji} \cap U_{jk}}} \\
U_{ij} \cap U_{ik} & \xrightarrow[\;\; e_{ik}|_{U_{ij} \cap U_{ik}} \;\;]{} & U_{ki} \cap U_{kj}
\end{array}
$$

for some η_{ijk}, where all three 1-morphisms are equivalences.

On the level of topological spaces, define the quotient topological space $Y = (\coprod_{i \in I} X_i)/ \sim$, where $\sim$ is the equivalence relation generated by $x_i \sim x_j$ if $i < j$, $x_i \in U_{ij} \subseteq X_i$ and $x_j \in U_{ji} \subseteq X_j$ with $e_{ij}(x_i) = x_j$. Suppose Y is Hausdorff and second countable. Then there exist a d-space $\mathbf{Y}$ and a 1-morphism $\mathbf{f}_i : \mathbf{X}_i \to \mathbf{Y}$ which is an equivalence with an open d-subspace $\hat{\mathbf{X}}_i \subseteq \mathbf{Y}$ for all $i \in I$, where $\mathbf{Y} = \bigcup_{i \in I} \hat{\mathbf{X}}_i$, such that $\mathbf{f}_i|_{U_{ij}}$ is an equivalence $\mathbf{U}_{ij} \to \hat{\mathbf{X}}_i \cap \hat{\mathbf{X}}_j$ for all $i < j$ in I, and there exists a 2-morphism $\eta_{ij} : \mathbf{f}_j \circ \mathbf{e}_{ij} \Rightarrow \mathbf{f}_i|_{U_{ij}}$. The d-space $\mathbf{Y}$ is unique up to equivalence, and is independent of choice of 2-morphisms η_{ijk}.

Suppose also that $\mathbf{Z}$ is a d-space, and $\mathbf{g}_i : \mathbf{X}_i \to \mathbf{Z}$ are 1-morphisms for all $i \in I$, and there exist 2-morphisms $\zeta_{ij} : \mathbf{g}_j \circ \mathbf{e}_{ij} \Rightarrow \mathbf{g}_i|_{U_{ij}}$ for all $i < j$ in I. Then there exist a 1-morphism $\mathbf{h} : \mathbf{Y} \to \mathbf{Z}$ and 2-morphisms $\zeta_i : \mathbf{h} \circ \mathbf{f}_i \Rightarrow \mathbf{g}_i$ for all $i \in I$. The 1-morphism $\mathbf{h}$ is unique up to 2-isomorphism, and is independent of the choice of 2-morphisms ζ_{ij}.

Remark 5.29. In Proposition 5.26, it is surprising that $\mathbf{h}$ is independent of η up to 2-isomorphism. It holds because of the existence of *partitions of unity* on nice C^∞-schemes, as in Proposition 5.8. Here is a sketch proof: suppose $\eta, \mathbf{h}, \zeta, \theta$ and $\eta', \mathbf{h}', \zeta', \theta'$ are alternative choices in Proposition 5.26. Then we have 2-morphisms $(\zeta')^{-1} \odot \zeta : \mathbf{h}|_U \Rightarrow \mathbf{h}'|_U$ and $(\theta')^{-1} \odot \theta : \mathbf{h}|_V \Rightarrow \mathbf{h}'|_V$. Choose a partition of unity $\{\alpha, 1 - \alpha\}$ on $\underline{X}$ subordinate to $\{\underline{U}, \underline{V}\}$, so that $\alpha : \underline{X} \to \mathbb{R}$ is smooth with α supported on $\underline{U} \subseteq \underline{X}$ and $1 - \alpha$ supported on $\underline{V} \subseteq \underline{X}$. Then $\alpha \cdot \big((\zeta')^{-1} \odot \zeta\big) + (1-\alpha) \cdot \big((\theta')^{-1} \odot \theta\big)$ is a 2-morphism $\mathbf{h} \Rightarrow \mathbf{h}'$, where $\alpha \cdot \big((\zeta')^{-1} \odot \zeta\big)$ makes sense on all of $\underline{X}$ (rather than just on $\underline{U}$ where $(\zeta')^{-1} \odot \zeta$ is defined) as α is supported on $\underline{U}$, so we extend by zero on $\underline{X} \setminus \underline{U}$.

Similarly, in Theorem 5.28, the compatibility conditions on the gluing data X_i, U_{ij}, e_{ij} are significantly weaker than you might expect, because of the existence of partitions of unity. The 2-morphisms η_{ijk} on overlaps $X_i \cap X_j \cap X_k$ are only required to exist, not to satisfy any further conditions. In particular, one might think that on overlaps $X_i \cap X_j \cap X_k \cap X_l$ we should require

$$\eta_{ikl} \odot (\mathrm{id}_{f_{kl}} * \eta_{ijk})|_{U_{ij} \cap U_{ik} \cap U_{il}} = \eta_{ijl} \odot (\eta_{jkl} * \mathrm{id}_{f_{ij}})|_{U_{ij} \cap U_{ik} \cap U_{il}}, \quad (5.4)$$

but we do not. Also, one might expect the ζ_{ij} should satisfy conditions on triple overlaps $X_i \cap X_j \cap X_k$, but they need not.

The moral is that constructing d-spaces by gluing together patches X_i is straightforward, as one only has to verify mild conditions on triple overlaps $X_i \cap X_j \cap X_k$. Again, this works because of the existence of partitions of unity on nice C^∞-schemes, which are used to construct the glued d-spaces $\mathbf{Z}$ and 1- and 2-morphisms in Theorems 5.27 and 5.28.

In contrast, for gluing d-stacks in [20, §9.4], we do need compatibility conditions of the form (5.4). The problem of gluing geometric spaces in an

∞-category $\mathcal{C}$ by equivalences, such as Spivak's derived manifolds [30], is discussed by Toën and Vezzosi [33, §1.3.4] and Lurie [24, §6.1.2]. It requires nontrivial conditions on overlaps $X_{i_1} \cap \cdots \cap X_{i_n}$ for all $n = 2, 3, \ldots$.

5.3.3 Fibre products in dSpa

Fibre products in 2-categories are explained in Appendix A. In [20, §2.5–§2.6] we discuss fibre products in **dSpa**, and their relation to transverse fibre products in **Man**.

Theorem 5.30. (a) *All fibre products exist in the 2-category* **dSpa**.
(b) *Let* $g : X \to Z$ *and* $h : Y \to Z$ *be smooth maps of manifolds without boundary, and write* $X = F_{\mathbf{Man}}^{\mathbf{dSpa}}(X)$, *and similarly for* Y, Z, g, h. *If* g, h *are transverse, so that a fibre product* $X \times_{g,Z,h} Y$ *exists in* **Man**, *then the fibre product* $X \times_{g,Z,h} Y$ *in* **dSpa** *is equivalent in* **dSpa** *to* $F_{\mathbf{Man}}^{\mathbf{dSpa}}(X \times_{g,Z,h} Y)$. *If* g, h *are not transverse then* $X \times_{g,Z,h} Y$ *exists in* **dSpa**, *but is not a manifold.*

To prove (a), given 1-morphisms $\boldsymbol{g} : \boldsymbol{X} \to \boldsymbol{Z}$ and $\boldsymbol{h} : \boldsymbol{Y} \to \boldsymbol{Z}$, we write down an explicit d-space $\boldsymbol{W} = (\underline{W}, \mathcal{O}'_W, \mathcal{E}_W, \imath_W, \jmath_W)$, 1-morphisms $\boldsymbol{e} = (\underline{e}, e', e'') : \boldsymbol{W} \to \boldsymbol{X}$ and $\boldsymbol{f} = (\underline{f}, f', f'') : \boldsymbol{W} \to \boldsymbol{Y}$ and a 2-morphism $\eta : \boldsymbol{g} \circ \boldsymbol{e} \Rightarrow \boldsymbol{h} \circ \boldsymbol{f}$, and verify the universal property for

$$
\begin{array}{ccc}
\boldsymbol{W} & \xrightarrow{\ \boldsymbol{f}\ } & \boldsymbol{Y} \\
{\scriptstyle \boldsymbol{e}}\downarrow & {\scriptstyle \eta}\nwarrow \ \ {\scriptstyle \boldsymbol{g}} & \downarrow{\scriptstyle \boldsymbol{h}} \\
\boldsymbol{X} & \xrightarrow[\ \boldsymbol{g}\ ]{} & \boldsymbol{Z}
\end{array}
$$

to be a 2-Cartesian square in **dSpa**. The underlying C^∞-scheme $\underline{W}$ is the fibre product $\underline{W} = \underline{X} \times_{\underline{g}, \underline{Z}, \underline{h}} \underline{Y}$ in $\mathbf{C}^\infty\mathbf{Sch}$, and $\underline{e} : \underline{W} \to \underline{X}$, $\underline{f} : \underline{W} \to \underline{Y}$ are the projections from the fibre product. The definitions of $\mathcal{O}'_W, \imath_W, \jmath_W, e', f'$ are complex, and we will not give them here. The remaining data $\mathcal{E}_W, e'', f'', \eta$, as well as the virtual cotangent sheaf $\phi_W : \mathcal{E}_W \to \mathcal{F}_W$, is characterized by the following commutative diagram in $\mathrm{qcoh}(\underline{W})$, with exact top row:

$$
\begin{array}{ccccccc}
 & & \begin{pmatrix} \underline{e}^*(g'') \\ -\underline{f}^*(h'') \\ (\underline{g}\circ\underline{e})^*(\phi_Z) \end{pmatrix} & & \begin{array}{c}\underline{e}^*(\mathcal{E}_X)\oplus \\ \underline{f}^*(\mathcal{E}_Y)\oplus \\ (\underline{g}\circ\underline{e})^*(\mathcal{F}_Z)\end{array} & \begin{pmatrix} e'' & f'' & \eta \end{pmatrix} & \\
(\underline{g}\circ\underline{e})^*(\mathcal{E}_Z) & \xrightarrow{\hspace{3cm}} & & & \xrightarrow{\hspace{2cm}} & & \mathcal{E}_W \longrightarrow 0 \\
 & \begin{pmatrix} -\underline{e}^*(\phi_X) & 0 \\ 0 & -\underline{f}^*(\phi_Y) \end{pmatrix} \searrow & & & \begin{pmatrix} \underline{e}^*(g^2) \\ -\underline{f}^*(h^2) \end{pmatrix}\Big\downarrow & & \Big\downarrow{\scriptstyle \phi_W} \\
 & & & & \begin{array}{c}\underline{e}^*(\mathcal{F}_X)\oplus \\ \underline{f}^*(\mathcal{F}_Y)\end{array} & \xrightarrow[\ \cong\ ]{\begin{pmatrix} e^2 & f^2 \end{pmatrix}} & \mathcal{F}_W.
\end{array}
$$

5.4 The 2-category of d-manifolds

Sections 5.4.1–5.4.8 survey the results of [20, §3–§4] on d-manifolds. Section 5.4.9 briefly describes extensions to d-manifolds with boundary, d-manifolds with corners, and d-orbifolds from [20, §6–§12], and §5.4.10 discusses d-manifold bordism and virtual classes for d-manifolds and d-orbifolds following [20, §13]. Section 5.4.11 explains the relationship between d-manifolds and d-orbifolds and other classes of geometric spaces, summarizing [20, §14].

5.4.1 The definition of d-manifolds

Definition 5.31. A d-space U is called a *principal d-manifold* if it is equivalent in **dSpa** to a fibre product $X \times_{g,Z,h} Y$ with $X, Y, Z \in \hat{\mathbf{Man}}$. That is,

$$U \simeq F_{\mathbf{Man}}^{\mathbf{dSpa}}(X) \times_{F_{\mathbf{Man}}^{\mathbf{dSpa}}(g), F_{\mathbf{Man}}^{\mathbf{dSpa}}(Z), F_{\mathbf{Man}}^{\mathbf{dSpa}}(h)} F_{\mathbf{Man}}^{\mathbf{dSpa}}(Y)$$

for manifolds X, Y, Z and smooth maps $g : X \to Z$ and $h : Y \to Z$. The *virtual dimension* vdim U of U is defined to be vdim $U = \dim X + \dim Y - \dim Z$. Proposition 5.40(b) below shows that if $U \neq \emptyset$ then vdim U depends only on the d-space U, and not on the choice of X, Y, Z, g, h, and so is well defined.

A d-space W is called a *d-manifold of virtual dimension* $n \in \mathbb{Z}$, written vdim $W = n$, if W can be covered by nonempty open d-subspaces U which are principal d-manifolds with vdim $U = n$.

Write **dMan** for the full 2-subcategory of d-manifolds in **dSpa**. If $X \in \hat{\mathbf{Man}}$ then $X \simeq X \times_* *$, so X is a principal d-manifold, and thus a d-manifold. Therefore $\hat{\mathbf{Man}}$ is a 2-subcategory of **dMan**. We say that a d-manifold X *is a manifold* if it lies in $\hat{\mathbf{Man}}$. The 2-functor $F_{\mathbf{Man}}^{\mathbf{dSpa}} : \mathbf{Man} \to \mathbf{dSpa}$ maps into **dMan**, and we will write $F_{\mathbf{Man}}^{\mathbf{dMan}} = F_{\mathbf{Man}}^{\mathbf{dSpa}} : \mathbf{Man} \to \mathbf{dMan}$.

Here [20, §3.2] are alternative descriptions of principal d-manifolds:

Proposition 5.32. *The following are equivalent characterizations of when a d-space W is a principal d-manifold:*

(a) $W \simeq X \times_{g,Z,h} Y$ *for $X, Y, Z \in \hat{\mathbf{Man}}$.*

(b) $W \simeq X \times_{i,Z,j} Y$, *where X, Y, Z are manifolds, $i : X \to Z$, $j : Y \to Z$ are embeddings, $X = F_{\mathbf{Man}}^{\mathbf{dSpa}}(X)$, and similarly for Y, Z, i, j. That is, W is an intersection of two submanifolds X, Y in Z, in the sense of d-spaces.*

(c) $W \simeq V \times_{s,E,0} V$, *where V is a manifold, $E \to V$ is a vector bundle, $s : V \to E$ is a smooth section, $0 : V \to E$ is the zero section, $V =$*

$F_{\mathbf{Man}}^{\mathbf{dSpa}}(V)$, *and similarly for* $E, s, 0$. *That is,* $\mathbf{W}$ *is the zeroes* $s^{-1}(0)$ *of a smooth section* s *of a vector bundle* E, *in the sense of d-spaces.*

5.4.2 'Standard model' d-manifolds, 1- and 2-morphisms

The next three examples, taken from [20, §3.2 & §3.4], give explicit models for principal d-manifolds in the form $V \times_{s,E,0} V$ from Proposition 5.32(c) and their 1- and 2-morphisms, which we call *standard models*.

Example 5.33. Let V be a manifold, $E \rightarrow V$ a vector bundle (which we sometimes call the *obstruction bundle*), and $s \in C^\infty(E)$. We will write down an explicit principal d-manifold $\mathbf{S} = (\underline{S}, \mathcal{O}'_S, \mathcal{E}_S, \iota_S, \jmath_S)$ which is equivalent to $V \times_{s,E,0} V$ in Proposition 5.32(c). We call $\mathbf{S}$ the *standard model* of (V, E, s), and also write it $\mathbf{S}_{V,E,s}$. Proposition 5.32 shows that every principal d-manifold $\mathbf{W}$ is equivalent to $\mathbf{S}_{V,E,s}$ for some V, E, s.

Write $C^\infty(V)$ for the C^∞-ring of smooth functions $c : V \rightarrow \mathbb{R}$, and $C^\infty(E), C^\infty(E^*)$ for the vector spaces of smooth sections of E, E^* over V. Then s lies in $C^\infty(E)$, and $C^\infty(E), C^\infty(E^*)$ are modules over $C^\infty(V)$, and there is a natural bilinear product $\cdot : C^\infty(E^*) \times C^\infty(E) \rightarrow C^\infty(V)$. Define $I_s \subseteq C^\infty(V)$ to be the ideal generated by s. That is,

$$I_s = \big\{ \alpha \cdot s : \alpha \in C^\infty(E^*) \big\} \subseteq C^\infty(V). \tag{5.5}$$

Let $I_s^2 = \langle fg : f, g \in I_s \rangle_\mathbb{R}$ be the square of I_s. Then I_s^2 is an ideal in $C^\infty(V)$, the ideal generated by $s \otimes s \in C^\infty(E \otimes E)$. That is,

$$I_s^2 = \big\{ \beta \cdot (s \otimes s) : \beta \in C^\infty(E^* \otimes E^*) \big\} \subseteq C^\infty(V).$$

Define C^∞-rings $\mathfrak{C} = C^\infty(V)/I_s$, $\mathfrak{C}' = C^\infty(V)/I_s^2$, and let $\pi : \mathfrak{C}' \rightarrow \mathfrak{C}$ be the natural projection from the inclusion $I_s^2 \subseteq I_s$. Define a topological space $S = \{v \in V : s(v) = 0\}$, as a subspace of V. Now $s(v) = 0$ if and only if $(s \otimes s)(v) = 0$. Thus S is the underlying topological space for both Spec $\mathfrak{C}$ and Spec $\mathfrak{C}'$. So Spec $\mathfrak{C} = \underline{S} = (S, \mathcal{O}_S)$, Spec $\mathfrak{C}' = \underline{S}' = (S, \mathcal{O}'_S)$, and Spec $\pi = \underline{\iota}_S = (\mathrm{id}_S, \iota_S) : \underline{S}' \rightarrow \underline{S}$, where $\underline{S}, \underline{S}'$ are fair affine C^∞-schemes, and $\mathcal{O}_S, \mathcal{O}'_S$ are sheaves of C^∞-rings on S, and $\iota_S : \mathcal{O}'_S \rightarrow \mathcal{O}_S$ is a morphism of sheaves of C^∞-rings. Since π is surjective with kernel the square zero ideal I_s/I_s^2, ι_S is surjective, with kernel $\mathcal{I}_S$ a sheaf of square zero ideals in $\mathcal{O}'_S$.

From (5.5) we have a surjective $C^\infty(V)$-module morphism $C^\infty(E^*) \rightarrow I_s$ mapping $\alpha \longmapsto \alpha \cdot s$. Applying $\otimes_{C^\infty(V)} \mathfrak{C}$ gives a surjective $\mathfrak{C}$-module morphism

$$\sigma : C^\infty(E^*)/(I_s \cdot C^\infty(E^*)) \longrightarrow I_s/I_s^2,$$
$$\sigma : \alpha + (I_s \cdot C^\infty(E^*)) \longmapsto \alpha \cdot s + I_s^2.$$

Define $\mathcal{E}_S = \mathrm{MSpec}\big(C^\infty(E^*)/(I_s \cdot C^\infty(E^*))\big)$. Also $\mathrm{MSpec}(I_s/I_s^2) = \mathcal{I}_S$, so $j_S = \mathrm{MSpec}\,\sigma$ is a surjective morphism $j_S : \mathcal{E}_S \to \mathcal{I}_S$ in $\mathrm{qcoh}(\underline{S})$. Therefore $\mathbf{S}_{V,E,s} = \mathbf{S} = (\underline{S}, \mathcal{O}'_{\underline{S}}, \mathcal{E}_S, \imath_S, j_S)$ is a d-space.

In fact $\mathcal{E}_S$ is a vector bundle on $\underline{S}$ naturally isomorphic to $\underline{E}^*|_{\underline{S}}$, where $\underline{E}$ is the vector bundle on $\underline{V} = F^{\mathbf{C}^\infty\mathbf{Sch}}_{\mathbf{Man}}(V)$ corresponding to $E \to V$. Also $\mathcal{F}_S \cong T^*\underline{V}|_{\underline{S}}$. The morphism $\phi_S : \mathcal{E}_S \to \mathcal{F}_S$ can be interpreted as follows: choose a connection ∇ on $E \to V$. Then $\nabla s \in C^\infty(E \otimes T^*V)$, so we can regard ∇s as a morphism of vector bundles $E^* \to T^*V$ on V. This lifts to a morphism of vector bundles $\hat\nabla s : \underline{E}^* \to T^*\underline{V}$ on the C^∞-scheme $\underline{V}$, and ϕ_S is identified with $\hat\nabla s|_{\underline{S}} : \underline{E}^*|_{\underline{S}} \to T^*\underline{V}|_{\underline{S}}$ under the isomorphisms $\mathcal{E}_S \cong \underline{E}^*|_{\underline{S}}$, $\mathcal{F}_S \cong T^*\underline{V}|_{\underline{S}}$.

Proposition 5.32 implies that every principal d-manifold $\mathbf{W}$ is equivalent to $\mathbf{S}_{V,E,s}$ for some V, E, s. The notation $O(s)$ and $O(s^2)$ used below should be interpreted as follows. Let V be a manifold, $E \to V$ a vector bundle, and $s \in C^\infty(E)$. If $F \to V$ is another vector bundle and $t \in C^\infty(F)$, then we write $t = O(s)$ if $t = \alpha \cdot s$ for some $\alpha \in C^\infty(F \otimes E^*)$, and $t = O(s^2)$ if $t = \beta \cdot (s \otimes s)$ for some $\beta \in C^\infty(F \otimes E^* \otimes E^*)$. Similarly, if W is a manifold and $f, g : V \to W$ are smooth then we write $f = g + O(s)$ if $c \circ f - c \circ g = O(s)$ for all smooth $c : W \to \mathbb{R}$, and $f = g + O(s^2)$ if $c \circ f - c \circ g = O(s^2)$ for all c.

Example 5.34. Let V, W be manifolds, $E \to V$, $F \to W$ be vector bundles, and $s \in C^\infty(E), t \in C^\infty(F)$. Write $X = \mathbf{S}_{V,E,s}, Y = \mathbf{S}_{W,F,t}$ for the 'standard model' principal d-manifolds from Example 5.33. Suppose $f : V \to W$ is a smooth map, and $\hat f : E \to f^*(F)$ is a morphism of vector bundles on V satisfying

$$\hat f \circ s = f^*(t) + O(s^2) \quad \text{in } C^\infty\big(f^*(F)\big). \tag{5.6}$$

We will define a 1-morphism $\mathbf{g} = (g, g', g'') : X \to Y$ in $\mathbf{dMan}$ using $f, \hat f$. We will also write $\mathbf{g} : X \to Y$ as $\mathbf{S}_{f,\hat f} : \mathbf{S}_{V,E,s} \to \mathbf{S}_{W,F,t}$, and call it a *standard model 1-morphism*. If $x \in X$ then $x \in V$ with $s(x) = 0$, so (5.6) implies that

$$t\big(f(x)\big) = \big(f^*(t)\big)(x) = \hat f\big(s(x)\big) + O\big(s(x)^2\big) = 0,$$

so $f(x) \in Y \subseteq W$. Thus $g := f|_X$ maps $X \to Y$.

Define morphisms of C^∞-rings

$$\phi : C^\infty(W)/I_t \longrightarrow C^\infty(V)/I_s, \quad \phi' : C^\infty(W)/I_t^2 \longrightarrow C^\infty(V)/I_s^2,$$

by $\quad \phi : c + I_t \longmapsto c \circ f + I_s, \quad \phi' : c + I_t^2 \longmapsto c \circ f + I_s^2.$

Here ϕ is well-defined since if $c \in I_t$ then $c = \gamma \cdot t$ for some $\gamma \in C^\infty(F^*)$, so

$$c \circ f = (\gamma \cdot t) \circ f = f^*(\gamma) \cdot f^*(t) = f^*(\gamma) \cdot \left(\hat{f} \circ s + O(s^2)\right)$$
$$= \left(\hat{f} \circ f^*(\gamma)\right) \cdot s + O(s^2) \in I_s.$$

Similarly if $c \in I_t^2$ then $c \circ f \in I_s^2$, so ϕ' is well-defined. Thus we have C^∞-scheme morphisms $\underline{g} = (g, g^\sharp) = \operatorname{Spec} \phi : \underline{X} \to \underline{Y}$, and $(g, g') = \operatorname{Spec} \phi' :$ $(X, \mathcal{O}'_X) \to (Y, \mathcal{O}'_Y)$, both with underlying map g. Hence $g^\sharp : g^{-1}(\mathcal{O}_Y) \to \mathcal{O}_X$ and $g' : g^{-1}(\mathcal{O}'_Y) \to \mathcal{O}'_X$ are morphisms of sheaves of C^∞-rings on X.

Since $\underline{g}^*(\mathcal{E}_Y) = \operatorname{MSpec}\!\left(C^\infty(f^*(F^*))/(I_s \cdot C^\infty(f^*(F^*)))\right)$, we may define $g'' : \underline{g}^*(\mathcal{E}_Y) \to \mathcal{E}_X$ by $g'' = \operatorname{MSpec}(G'')$, where

$$G'' : C^\infty(f^*(F^*))/(I_s \cdot C^\infty(f^*(F^*))) \longrightarrow C^\infty(E^*)/(I_s \cdot C^\infty(E^*))$$

is defined by $\quad G'' : \gamma + I_s \cdot C^\infty(f^*(F^*)) \longmapsto \gamma \circ \hat{f} + I_s \cdot C^\infty(E^*).$

This defines $\boldsymbol{g} = (\underline{g}, g', g'')$. One can show it is a 1-morphism $\boldsymbol{g} : \boldsymbol{X} \to \boldsymbol{Y}$ in **dSpa**, which we also write as $\boldsymbol{S}_{f,\hat{f}} : \boldsymbol{S}_{V,E,s} \to \boldsymbol{S}_{W,F,t}$.

Now suppose $\widetilde{V}$ is an open neighbourhood of $s^{-1}(0)$ in V, and let $\widetilde{E} = E|_{\widetilde{V}}$ and $\widetilde{s} = s|_{\widetilde{V}}$. Write $i_{\widetilde{V}} : \widetilde{V} \to V$ for the inclusion. Then $i_{\widetilde{V}}^*(E) = \widetilde{E}$, and $\operatorname{id}_{\widetilde{E}} \circ \widetilde{s} = \widetilde{s} = i_{\widetilde{V}}^*(s)$. Thus we have a 1-morphism $\boldsymbol{i}_{\widetilde{V},V} = \boldsymbol{S}_{i_{\widetilde{V}},\operatorname{id}_{\widetilde{E}}} :$ $\boldsymbol{S}_{\widetilde{V},\widetilde{E},\widetilde{s}} \to \boldsymbol{S}_{V,E,s}$. It is easy to show that $\boldsymbol{i}_{\widetilde{V},V}$ is a 1-*isomorphism*, with an inverse $\boldsymbol{i}_{\widetilde{V},V}^{-1}$. That is, making V smaller without making $s^{-1}(0)$ smaller does not really change $\boldsymbol{S}_{V,E,s}$; the d-manifold $\boldsymbol{S}_{V,E,s}$ depends only on E, s on an arbitrarily small open neighbourhood of $s^{-1}(0)$ in V.

Example 5.35. Let V, W be manifolds, $E \to V$, $F \to W$ be vector bundles, and $s \in C^\infty(E)$, $t \in C^\infty(F)$. Suppose $f, g : V \to W$ are smooth and $\hat{f} : E \to f^*(F)$, $\hat{g} : E \to g^*(F)$ are vector bundle morphisms with $\hat{f} \circ s = f^*(t) + O(s^2)$ and $\hat{g} \circ s = g^*(t) + O(s^2)$, so we have 1-morphisms $\boldsymbol{S}_{f,\hat{f}}, \boldsymbol{S}_{g,\hat{g}} : \boldsymbol{S}_{V,E,s} \to \boldsymbol{S}_{W,F,t}$. It is easy to show that $\boldsymbol{S}_{f,\hat{f}} = \boldsymbol{S}_{g,\hat{g}}$ if and only if $g = f + O(s^2)$ and $\hat{g} = \hat{f} + O(s)$.

Now suppose $\Lambda : E \to f^*(TW)$ is a morphism of vector bundles on V. Taking the dual of Λ and lifting to $\underline{V}$ gives $\Lambda^* : \underline{f}^*(T^*\underline{W}) \to \mathcal{E}^*$. Restricting to the C^∞-subscheme $\underline{X} = \underline{s}^{-1}(0)$ in $\underline{V}$ gives $\lambda = \Lambda^*|_{\underline{X}} :$ $\underline{f}^*(\mathcal{F}_Y) \cong \underline{f}^*(T^*\underline{W})|_{\underline{X}} \to \mathcal{E}^*|_{\underline{X}} = \mathcal{E}_X$. One can show that λ is a 2-morphism $\boldsymbol{S}_{f,\hat{f}} \Rightarrow \boldsymbol{S}_{g,\hat{g}}$ if and only if

$$g = f + \Lambda \circ s + O(s^2) \quad \text{and} \quad \hat{g} = \hat{f} + f^*(\mathrm{d}t) \circ \Lambda + O(s).$$

We write λ as $S_\Lambda : S_{f,\hat{f}} \Rightarrow S_{g,\hat{g}}$, and call it a *standard model 2-morphism*. Every 2-morphism $\eta : S_{f,\hat{f}} \Rightarrow S_{g,\hat{g}}$ is S_Λ for some Λ. Two vector bundle morphisms $\Lambda, \Lambda' : E \to f^*(TW)$ have $S_\Lambda = S_{\Lambda'}$ if and only if $\Lambda = \Lambda' + O(s)$.

If X is a d-manifold and $x \in X$ then x has an open neighbourhood U in X equivalent in **dSpa** to $S_{V,E,s}$ for some manifold V, vector bundle $E \to V$ and $s \in C^\infty(E)$. In [20, §3.3] we investigate the extent to which X determines V, E, s near a point in X and V, and prove:

Theorem 5.36. *Let X be a d-manifold, and $x \in X$. Then there exists an open neighbourhood U of x in X and an equivalence $U \simeq S_{V,E,s}$ in **dMan** for some manifold V, vector bundle $E \to V$ and $s \in C^\infty(E)$ which identifies $x \in U$ with a point $v \in V$ such that $s(v) = \mathrm{d}s(v) = 0$, where $S_{V,E,s}$ is as in Example 5.33. These V, E, s are determined up to non-canonical isomorphism near v by X near x, and in fact they depend only on the underlying C^∞-scheme $\underline{X}$ and the integer* vdim X.

Thus, if we impose the extra condition $\mathrm{d}s(v) = 0$, which is in fact equivalent to choosing V, E, s with $\dim V$ as small as possible, then V, E, s are determined uniquely near v by X near x (that is, V, E, s are determined locally up to isomorphism, but not up to canonical isomorphism). If we drop the condition $\mathrm{d}s(v) = 0$ then V, E, s are determined uniquely near v by X near x and $\dim V$.

Theorem 5.36 shows that any d-manifold $X = (\underline{X}, \mathcal{O}'_X, \mathcal{E}_X, \iota_X, \jmath_X)$ is determined up to equivalence in **dSpa** near any point $x \in X$ by the 'classical' underlying C^∞-scheme $\underline{X}$ and the integer vdim X. So we can ask: what extra information about X is contained in the 'derived' data $\mathcal{O}'_X, \mathcal{E}_X, \iota_X, \jmath_X$? One can think of this extra information as like a vector bundle $\mathcal{E}$ over $\underline{X}$. The only local information in a vector bundle $\mathcal{E}$ is rank $\mathcal{E} \in \mathbb{Z}$, but globally it also contains nontrivial algebraic-topological information.

Suppose now that $f : X \to Y$ is a 1-morphism in **dMan**, and $x \in X$ with $f(x) = y \in Y$. Then by Theorem 5.36 we have $X \simeq S_{V,E,s}$ near x and $Y \simeq S_{W,F,t}$ near y. So up to composition with equivalences, we can identify f near x with a 1-morphism $g : S_{V,E,s} \to S_{W,F,t}$. Thus, to understand arbitrary 1-morphisms f in **dMan** near a point, it is enough to study 1-morphisms $g : S_{V,E,s} \to S_{W,F,t}$. Our next theorem, proved in [20, §3.4], shows that after making V smaller, every 1-morphism $g : S_{V,E,s} \to S_{W,F,t}$ is of the form $S_{f,\hat{f}}$.

Theorem 5.37. *Let V, W be manifolds, $E \to V$, $F \to W$ be vector bundles, and $s \in C^\infty(E)$, $t \in C^\infty(F)$. Define principal d-manifolds $X = S_{V,E,s}$, $Y = S_{W,F,t}$, with topological spaces $X = \{v \in V : s(v) = 0\}$ and $Y = \{w \in W : t(w) = 0\}$. Suppose $\boldsymbol{g} : \boldsymbol{X} \to \boldsymbol{Y}$ is a 1-morphism. Then there exist an open neighbourhood $\widetilde{V}$ of X in V, a smooth map $f : \widetilde{V} \to W$, and a morphism of vector bundles $\hat{f} : \widetilde{E} \to f^*(F)$ with $\hat{f} \circ \widetilde{s} = f^*(t)$, where $\widetilde{E} = E|_{\widetilde{V}}$, $\widetilde{s} = s|_{\widetilde{V}}$, such that $\boldsymbol{g} = \boldsymbol{S}_{f,\hat{f}} \circ \boldsymbol{i}_{\widetilde{V},V}^{-1}$, where $\boldsymbol{i}_{\widetilde{V},V} = \boldsymbol{S}_{\mathrm{id}_{\widetilde{V}},\mathrm{id}_{\widetilde{E}}} : \boldsymbol{S}_{\widetilde{V},\widetilde{E},\widetilde{s}} \to \boldsymbol{S}_{V,E,s}$ is a 1-isomorphism, and $\boldsymbol{S}_{f,\hat{f}} : \boldsymbol{S}_{\widetilde{V},\widetilde{E},\widetilde{s}} \to \boldsymbol{S}_{W,F,t}$.*

These results give a good differential-geometric picture of d-manifolds and their 1- and 2-morphisms near a point. The $O(s)$ and $O(s^2)$ notation helps keep track of what information from V, E, s and $f, \hat{f}$ and Λ is remembered and what forgotten by the d-manifolds $\boldsymbol{S}_{V,E,s}$, 1-morphisms $\boldsymbol{S}_{f,\hat{f}}$ and 2-morphisms $\boldsymbol{S}_\Lambda$.

5.4.3 The 2-category of virtual vector bundles

In our theory of derived differential geometry, it is a general principle that 1-categories in classical differential geometry should often be replaced by 2-categories, and classical concepts be replaced by 2-categorical analogues.

In classical differential geometry, if X is a manifold, the vector bundles $E \to X$ and their morphisms form a category $\mathrm{vect}(X)$. The cotangent bundle T^*X is an important example of a vector bundle. If $f : X \to Y$ is smooth then pullback $f^* : \mathrm{vect}(Y) \to \mathrm{vect}(X)$ is a functor. There is a natural morphism $\mathrm{d}f^* : f^*(T^*Y) \to T^*X$. We now explain 2-categorical analogues of all this for d-manifolds, following [20, §3.1–§3.2].

Definition 5.38. Let $\underline{X}$ be a C^∞-scheme, which will usually be the C^∞-scheme underlying a d-manifold X. We will define a 2-category $\mathrm{vqcoh}(\underline{X})$ of *virtual quasicoherent sheaves* on $\underline{X}$. *Objects* of $\mathrm{vqcoh}(\underline{X})$ are morphisms $\phi : \mathcal{E}^1 \to \mathcal{E}^2$ in $\mathrm{qcoh}(\underline{X})$, which we also may write as $(\mathcal{E}^1, \mathcal{E}^2, \phi)$ or $(\mathcal{E}^\bullet, \phi)$. Given objects $\phi : \mathcal{E}^1 \to \mathcal{E}^2$ and $\psi : \mathcal{F}^1 \to \mathcal{F}^2$, a 1-*morphism* $(f^1, f^2) : (\mathcal{E}^\bullet, \phi) \to (\mathcal{F}^\bullet, \psi)$ is a pair of morphisms $f^1 : \mathcal{E}^1 \to \mathcal{F}^1$, $f^2 : \mathcal{E}^2 \to \mathcal{F}^2$ in $\mathrm{qcoh}(\underline{X})$ such that $\psi \circ f^1 = f^2 \circ \phi$. We write $f^\bullet$ for (f^1, f^2).

The *identity* 1-*morphism* of $(\mathcal{E}^\bullet, \phi)$ is $(\mathrm{id}_{\mathcal{E}^1}, \mathrm{id}_{\mathcal{E}^2})$. The *composition* of 1-morphisms $f^\bullet : (\mathcal{E}^\bullet, \phi) \to (\mathcal{F}^\bullet, \psi)$ and $g^\bullet : (\mathcal{F}^\bullet, \psi) \to (\mathcal{G}^\bullet, \xi)$ is $g^\bullet \circ f^\bullet = (g^1 \circ f^1, g^2 \circ f^2) : (\mathcal{E}^\bullet, \phi) \to (\mathcal{G}^\bullet, \xi)$.

Given $f^\bullet, g^\bullet : (\mathcal{E}^\bullet, \phi) \to (\mathcal{F}^\bullet, \psi)$, a 2-*morphism* $\eta : f^\bullet \Rightarrow g^\bullet$ is a morphism $\eta : \mathcal{E}^2 \to \mathcal{F}^1$ in $\mathrm{qcoh}(\underline{X})$ such that $g^1 = f^1 + \eta \circ \phi$ and $g^2 = f^2 + \psi \circ \eta$. The *identity* 2-*morphism* for $f^\bullet$ is $\mathrm{id}_{f^\bullet} = 0$. If $f^\bullet, g^\bullet, h^\bullet : (\mathcal{E}^\bullet, \phi) \to (\mathcal{F}^\bullet, \psi)$

are 1-morphisms and $\eta : f^\bullet \Rightarrow g^\bullet, \zeta : g^\bullet \Rightarrow h^\bullet$ are 2-morphisms, the *vertical composition of 2-morphisms* $\zeta \odot \eta : f^\bullet \Rightarrow h^\bullet$ is $\zeta \odot \eta = \zeta + \eta$. If $f^\bullet, \widetilde{f}^\bullet :$ $(\mathcal{E}^\bullet, \phi) \to (\mathcal{F}^\bullet, \psi)$ and $g^\bullet, \widetilde{g}^\bullet : (\mathcal{F}^\bullet, \psi) \to (\mathcal{G}^\bullet, \xi)$ are 1-morphisms and $\eta : f^\bullet \Rightarrow \widetilde{f}^\bullet, \zeta : g^\bullet \Rightarrow \widetilde{g}^\bullet$ are 2-morphisms, the *horizontal composition of 2-morphisms* $\zeta * \eta : g^\bullet \circ f^\bullet \to \widetilde{g}^\bullet \circ \widetilde{f}^\bullet$ is $\zeta * \eta = g^1 \circ \eta + \zeta \circ f^2 + \zeta \circ \psi \circ \eta$. This defines a strict 2-category vqcoh($\underline{X}$), the obvious 2-category of 2-term complexes in qcoh($\underline{X}$).

If $\underline{U} \subseteq \underline{X}$ is an open C^∞-subscheme then restriction from $\underline{X}$ to $\underline{U}$ defines a strict 2-functor $|_{\underline{U}} :$ vqcoh($\underline{X}$) $\to$ vqcoh($\underline{U}$). An object $(\mathcal{E}^\bullet, \phi)$ in vqcoh($\underline{X}$) is called a *virtual vector bundle of rank* $d \in \mathbb{Z}$ if $\underline{X}$ may be covered by open $\underline{U} \subseteq \underline{X}$ such that $(\mathcal{E}^\bullet, \phi)|_{\underline{U}}$ is equivalent in vqcoh($\underline{U}$) to some $(\mathcal{F}^\bullet, \psi)$ for $\mathcal{F}^1, \mathcal{F}^2$ vector bundles on $\underline{U}$ with rank $\mathcal{F}^2 -$ rank $\mathcal{F}^1 = d$. We write rank($\mathcal{E}^\bullet, \phi$) $= d$. If $\underline{X} \neq \emptyset$ then rank($\mathcal{E}^\bullet, \phi$) depends only on $\mathcal{E}^1, \mathcal{E}^2, \phi$, so it is well-defined. Write vvect($\underline{X}$) for the full 2-subcategory of virtual vector bundles in vqcoh($\underline{X}$).

If $\underline{f} : \underline{X} \to \underline{Y}$ is a C^∞-scheme morphism then pullback gives a strict 2-functor $\underline{f}^* :$ vqcoh($\underline{Y}$) $\to$ vqcoh($\underline{X}$), which maps vvect($\underline{Y}$) $\to$ vvect($\underline{X}$).

We apply these ideas to d-spaces.

Definition 5.39. Let $X = (\underline{X}, \mathcal{O}'_X, \mathcal{E}_X, \iota_X, \jmath_X)$ be a d-space. Define the *virtual cotangent sheaf* T^*X of X to be the morphism $\phi_X : \mathcal{E}_X \to \mathcal{F}_X$ in qcoh($\underline{X}$) from Definition 5.22, regarded as a virtual quasicoherent sheaf on $\underline{X}$.

Let $\boldsymbol{f} = (\underline{f}, f', f'') : X \to Y$ be a 1-morphism in **dSpa**. Then $T^*X = (\mathcal{E}_X, \mathcal{F}_X, \phi_X)$ and $\underline{f}^*(T^*Y) = (\underline{f}^*(\mathcal{E}_Y), \underline{f}^*(\mathcal{F}_Y), \underline{f}^*(\phi_Y))$ are virtual quasicoherent sheaves on $\underline{X}$, and $\Omega_{\boldsymbol{f}} := (f'', f^2)$ is a 1-morphism $\underline{f}^*(T^*Y) \to T^*X$ in vqcoh($\underline{X}$), as (5.2) commutes.

Let $\boldsymbol{f}, \boldsymbol{g} : X \to Y$ be 1-morphisms in **dSpa**, and $\eta : \boldsymbol{f} \Rightarrow \boldsymbol{g}$ a 2-morphism. Then $\eta : \underline{f}^*(\mathcal{F}_Y) \to \mathcal{E}_X$ with $g'' = f'' + \eta \circ \underline{f}^*(\phi_Y)$ and $g^2 = f^2 + \phi_X \circ \eta$, as in (5.3). It follows that η is a 2-morphism $\Omega_{\boldsymbol{f}} \Rightarrow \Omega_{\boldsymbol{g}}$ in vqcoh($\underline{X}$). Thus, objects, 1-morphisms and 2-morphisms in **dSpa** lift to objects, 1-morphisms and 2-morphisms in vqcoh($\underline{X}$).

The next proposition justifies the definition of virtual vector bundle. Because of part (b), if W is a d-manifold we call T^*W the *virtual cotangent bundle* of W, rather than the virtual cotangent sheaf.

Proposition 5.40. (a) *Let* V *be a manifold,* $E \to V$ *a vector bundle, and* $s \in C^\infty(E)$. *Then Example 5.33 defines a principal d-manifold* $S_{V,E,s}$.

Its cotangent bundle $T^*S_{V,E,s}$ is a virtual vector bundle on $\underline{S}_{V,E,s}$ of rank dim V − rank E.

(b) *Let W be a d-manifold. Then T^*W is a virtual vector bundle on $\underline{W}$ of rank* vdim W. *Hence if $W \neq \emptyset$ then* vdim W *is well-defined.*

The virtual cotangent bundle T^*X of a d-manifold X contains only a fraction of the information in $X = (\underline{X}, \mathcal{O}'_X, \mathcal{E}_X, \imath_X, \jmath_X)$, but many interesting properties of d-manifolds X and 1-morphisms $f : X \to Y$ can be expressed solely in terms of virtual cotangent bundles T^*X, T^*Y and 1-morphisms $\Omega_f : f^*(T^*Y) \to T^*X$. Here is an example of this.

Definition 5.41. Let $\underline{X}$ be a C^∞-scheme. We say that a virtual vector bundle $(\mathcal{E}^1, \mathcal{E}^2, \phi)$ on $\underline{X}$ *is a vector bundle* if it is equivalent in vvect($\underline{X}$) to $(0, \mathcal{E}, 0)$ for some vector bundle $\mathcal{E}$ on $\underline{X}$. One can show $(\mathcal{E}^1, \mathcal{E}^2, \phi)$ is a vector bundle if and only if ϕ has a left inverse in qcoh($\underline{X}$).

Proposition 5.42. *Let X be a d-manifold. Then X is a manifold (that is, $X \in \hat{\textbf{Man}}$) if and only if T^*X is a vector bundle, or equivalently, if $\phi_X : \mathcal{E}_X \to \mathcal{F}_X$ has a left inverse in* qcoh($\underline{X}$).

5.4.4   Equivalences in dMan, and gluing by equivalences

Equivalences in a 2-category are defined in Appendix A. Equivalences in **dMan** are the best derived analogue of isomorphisms in **Man**, that is, of diffeomorphisms of manifolds. A smooth map of manifolds $f : X \to Y$ is called *étale* if it is a local diffeomorphism. Here is the derived analogue.

Definition 5.43. Let $f : X \to Y$ be a 1-morphism in **dMan**. We call f *étale* if it is a *local equivalence*, that is, if for each $x \in X$ there exist open $x \in U \subseteq X$ and $f(x) \in V \subseteq Y$ such that $f(U) = V$ and $f|_U : U \to V$ is an equivalence.

If $f : X \to Y$ is a smooth map of manifolds, then f is étale if and only if $df^* : f^*(T^*Y) \to T^*X$ is an isomorphism of vector bundles. (The analogue is false for schemes.) In [20, §3.5] we prove a version of this for d-manifolds:

Theorem 5.44. *Suppose $f : X \to Y$ is a 1-morphism of d-manifolds. Then the following are equivalent:*

  (i) *f is étale;*

  (ii) *$\Omega_f : \underline{f}^*(T^*Y) \to T^*X$ is an equivalence in* vqcoh($\underline{X}$); *and*

(iii) *the following is a split short exact sequence in* $\mathrm{qcoh}(\underline{X})$:

$$0 \longrightarrow \underline{f}^*(\mathcal{E}_Y) \xrightarrow{\ f'' \oplus -\underline{f}^*(\phi_Y)\ } \mathcal{E}_X \oplus \underline{f}^*(\mathcal{F}_Y) \xrightarrow{\ \phi_X \oplus f^2\ } \mathcal{F}_X \longrightarrow 0.$$

If in addition $f : X \to Y$ *is a bijection, then* $\boldsymbol{f}$ *is an equivalence in* **dMan**.

The analogue of Theorem 5.44 for d-spaces is false. When $\boldsymbol{f} : \boldsymbol{X} \to \boldsymbol{Y}$ is a 'standard model' 1-morphism $\boldsymbol{S}_{f,\hat{f}} : \boldsymbol{S}_{V,E,s} \to \boldsymbol{S}_{W,F,t}$, as in §5.4.2, we can express the conditions for $\boldsymbol{S}_{f,\hat{f}}$ to be étale or an equivalence in terms of $f, \hat{f}$.

Theorem 5.45. *Let* V, W *be manifolds,* $E \to V$, $F \to W$ *be vector bundles,* $s \in C^\infty(E)$, $t \in C^\infty(F)$, $f : V \to W$ *be smooth, and* $\hat{f} : E \to f^*(F)$ *be a morphism of vector bundles on* V *with* $\hat{f} \circ s = f^*(t) + O(s^2)$. *Then Example 5.34 defines a 1-morphism* $\boldsymbol{S}_{f,\hat{f}} : \boldsymbol{S}_{V,E,s} \to \boldsymbol{S}_{W,F,t}$ *in* **dMan**. *This* $\boldsymbol{S}_{f,\hat{f}}$ *is étale if and only if for each* $v \in V$ *with* $s(v) = 0$ *and* $w = f(v) \in W$, *the following sequence of vector spaces is exact:*

$$0 \longrightarrow T_v V \xrightarrow{\ ds(v) \oplus df(v)\ } E_v \oplus T_w W \xrightarrow{\ \hat{f}(v) \oplus -dt(w)\ } F_w \longrightarrow 0.$$

Also $\boldsymbol{S}_{f,\hat{f}}$ *is an equivalence if and only if in addition* $f|_{s^{-1}(0)} : s^{-1}(0) \to t^{-1}(0)$ *is a bijection, where* $s^{-1}(0) = \{v \in V : s(v) = 0\}$, $t^{-1}(0) = \{w \in W : t(w) = 0\}$.

Section 5.3.2 discussed gluing d-spaces by equivalences on open d-subspaces. It generalizes immediately to d-manifolds: if in Theorem 5.28 we fix $n \in \mathbb{Z}$ and take the initial d-spaces $\boldsymbol{X}_i$ to be d-manifolds with $\mathrm{vdim}\, \boldsymbol{X}_i = n$, then the glued d-space $\boldsymbol{Y}$ is also a d-manifold with $\mathrm{vdim}\, \boldsymbol{Y} = n$.

Here is an analogue of Theorem 5.28, taken from [20, §3.6], in which we take the d-spaces $\boldsymbol{X}_i$ to be 'standard model' d-manifolds $\boldsymbol{S}_{V_i,E_i,s_i}$, and the 1-morphisms $\boldsymbol{e}_{ij}$ to be 'standard model' 1-morphisms $\boldsymbol{S}_{e_{ij},\hat{e}_{ij}}$. We also use Theorem 5.45 in (iii) to characterize when $\boldsymbol{e}_{ij}$ is an equivalence.

Theorem 5.46. *Suppose we are given the following data:*

(a) *an integer* n;

(b) *a Hausdorff, second countable topological space* X;

(c) *an indexing set* I, *and a total order* $<$ *on* I;

(d) *for each* i *in* I, *a manifold* V_i, *a vector bundle* $E_i \to V_i$ *with* $\dim V_i -$ rank $E_i = n$, *a smooth section* $s_i : V_i \to E_i$, *and a homeomorphism* $\psi_i : X_i \to \hat{X}_i$, *where* $X_i = \{v_i \in V_i : s_i(v_i) = 0\}$ *and* $\hat{X}_i \subseteq X$ *is open; and*

(e) *for all $i < j$ in I, an open submanifold $V_{ij} \subseteq V_i$, a smooth map $e_{ij} :$ $V_{ij} \to V_j$, and a morphism of vector bundles $\hat{e}_{ij} : E_i|_{V_{ij}} \to e_{ij}^*(E_j)$.*

Using notation $O(s_i)$, $O(s_i^2)$ as in §5.4.2, let this data satisfy the conditions:

(i) $X = \bigcup_{i \in I} \hat{X}_i$;

(ii) *if $i < j$ in I then $\hat{e}_{ij} \circ s_i|_{V_{ij}} = e_{ij}^*(s_j) + O(s_i^2)$, $\psi_i(X_i \cap V_{ij}) = \hat{X}_i \cap \hat{X}_j$, and $\psi_i|_{X_i \cap V_{ij}} = \psi_j \circ e_{ij}|_{X_i \cap V_{ij}}$, and if $v_i \in V_{ij}$ with $s_i(v_i) = 0$ and $v_j = e_{ij}(v_i)$ then the following is exact:*

$$0 \longrightarrow T_{v_i} V_i \xrightarrow{\mathrm{d}s_i(v_i) \oplus \mathrm{d}e_{ij}(v_i)} E_i|_{v_i} \oplus T_{v_j} V_j \xrightarrow{\hat{e}_{ij}(v_i) \oplus -\mathrm{d}s_j(v_j)} E_j|_{v_j} \longrightarrow 0;$$

(iii) *if $i < j < k$ in I then*

$$e_{ik}|_{V_{ij} \cap V_{ik}} = e_{jk} \circ e_{ij}|_{V_{ij} \cap V_{ik}} + O(s_i^2) \qquad and$$

$$\hat{e}_{ik}|_{V_{ij} \cap V_{ik}} = e_{ij}|_{V_{ij} \cap V_{ik}}^*(\hat{e}_{jk}) \circ \hat{e}_{ij}|_{V_{ij} \cap V_{ik}} + O(s_i).$$

Then there exist a d-manifold X with $\mathrm{vdim}\, X = n$ and underlying topological space X, and a 1-morphism $\boldsymbol{\psi}_i : \boldsymbol{S}_{V_i, E_i, s_i} \to X$ with underlying continuous map ψ_i which is an equivalence with the open d-submanifold $\hat{X}_i \subseteq X$ corresponding to $\hat{X}_i \subseteq X$ for all $i \in I$, such that for all $i < j$ in I there exists a 2-morphism $\eta_{ij} : \boldsymbol{\psi}_j \circ \boldsymbol{S}_{e_{ij}, \hat{e}_{ij}} \Rightarrow \boldsymbol{\psi}_i \circ \boldsymbol{i}_{V_{ij}, V_i}$, where $\boldsymbol{S}_{e_{ij}, \hat{e}_{ij}} : \boldsymbol{S}_{V_{ij}, E_i|_{V_{ij}}, s_i|_{V_{ij}}} \to \boldsymbol{S}_{V_j, E_j, s_j}$ and $\boldsymbol{i}_{V_{ij}, V_i} : \boldsymbol{S}_{V_{ij}, E_i|_{V_{ij}}, s_i|_{V_{ij}}} \to \boldsymbol{S}_{V_i, E_i, s_i}$. This d-manifold X is unique up to equivalence in $\mathbf{dMan}$.

Suppose also that Y is a manifold, and $g_i : V_i \to Y$ are smooth maps for all $i \in I$, and $g_j \circ e_{ij} = g_i|_{V_{ij}} + O(s_i)$ for all $i < j$ in I. Then there exists a 1-morphism $\boldsymbol{h} : X \to Y$ unique up to 2-isomorphism, where $Y = F_{\mathbf{Man}}^{\mathbf{dMan}}(Y) = \boldsymbol{S}_{Y, 0, 0}$, and 2-morphisms $\zeta_i : \boldsymbol{h} \circ \boldsymbol{\psi}_i \Rightarrow \boldsymbol{S}_{g_i, 0}$ for all $i \in I$. Here $\boldsymbol{S}_{Y, 0, 0}$ is from Example 5.33 with vector bundle E and section s both zero, and $\boldsymbol{S}_{g_i, 0} : \boldsymbol{S}_{V_i, E_i, s_i} \to \boldsymbol{S}_{Y, 0, 0} = Y$ is from Example 5.34 with $\hat{g}_i = 0$.

The hypotheses of Theorem 5.46 are similar to the notion of *good coordinate system* in the theory of Kuranishi spaces of Fukaya and Ono [10, Def. 6.1]. The importance of Theorem 5.46 is that all the ingredients are described wholly in differential-geometric or topological terms. So we can use the theorem as a tool to prove the existence of d-manifold structures on spaces coming from other areas of geometry, for instance, on moduli spaces.

5.4.5 Submersions, immersions and embeddings

Let $f : X \to Y$ be a smooth map of manifolds. Then $\mathrm{d}f^* : f^*(T^*Y) \to T^*X$ is a morphism of vector bundles on X, and f is a *submersion* if $\mathrm{d}f^*$ is injective, and f is an *immersion* if df^* is surjective. Here the appropriate notions of injective and surjective for morphisms of vector bundles are stronger than the corresponding notions for sheaves: df^* is *injective* if it has a left inverse, and *surjective* if it has a right inverse.

In a similar way, if $f : X \to Y$ is a 1-morphism of d-manifolds, we would like to define f to be a submersion or immersion if the 1-morphism $\Omega_f : \underline{f^*(T^*Y)} \to T^*X$ in vvect($\underline{X}$) is injective or surjective in some suitable sense. It turns out that there are two different notions of injective and surjective 1-morphisms in the 2-category vvect($\underline{X}$), a weak and a strong:

Definition 5.47. Let $\underline{X}$ be a C^∞-scheme, $(\mathcal{E}^1, \mathcal{E}^2, \phi)$ and $(\mathcal{F}^1, \mathcal{F}^2, \psi)$ be virtual vector bundles on $\underline{X}$, and $(f^1, f^2) : (\mathcal{E}^\bullet, \phi) \to (\mathcal{F}^\bullet, \psi)$ be a 1-morphism in vvect($\underline{X}$). Then we have a complex in qcoh($\underline{X}$):

$$0 \longrightarrow \mathcal{E}^1 \underset{\gamma}{\overset{f^1 \oplus -\phi}{\rightleftarrows}} \mathcal{F}^1 \oplus \mathcal{E}^2 \underset{\delta}{\overset{\psi \oplus f^2}{\rightleftarrows}} \mathcal{F}^2 \longrightarrow 0. \qquad (5.7)$$

One can show that $f^\bullet$ is an equivalence in vvect($\underline{X}$) if and only if (5.7) is a *split short exact sequence* in qcoh($\underline{X}$). That is, $f^\bullet$ is an equivalence if and only if there exist morphisms γ, δ as shown in (5.7) satisfying the conditions:

$$\gamma \circ \delta = 0, \qquad \gamma \circ (f^1 \oplus -\phi) = \mathrm{id}_{\mathcal{E}^1},$$
$$(f^1 \oplus -\phi) \circ \gamma + \delta \circ (\psi \oplus f^2) = \mathrm{id}_{\mathcal{F}^1 \oplus \mathcal{E}^2}, \qquad (\psi \oplus f^2) \circ \delta = \mathrm{id}_{\mathcal{F}^2}.$$
$$(5.8)$$

Our notions of $f^\bullet$ injective or surjective impose some but not all of (5.8):

(a) We call $f^\bullet$ *weakly injective* if there exists $\gamma : \mathcal{F}^1 \oplus \mathcal{E}^2 \to \mathcal{E}^1$ in qcoh($\underline{X}$) with $\gamma \circ (f^1 \oplus -\phi) = \mathrm{id}_{\mathcal{E}^1}$.
(b) We call $f^\bullet$ *injective* if there exist $\gamma : \mathcal{F}^1 \oplus \mathcal{E}^2 \to \mathcal{E}^1$ and $\delta : \mathcal{F}^2 \to \mathcal{F}^1 \oplus \mathcal{E}^2$ with $\gamma \circ \delta = 0$, $\gamma \circ (f^1 \oplus -\phi) = \mathrm{id}_{\mathcal{E}^1}$ and $(f^1 \oplus -\phi) \circ \gamma + \delta \circ (\psi \oplus f^2) = \mathrm{id}_{\mathcal{F}^1 \oplus \mathcal{E}^2}$.
(c) We call $f^\bullet$ *weakly surjective* if there exists $\delta : \mathcal{F}^2 \to \mathcal{F}^1 \oplus \mathcal{E}^2$ in qcoh($\underline{X}$) with $(\psi \oplus f^2) \circ \delta = \mathrm{id}_{\mathcal{F}^2}$.
(d) We call $f^\bullet$ *surjective* if there exist $\gamma : \mathcal{F}^1 \oplus \mathcal{E}^2 \to \mathcal{E}^1$ and $\delta : \mathcal{F}^2 \to \mathcal{F}^1 \oplus \mathcal{E}^2$ with $\gamma \circ \delta = 0$, $\gamma \circ (f^1 \oplus -\phi) = \mathrm{id}_{\mathcal{E}^1}$ and $(\psi \oplus f^2) \circ \delta = \mathrm{id}_{\mathcal{F}^2}$.

Using these we define weak and strong forms of submersions, immersions, and embeddings for d-manifolds.

Definition 5.48. Let $f : X \to Y$ be a 1-morphism of d-manifolds. Definition 5.39 defines a 1-morphism $\Omega_f : \underline{f^*(T^*Y)} \to T^*X$ in vvect($\underline{X}$). Then:

(a) We call f a *w-submersion* if Ω_f is weakly injective.
(b) We call f a *submersion* if Ω_f is injective.
(c) We call f a *w-immersion* if Ω_f is weakly surjective.
(d) We call f an *immersion* if Ω_f is surjective.
(e) We call f a *w-embedding* if it is a w-immersion and $f : X \to f(X)$ is a homeomorphism, so in particular f is injective.
(f) We call f an *embedding* if it is an immersion and f is a homeomorphism with its image.

Here w-submersion is short for *weak submersion*, etc. Conditions (a)–(d) all concern the existence of morphisms γ, δ in the next equation satisfying identities.

$$0 \longrightarrow \underline{f}^*(\mathcal{E}_Y) \underset{\gamma}{\overset{f'' \oplus -\underline{f}^*(\phi_Y)}{\underset{\longleftarrow}{\longrightarrow}}} \mathcal{E}_X \oplus \underline{f}^*(\mathcal{F}_Y) \underset{\delta}{\overset{\phi_X \oplus f^2}{\underset{\longleftarrow}{\longrightarrow}}} \mathcal{F}_X \longrightarrow 0.$$

Parts (c)–(f) enable us to define *d-submanifolds* of d-manifolds. *Open d-submanifolds* are open d-subspaces of a d-manifold. More generally, we call $i : X \to Y$ a *w-immersed*, or *immersed*, or *w-embedded*, or *embedded d-submanifold*, of Y, if X, Y are d-manifolds and i is a w-immersion, immersion, w-embedding, or embedding, respectively.

Here are some properties of these, taken from [20, §4.1–§4.2]:

Theorem 5.49. **(i)** *Any equivalence of d-manifolds is a w-submersion, submersion, w-immersion, immersion, w-embedding and embedding.*
(ii) *If $f, g : X \to Y$ are 2-isomorphic 1-morphisms of d-manifolds then f is a w-submersion, submersion, ..., embedding, if and only if g is.*
(iii) *Compositions of w-submersions, submersions, w-immersions, immersions, w-embeddings, and embeddings are 1-morphisms of the same kind.*
(iv) *The conditions that a 1-morphism of d-manifolds $f : X \to Y$ is a w-submersion, submersion, w-immersion or immersion are local in X and Y. That is, for each $x \in X$ with $f(x) = y \in Y$, it suffices to check the conditions for $f|_U : U \to V$ with V an open neighbourhood of y in Y, and U an open neighbourhood of x in $f^{-1}(V) \subseteq X$.*
(v) *Let $f : X \to Y$ be a submersion of d-manifolds. Then $\operatorname{vdim} X \geqslant \operatorname{vdim} Y$, and if $\operatorname{vdim} X = \operatorname{vdim} Y$ then f is étale.*
(vi) *Let $f : X \to Y$ be an immersion of d-manifolds. Then $\operatorname{vdim} X \leqslant \operatorname{vdim} Y$, and if $\operatorname{vdim} X = \operatorname{vdim} Y$ then f is étale.*
(vii) *Let $f : X \to Y$ be a smooth map of manifolds, and $\boldsymbol{f} = F_{\mathbf{Man}}^{\mathbf{dMan}}(f)$. Then $\boldsymbol{f}$ is a submersion, immersion, or embedding in $\mathbf{dMan}$ if and only if f*

is a submersion, immersion, or embedding in **Man**, *respectively. Also f is a w-immersion or w-embedding if and only if f is an immersion or embedding.*

(viii) *Let $f : X \to Y$ be a 1-morphism of d-manifolds, with Y a manifold. Then f is a w-submersion.*

(ix) *Let X, Y be d-manifolds, with Y a manifold. Then $\pi_X : X \times Y \to X$ is a submersion.*

(x) *Let $f : X \to Y$ be a submersion of d-manifolds, and $x \in X$ with $f(x) = y \in Y$. Then there exist open $x \in U \subseteq X$ and $y \in V \subseteq Y$ with $f(U) = V$, a manifold Z, and an equivalence $i : U \to V \times Z$, such that $f|_U : U \to V$ is 2-isomorphic to $\pi_V \circ i$, where $\pi_V : V \times Z \to V$ is the projection.*

(xi) *Let $f : X \to Y$ be a submersion of d-manifolds with Y a manifold. Then X is a manifold.*

5.4.6 D-transversality and fibre products

From §5.3.3, if $g : X \to Z$ and $h : Y \to Z$ are 1-morphisms of d-manifolds then a fibre product $W = X_{g,Z,h} Y$ exists in **dSpa**, and is unique up to equivalence. We want to know whether W is a d-manifold. We will define when g, h are *d-transverse*, which is a sufficient condition for W to be a d-manifold.

Recall that if $g : X \to Z$, $h : Y \to Z$ are smooth maps of manifolds, then a fibre product $W = X \times_{g,Z,h} Y$ in **Man** exists if g, h are *transverse*, that is, if $T_z Z = \mathrm{d}g|_x(T_x X) + \mathrm{d}h|_y(T_y Y)$ for all $x \in X$ and $y \in Y$ with $g(x) = h(y) = z \in Z$. Equivalently, $\mathrm{d}g|_x^* \oplus \mathrm{d}h|_y^* : T_z Z^* \to T_x^* X \oplus T_y^* Y$ should be injective. Writing $W = X \times_Z Y$ for the topological fibre product and $e : W \to X$, $f : W \to Y$ for the projections, with $g \circ e = h \circ f$, we see that g, h are transverse if and only if

$$e^*(\mathrm{d}g^*) \oplus f^*(\mathrm{d}h^*) : (g \circ e)^*(T^*Z) \to e^*(T^*X) \oplus f^*(T^*Y) \qquad (5.9)$$

is an injective morphism of vector bundles on the topological space W, that is, it has a left inverse. The condition that (5.10) has a left inverse is an analogue of this, but on (dual) obstruction rather than cotangent bundles.

Definition 5.50. Let X, Y, Z be d-manifolds and $g : X \to Z$, $h : Y \to Z$ be 1-morphisms. Let $\underline{W} = \underline{X} \times_{g,\underline{Z},h} \underline{Y}$ be the C^∞-scheme fibre product, and write $\underline{e} : \underline{W} \to \underline{X}$, $\underline{f} : \underline{W} \to \underline{Y}$ for the projections. Consider the morphism

$$\alpha = \underline{e}^*(g'') \oplus -\underline{f}^*(h'') \oplus (g \circ \underline{e})^*(\phi_Z) : (g \circ \underline{e})^*(\mathcal{E}_Z) \longrightarrow$$
$$\underline{e}^*(\mathcal{E}_X) \oplus \underline{f}^*(\mathcal{E}_Y) \oplus (g \circ \underline{e})^*(\mathcal{F}_Z) \qquad (5.10)$$

in $\mathrm{qcoh}(\underline{W})$. We call g, h *d-transverse* if α has a left inverse.

In the notation of §5.4.3 and §5.4.5, we have 1-morphisms Ω_g : $\underline{g}^*(T^*Z) \rightarrow T^*X$ in vvect($\underline{X}$) and Ω_h : $\underline{h}^*(T^*Z) \rightarrow T^*Y$ in vvect($\underline{Y}$). Pulling these back to vvect($\underline{W}$) using $\underline{e}^*, \underline{f}^*$ we form the 1-morphism in vvect($\underline{W}$):

$$\underline{e}^*(\Omega_g) \oplus \underline{f}^*(\Omega_h) : (\underline{g} \circ \underline{e})^*(T^*Z) \longrightarrow \underline{e}^*(T^*X) \oplus \underline{f}^*(T^*Y). \qquad (5.11)$$

For (5.10) to have a left inverse is equivalent to (5.11) being weakly injective, as in Definition 5.47. This is the d-manifold analogue of (5.9) being injective.

Here are the main results of [20, §4.3]:

Theorem 5.51. *Suppose X, Y, Z are d-manifolds and $g : X \rightarrow Z$, $h : Y \rightarrow Z$ are d-transverse 1-morphisms, and let $W = X \times_{g,Z,h} Y$ be the d-space fibre product. Then W is a d-manifold, with*

$$\text{vdim } W = \text{vdim } X + \text{vdim } Y - \text{vdim } Z. \qquad (5.12)$$

Theorem 5.52. *Suppose $g : X \rightarrow Z$, $h : Y \rightarrow Z$ are 1-morphisms of d-manifolds. The following are sufficient conditions for g, h to be d-transverse, so that $W = X \times_{g,Z,h} Y$ is a d-manifold of virtual dimension (5.12):*

(a) *Z is a manifold, that is, $Z \in \hat{\textbf{Man}}$; or*
(b) *g or h is a w-submersion.*

The point here is that roughly speaking, g, h are d-transverse if they map the direct sum of the obstruction spaces of X, Y surjectively onto the obstruction spaces of Z. If Z is a manifold its obstruction spaces are zero. If g is a w-submersion it maps the obstruction spaces of X surjectively onto the obstruction spaces of Z. In both cases, d-transversality follows. See [30, Th. 8.15] for the analogue of Theorem 5.52(a) for Spivak's derived manifolds.

Theorem 5.53. *Let X, Z be d-manifolds, Y a manifold, and $g : X \rightarrow Z$, $h : Y \rightarrow Z$ be 1-morphisms with g a submersion. Then $W = X \times_{g,Z,h} Y$ is a manifold, with $\dim W = \text{vdim } X + \dim Y - \text{vdim } Z$.*

Theorem 5.53 shows that we may think of submersions as *representable 1-morphisms* in **dMan**. We can locally characterize embeddings and immersions in **dMan** in terms of fibre products with $\mathbb{R}^n$ in **dMan**.

Theorem 5.54. **(a)** *Let X be a d-manifold and $g : X \rightarrow \mathbb{R}^n$ a 1-morphism in* **dMan***. Then the fibre product $W = X \times_{g,\mathbb{R}^n,0} *$ exists in* **dMan** *by Theorem 5.52(a), and the projection $e : W \rightarrow X$ is an embedding.*

(b) *Suppose* $f : X \to Y$ *is an immersion of d-manifolds, and* $x \in X$ *with* $f(x) = y \in Y$. *Then there exist open d-submanifolds* $x \in U \subseteq X$ *and* $y \in V \subseteq Y$ *with* $f(U) \subseteq V$, *and a 1-morphism* $g : V \to \mathbb{R}^n$ *with* $g(y) = 0$, *where* $n = \mathrm{vdim}\,Y - \mathrm{vdim}\,X \geqslant 0$, *fitting into a 2-Cartesian square in* **dMan** :

$$
\begin{array}{ccc}
U & \longrightarrow & * \\
{\scriptstyle f|_U}\big\downarrow & \qquad\Uparrow & \big\downarrow{\scriptstyle 0} \\
V & \underset{g}{\longrightarrow} & \mathbb{R}^n .
\end{array}
$$

If f *is an embedding we may take* $U = f^{-1}(V)$.

Remark 5.55. For the applications the author has in mind, it will be crucial that if $g : X \to Z$ and $h : Y \to Z$ are 1-morphisms with X, Y d-manifolds and Z a manifold then $W = X \times_Z Y$ is a d-manifold, with $\mathrm{vdim}\,W = \mathrm{vdim}\,X + \mathrm{vdim}\,Y - \dim Z$, as in Theorem 5.52(a). We will show by example, following Spivak [30, Prop. 1.7], that if d-manifolds **dMan** were an ordinary category containing manifolds as a full subcategory, then this would be false.

Consider the fibre product $* \times_{0,\mathbb{R},0} *$ in **dMan**. If **dMan** were a 1-category then as $*$ is a terminal object, the fibre product would be $*$. But then

$$
\mathrm{vdim}(* \times_{0,\mathbb{R},0} *) = \mathrm{vdim}\,* = 0 \neq -1 = \mathrm{vdim}\,* + \mathrm{vdim}\,* - \mathrm{vdim}\,\mathbb{R},
$$

so equation (5.12) and Theorem 5.52(a) would be false. Thus, if we want fibre products of d-manifolds over manifolds to be well behaved, then **dMan** must be at least a 2-category. It could be an ∞-category, as for Spivak's derived manifolds [30], or some other kind of higher category. Making d-manifolds into a 2-category, as we have done, is the simplest of the available options.

5.4.7 Embedding d-manifolds into manifolds

Let V be a manifold, $E \to V$ a vector bundle, and $s \in C^\infty(E)$. Then Example 5.33 defines a 'standard model' principal d-manifold $S_{V,E,s}$. When E and s are zero, we have $S_{V,0,0} = V = F_{\mathbf{Man}}^{\mathbf{dMan}}(V)$, so that $S_{V,0,0}$ is a manifold. For general V, E, s, taking $f = \mathrm{id}_V : V \to V$ and $\hat{f} = 0 : 0 \to E$ in Example 5.34 gives a 'standard model' 1-morphism $S_{\mathrm{id}_V,0} : S_{V,E,s} \to S_{V,0,0} = V$. One can show $S_{\mathrm{id}_V,0}$ is an embedding, in the sense of Definition 5.48. Any principal d-manifold U is equivalent to some $S_{V,E,s}$. Thus we deduce:

Lemma 5.56. *Any principal d-manifold* U *admits an embedding* $i : U \to V$ *into a manifold* V.

Theorem 5.61 below is a converse to this: if a d-manifold X can be embedded into a manifold Y, then X is principal. So it will be useful to study embeddings of d-manifolds into manifolds. The following facts are due to Whitney [34].

Theorem 5.57. **(a)** *Let X be an m-manifold and $n \geqslant 2m$. Then a generic smooth map $f : X \to \mathbb{R}^n$ is an immersion.*
(b) *Let X be an m-manifold and $n \geqslant 2m + 1$. Then there exists an embedding $f : X \to \mathbb{R}^n$, and we can choose such f with $f(X)$ closed in $\mathbb{R}^n$. Generic smooth maps $f : X \to \mathbb{R}^n$ are embeddings.*

In [20, §4.4] we generalize Theorem 5.57 to d-manifolds.

Theorem 5.58. *Let X be a d-manifold. Then there exist immersions and/or embeddings $\boldsymbol{f} : \boldsymbol{X} \to \mathbb{R}^n$ for some $n \gg 0$ if and only if there is an upper bound for $\dim T_x^* X$ for all $x \in X$. If there is such an upper bound, then immersions $\boldsymbol{f} : \boldsymbol{X} \to \mathbb{R}^n$ exist provided $n \geqslant 2 \dim T_x^* X$ for all $x \in X$, and embeddings $\boldsymbol{f} : \boldsymbol{X} \to \mathbb{R}^n$ exist provided $n \geqslant 2 \dim T_x^* X + 1$ for all $x \in X$. For embeddings we may also choose $\boldsymbol{f}$ with $f(X)$ closed in $\mathbb{R}^n$.*

Here is an example in which the condition does not hold.

Example 5.59. $\mathbb{R}^k \times_{0, \mathbb{R}^k, 0} *$ is a principal d-manifold of virtual dimension 0, with C^∞-scheme $\underline{\mathbb{R}}^k$, and obstruction bundle $\mathbb{R}^k$. Thus $\boldsymbol{X} = \coprod_{k \geqslant 0} \mathbb{R}^k \times_{0, \mathbb{R}^k, 0} *$ is a d-manifold of virtual dimension 0, with C^∞-scheme $\underline{X} = \coprod_{k \geqslant 0} \underline{\mathbb{R}}^k$. Since $T_x^* \underline{X} \cong \mathbb{R}^n$ for $x \in \mathbb{R}^n \subset \coprod_{k \geqslant 0} \mathbb{R}^k$, $\dim T_x^* \underline{X}$ realizes all values $n \geqslant 0$. Hence there cannot exist immersions or embeddings $\boldsymbol{f} : \boldsymbol{X} \to \mathbb{R}^n$ for any $n \geqslant 0$.

As $x \longmapsto \dim T_x^* X$ is an upper semicontinuous map $X \to \mathbb{N}$, if X is compact then $\dim T_x^* X$ is bounded above, giving:

Corollary 5.60. *Let X be a compact d-manifold. Then there exists an embedding $\boldsymbol{f} : \boldsymbol{X} \to \mathbb{R}^n$ for some $n \gg 0$.*

If a d-manifold X can be embedded into a manifold Y, we show in [20, §4.4] that we can write X as the zeroes of a section of a vector bundle over Y near its image. See [30, Prop. 9.5] for the analogue for Spivak's derived manifolds.

Theorem 5.61. *Suppose X is a d-manifold, Y a manifold, and $\boldsymbol{f} : \boldsymbol{X} \to \boldsymbol{Y}$ an embedding, in the sense of Definition 5.48. Then there exist an open subset V in Y with $\boldsymbol{f}(\boldsymbol{X}) \subseteq \boldsymbol{V}$, a vector bundle $E \to V$, and $s \in C^\infty(E)$ fitting into a 2-Cartesian diagram in* **dSpa** :

$$
\begin{array}{ccc}
X & \xrightarrow{\quad\quad} & V \\
{\scriptstyle f}\downarrow\; {\scriptstyle f} & \eta\!\!\nearrow & \downarrow{\scriptstyle 0} \\
V & \xrightarrow[\;\;s\;\;]{} & E.
\end{array}
$$

Here $\boldsymbol{Y} = F_{\mathbf{Man}}^{\mathbf{dMan}}(Y)$, and similarly for $\boldsymbol{V}, \boldsymbol{E}, \boldsymbol{s}, \boldsymbol{0}$, with $\boldsymbol{0} : \boldsymbol{V} \to \boldsymbol{E}$ the zero section. Hence $\boldsymbol{X}$ is equivalent to the 'standard model' d-manifold $\boldsymbol{S}_{V,E,s}$ of Example 5.33, and is a principal d-manifold.

Combining Theorems 5.58 and 5.61, Lemma 5.56, and Corollary 5.60 yields:

Corollary 5.62. *Let $\boldsymbol{X}$ be a d-manifold. Then $\boldsymbol{X}$ is a principal d-manifold if and only if $\dim T_x^*\underline{X}$ is bounded above for all $x \in \underline{X}$. In particular, if $\boldsymbol{X}$ is compact, then $\boldsymbol{X}$ is principal.*

Corollary 5.62 suggests that most interesting d-manifolds are principal, in a similar way to most interesting C^∞-schemes being affine in Remark 5.9(ii). Example 5.59 gives a d-manifold which is not principal.

5.4.8 Orientations on d-manifolds

Let X be an n-manifold. Then T^*X is a rank n vector bundle on X, so its top exterior power $\Lambda^n T^*X$ is a line bundle (rank 1 vector bundle) on X. In algebraic geometry, $\Lambda^n T^*X$ would be called the canonical bundle of X. We define an *orientation* ω on X to be an *orientation on the fibres of* $\Lambda^n T^*X$. That is, ω is an equivalence class $[\tau]$ of isomorphisms of line bundles $\tau :$ $O_X \to \Lambda^n T^*X$, where O_X is the trivial line bundle $\mathbb{R} \times X \to X$, and τ, τ' are equivalent if $\tau' = \tau \cdot c$ for some smooth $c : X \to (0, \infty)$.

To generalize all this to d-manifolds, we will need a notion of the 'top exterior power' $\mathcal{L}_{(\mathcal{E}^\bullet, \phi)}$ of a virtual vector bundle $(\mathcal{E}^\bullet, \phi)$ in §5.4.3. As the definition is long and complicated, we will not give it, but just state its important properties.

Theorem 5.63. *Let $\underline{X}$ be a C^∞-scheme, and $(\mathcal{E}^\bullet, \phi)$ a virtual vector bundle on $\underline{X}$. Then in [20, §4.5] we define a line bundle (rank 1 vector bundle) $\mathcal{L}_{(\mathcal{E}^\bullet, \phi)}$ on $\underline{X}$, which we call the **orientation line bundle** of $(\mathcal{E}^\bullet, \phi)$. This satisfies:*

(a) *Suppose $\mathcal{E}^1, \mathcal{E}^2$ are vector bundles on $\underline{X}$ with ranks k_1, k_2, and $\phi : \mathcal{E}^1 \rightarrow \mathcal{E}^2$ is a morphism. Then $(\mathcal{E}^\bullet, \phi)$ is a virtual vector bundle of rank $k_2 - k_1$, and there is a canonical isomorphism $\mathcal{L}_{(\mathcal{E}^\bullet, \phi)} \cong \Lambda^{k_1}(\mathcal{E}^1)^* \otimes \Lambda^{k_2}\mathcal{E}^2$.*

(b) *Let $f^\bullet : (\mathcal{E}^\bullet, \phi) \rightarrow (\mathcal{F}^\bullet, \psi)$ be an equivalence in $\mathrm{vvect}(\underline{X})$. Then there is a canonical isomorphism $\mathcal{L}_{f^\bullet} : \mathcal{L}_{(\mathcal{E}^\bullet, \phi)} \rightarrow \mathcal{L}_{(\mathcal{F}^\bullet, \psi)}$ in $\mathrm{qcoh}(\underline{X})$.*

(c) *If $(\mathcal{E}^\bullet, \phi) \in \mathrm{vvect}(\underline{X})$ then $\mathcal{L}_{\mathrm{id}_\phi} = \mathrm{id}_{\mathcal{L}_{(\mathcal{E}^\bullet, \phi)}} : \mathcal{L}_{(\mathcal{E}^\bullet, \phi)} \rightarrow \mathcal{L}_{(\mathcal{E}^\bullet, \phi)}$.*

(d) *If $f^\bullet : (\mathcal{E}^\bullet, \phi) \rightarrow (\mathcal{F}^\bullet, \psi)$ and $g^\bullet : (\mathcal{F}^\bullet, \psi) \rightarrow (\mathcal{G}^\bullet, \xi)$ are equivalences in $\mathrm{vvect}(\underline{X})$ then $\mathcal{L}_{g^\bullet \circ f^\bullet} = \mathcal{L}_{g^\bullet} \circ \mathcal{L}_{f^\bullet} : \mathcal{L}_{(\mathcal{E}^\bullet, \phi)} \rightarrow \mathcal{L}_{(\mathcal{G}^\bullet, \xi)}$.*

(e) *If $f^\bullet, g^\bullet : (\mathcal{E}^\bullet, \phi) \rightarrow (\mathcal{F}^\bullet, \psi)$ are 2-isomorphic equivalences in $\mathrm{vvect}(\underline{X})$ then $\mathcal{L}_{f^\bullet} = \mathcal{L}_{g^\bullet} : \mathcal{L}_{(\mathcal{E}^\bullet, \phi)} \rightarrow \mathcal{L}_{(\mathcal{F}^\bullet, \psi)}$.*

(f) *Let $\underline{f} : \underline{X} \rightarrow \underline{Y}$ be a morphism of C^∞-schemes, and $(\mathcal{E}^\bullet, \phi) \in \mathrm{vvect}(\underline{Y})$. Then there is a canonical isomorphism $I_{\underline{f}, (\mathcal{E}^\bullet, \phi)} : \underline{f}^*(\mathcal{L}_{(\mathcal{E}^\bullet, \phi)}) \rightarrow \mathcal{L}_{\underline{f}^*(\mathcal{E}^\bullet, \phi)}$.*

Now we can define orientations on d-manifolds.

Definition 5.64. Let X be a d-manifold. Then the virtual cotangent bundle T^*X is a virtual vector bundle on $\underline{X}$ by Proposition 5.40(b), so Theorem 5.63 gives a line bundle $\mathcal{L}_{T^*X}$ on $\underline{X}$. We call $\mathcal{L}_{T^*X}$ the *orientation line bundle* of X.

An *orientation* ω on X is an orientation on $\mathcal{L}_{T^*X}$. That is, ω is an equivalence class $[\tau]$ of isomorphisms $\tau : \mathcal{O}_X \rightarrow \mathcal{L}_{T^*X}$ in $\mathrm{qcoh}(\underline{X})$, where τ, τ' are equivalent if they are proportional by a smooth positive function on $\underline{X}$.

If $\omega = [\tau]$ is an orientation on X, the *opposite orientation* is $-\omega = [-\tau]$, which changes the sign of the isomorphism $\tau : \mathcal{O}_X \rightarrow \mathcal{L}_{T^*X}$. When we refer to X as an oriented d-manifold, $-X$ will mean X with the opposite orientation, that is, X is short for (X, ω) and $-X$ is short for $(X, -\omega)$.

Example 5.65. (a) Let X be an n-manifold, and $\mathbf{X} = F^{\mathbf{dMan}}_{\mathbf{Man}}(X)$ the associated d-manifold. Then $\underline{X} = F^{\mathbf{C}^\infty\mathbf{Sch}}_{\mathbf{Man}}(X)$, $\mathcal{E}_X = 0$ and $\mathcal{F}_X = T^*\underline{X}$. So $\mathcal{E}_X, \mathcal{F}_X$ are vector bundles of ranks $0, n$. As $\Lambda^0 \mathcal{E}_X \cong \mathcal{O}_X$, Theorem 5.63(a) gives a canonical isomorphism $\mathcal{L}_{T^*X} \cong \Lambda^n T^*\underline{X}$. That is, $\mathcal{L}_{T^*X}$ is isomorphic to the lift to C^∞-schemes of the line bundle $\Lambda^n T^*X$ on the manifold X.

As above, an orientation on $\mathbf{X}$ is an orientation on the line bundle $\Lambda^n T^*X$. Hence orientations on the d-manifold $\mathbf{X} = F^{\mathbf{dMan}}_{\mathbf{Man}}(X)$ in the sense of Definition 5.64 are equivalent to orientations on the manifold X in the usual sense.

(b) Let V be an n-manifold, $E \rightarrow V$ a vector bundle of rank k, and $s \in C^\infty(E)$. Then Example 5.33 defines a 'standard model' principal d-manifold $\mathbf{S} = \mathbf{S}_{V,E,s}$, which has $\mathcal{E}_S \cong \underline{E}^*|_{\underline{S}}$, $\mathcal{F}_S \cong T^*\underline{V}|_{\underline{S}}$, where $\underline{E}, T^*\underline{V}$ are the lifts of the vector bundles E, T^*V on V to $\underline{V}$. Hence $\mathcal{E}_S$, $\mathcal{F}_S$ are vector bundles on $\underline{S}_{V,E,s}$ of ranks k, n, so Theorem 5.63(a) gives an isomorphism $\mathcal{L}_{T^*S_{V,E,s}} \cong (\Lambda^k \underline{E} \otimes \Lambda^n T^*\underline{V})|_{\underline{S}}$.

Thus $\mathcal{L}_{T^*S_{V,E,s}}$ is the lift to $\underline{S}_{V,E,s}$ of the line bundle $\Lambda^k E \otimes \Lambda^n T^*V$ over the manifold V. Therefore we may induce an orientation on the d-manifold $S_{V,E,s}$ from an orientation on the line bundle $\Lambda^k E \otimes \Lambda^n T^*V$ over V. Equivalently, we can induce an orientation on $S_{V,E,s}$ from an orientation on the total space of the vector bundle E^* over V, or from an orientation on the total space of E.

We can construct orientations on d-transverse fibre products of oriented d-manifolds. Note that (5.13) depends on an *orientation convention*: a different choice would change (5.13) by a sign depending on vdim X, vdim Y, vdim Z. Our conventions follow those of Fukaya et al. [9, §8.2] for Kuranishi spaces.

Theorem 5.66. *Work in the situation of Theorem 5.51, so that W, X, Y, Z are d-manifolds with $W = X \times_{g,Z,h} Y$ for g, h d-transverse, where $e : W \to X$, $f : W \to Y$ are the projections. Then we have orientation line bundles $\mathcal{L}_{T^*W}, \ldots, \mathcal{L}_{T^*Z}$ on $\underline{W}, \ldots, \underline{Z}$, so $\mathcal{L}_{T^*W}$, $\underline{e}^*(\mathcal{L}_{T^*X})$, $\underline{f}^*(\mathcal{L}_{T^*Y})$, $(\underline{g}\circ\underline{e})^*(\mathcal{L}_{T^*Z})$ are line bundles on $\underline{W}$. With a suitable choice of orientation convention, there is a canonical isomorphism*

$$\Phi : \mathcal{L}_{T^*W} \longrightarrow \underline{e}^*(\mathcal{L}_{T^*X}) \otimes_{\mathcal{O}_W} \underline{f}^*(\mathcal{L}_{T^*Y}) \otimes_{\mathcal{O}_W} (\underline{g}\circ\underline{e})^*(\mathcal{L}_{T^*Z})^*. \quad (5.13)$$

Hence, if X, Y, Z are oriented d-manifolds, then W also has a natural orientation, since trivializations of $\mathcal{L}_{T^*X}$, $\mathcal{L}_{T^*Y}$, $\mathcal{L}_{T^*Z}$ induce a trivialization of $\mathcal{L}_{T^*W}$ by (5.13).

Fibre products have natural commutativity and associativity properties. When we include orientations, the orientations differ by some sign. Here is an analogue of results of Fukaya et al. [9, Lem. 8.2.3] for Kuranishi spaces.

Proposition 5.67. *Suppose $V, \ldots, Z$ are oriented d-manifolds, $e, \ldots, h$ are 1-morphisms, and all fibre products below are d-transverse. Then the following hold, in oriented d-manifolds:*

(a) *For $g : X \to Z$ and $h : Y \to Z$ we have*

$$X \times_{g,Z,h} Y \simeq (-1)^{(\text{vdim } X - \text{vdim } Z)(\text{vdim } Y - \text{vdim } Z)} Y \times_{h,Z,g} X.$$

In particular, when $Z = *$ *so that* $X \times_Z Y = X \times Y$ *we have*

$$X \times Y \simeq (-1)^{\text{vdim} X \text{ vdim} Y} Y \times X.$$

(b) *For* $e : V \to Y$, $f : W \to Y$, $g : W \to Z$, *and* $h : X \to Z$ *we have*

$$V \times_{e,Y,f \circ \pi_W} (W \times_{g,Z,h} X) \simeq (V \times_{e,Y,f} W) \times_{g \circ \pi_W, Z, h} X.$$

(c) *For* $e : V \to Y$, $f : V \to Z$, $g : W \to Y$, *and* $h : X \to Z$ *we have*

$$V \times_{(e,f),Y \times Z, g \times h} (W \times X) \simeq$$
$$(-1)^{\text{vdim} Z(\text{vdim} Y + \text{vdim} W)} (V \times_{e,Y,g} W) \times_{f \circ \pi_V, Z, h} X.$$

5.4.9 D-manifolds with boundary and corners, d-orbifolds

For brevity, this section will give much less detail than §5.4.1–§5.4.8. So far we have discussed only manifolds *without boundary* (locally modelled on $\mathbb{R}^n$). One can also consider *manifolds with boundary* (locally modelled on $[0, \infty) \times \mathbb{R}^{n-1}$) and *manifolds with corners* (locally modelled on $[0, \infty)^k \times \mathbb{R}^{n-k}$). In [17] the author studied manifolds with boundary and with corners, giving a new definition of *smooth map* $f : X \to Y$ between manifolds with corners X, Y, satisfying extra conditions over $\partial^k X, \partial^l Y$. This yields categories $\mathbf{Man^b}$, $\mathbf{Man^c}$ of manifolds with boundary and with corners with good properties *as categories*.

In [20, §6–§7], the author defined 2-categories $\mathbf{dSpa^b}$, $\mathbf{dSpa^c}$ of *d-spaces with boundary* and *with corners*, and 2-subcategories $\mathbf{dMan^b}$, $\mathbf{dMan^c}$ of *d-manifolds with boundary* and *with corners*. Objects in $\mathbf{dSpa^b}$, $\mathbf{dSpa^c}$, $\mathbf{dMan^b}$, $\mathbf{dMan^c}$ are quadruples $\mathbf{X} = (X, \partial X, i_{\mathbf{X}}, \omega_{\mathbf{X}})$, where $X, \partial X$ are d-spaces, and $i_{\mathbf{X}} : \partial X \to X$ is a 1-morphism, such that ∂X is locally equivalent to a fibre product $X \times_{[0,\infty)} *$ in $\mathbf{dSpa}$, in a similar way to Theorem 5.54(b). This implies that the 'conormal bundle' $\mathcal{N}_{\mathbf{X}}$ of ∂X in X is a line bundle on $\underline{\partial X}$. The final piece of data $\omega_{\mathbf{X}}$ is an orientation on $\mathcal{N}_{\mathbf{X}}$, giving a notion of 'outward-pointing normal vectors' to $\partial \mathbf{X}$ in $\mathbf{X}$. Here are some properties of these:

Theorem 5.68. *In* [20, §7] *we define strict 2-categories* $\mathbf{dMan^b}$, $\mathbf{dMan^c}$ *of* **d-manifolds with boundary** *and* **d-manifolds with corners**. *These have the following properties:*

(a) $\mathbf{dMan^b}$ *is a full 2-subcategory of* $\mathbf{dMan^c}$. *There is a full and faithful 2-functor* $F_{\mathbf{dMan}}^{\mathbf{dMan^c}} : \mathbf{dMan} \hookrightarrow \mathbf{dMan^c}$ *whose image is a full 2-subcategory* $\overline{\mathbf{dMan}}$ *in* $\mathbf{dMan^b}$, *so that* $\overline{\mathbf{dMan}} \subset \mathbf{dMan^b} \subset \mathbf{dMan^c}$.

(b) *There are full and faithful 2-functors* $F_{\mathbf{Man^b}}^{\mathbf{dMan^b}} : \mathbf{Man^b} \to \mathbf{dMan^b}$ *and* $F_{\mathbf{Man^c}}^{\mathbf{dMan^c}} : \mathbf{Man^c} \to \mathbf{dMan^c}$. *We write* $\bar{\mathbf{Man}}^{\mathbf{b}}, \bar{\mathbf{Man}}^{\mathbf{c}}$ *for the full 2-subcategories of objects in* $\mathbf{dMan^b}, \mathbf{dMan^c}$ *equivalent to objects in the images of* $F_{\mathbf{Man^b}}^{\mathbf{dMan^b}}, F_{\mathbf{Man^c}}^{\mathbf{dMan^c}}$.

(c) *Each object* $\mathbf{X} = (X, \partial X, i_{\mathbf{X}}, \omega_{\mathbf{X}})$ *in* $\mathbf{dMan^b}$ *or* $\mathbf{dMan^c}$ *has a **virtual dimension*** $\operatorname{vdim} \mathbf{X} \in \mathbb{Z}$. *The virtual cotangent sheaf* T^*X *of the underlying d-space* X *is a virtual vector bundle on* $\underline{X}$ *with rank* $\operatorname{vdim} \mathbf{X}$.

(d) *Each d-manifold with corners* $\mathbf{X}$ *has a **boundary*** $\partial \mathbf{X}$, *which is another d-manifold with corners with* $\operatorname{vdim} \partial \mathbf{X} = \operatorname{vdim} \mathbf{X} - 1$. *The d-space 1-morphism* $i_{\mathbf{X}}$ *in* $\mathbf{X}$ *is also a 1-morphism* $i_{\mathbf{X}} : \partial \mathbf{X} \to \mathbf{X}$ *in* $\mathbf{dMan^c}$. *If* $\mathbf{X} \in \mathbf{dMan^b}$ *then* $\partial \mathbf{X} \in \bar{\mathbf{dMan}}$, *and if* $\mathbf{X} \in \bar{\mathbf{dMan}}$ *then* $\partial \mathbf{X} = \emptyset$.

(e) *Boundaries in* $\mathbf{dMan^b}, \mathbf{dMan^c}$ *have strong functorial properties. For instance, there is an interesting class of **simple** 1-morphisms* $\boldsymbol{f} : \mathbf{X} \to \mathbf{Y}$ *in* $\mathbf{dMan^b}$ *and* $\mathbf{dMan^c}$, *which satisfy a discrete condition broadly saying that* $\boldsymbol{f}$ *maps* $\partial^k \mathbf{X}$ *to* $\partial^k \mathbf{Y}$ *for all* k. *These have the property that for all simple* $\boldsymbol{f} : \mathbf{X} \to \mathbf{Y}$ *there is a unique simple 1-morphism* $\boldsymbol{f}_{-} : \partial \mathbf{X} \to \partial \mathbf{Y}$ *with* $\boldsymbol{f} \circ i_{\mathbf{X}} = i_{\mathbf{Y}} \circ \boldsymbol{f}_{-}$, *and the following diagram is 2-Cartesian in* $\mathbf{dMan^c}$

$$
\begin{array}{ccc}
\partial\mathbf{X} & \xrightarrow{\;\;\boldsymbol{f}_{-}\;\;} & \partial\mathbf{Y} \\
{\scriptstyle i_{\mathbf{X}}}\big\downarrow & {\scriptstyle \mathrm{id}_{\boldsymbol{f} \circ i_{\mathbf{X}}} \Uparrow} & \big\downarrow{\scriptstyle i_{\mathbf{Y}}} \\
\mathbf{X} & \xrightarrow[\;\;\boldsymbol{f}\;\;]{} & \mathbf{Y},
\end{array}
$$

so that $\partial \mathbf{X} \simeq \mathbf{X} \times_{\boldsymbol{f}, \mathbf{Y}, i_{\mathbf{Y}}} \partial \mathbf{Y}$ *in* $\mathbf{dMan^c}$. *If* $\boldsymbol{f}, \boldsymbol{g} : \mathbf{X} \to \mathbf{Y}$ *are simple 1-morphisms and* $\eta : \boldsymbol{f} \Rightarrow \boldsymbol{g}$ *is a 2-morphism in* $\mathbf{dMan^c}$ *then there is a natural 2-morphism* $\eta_{-} : \boldsymbol{f}_{-} \Rightarrow \boldsymbol{g}_{-}$ *in* $\mathbf{dMan^c}$.

(f) *An **orientation** on a d-manifold with corners* $\mathbf{X} = (X, \partial X, i_{\mathbf{X}}, \omega_{\mathbf{X}})$ *is an orientation on the line bundle* $\mathcal{L}_{T^*X}$ *on* $\underline{X}$. *If* $\mathbf{X}$ *is an oriented d-manifold with corners, there is a natural orientation on* $\partial \mathbf{X}$, *constructed using the orientation on* $\mathbf{X}$ *and the data* $\omega_{\mathbf{X}}$ *in* $\mathbf{X}$.

(g) *Almost all the results of* §5.4.1–§5.4.8 *on d-manifolds without boundary extend to d-manifolds with boundary and with corners, with some changes.*

One moral of [17] and [20, §5–§7] is that doing 'things with corners' properly is a great deal more complicated, but also more interesting, than you would believe if you had not thought about the issues involved.

Example 5.69. (i) Let $\mathbf{X}$ be the fibre product $[0, \infty) \times_{i, \mathbb{R}, 0} *$ in $\mathbf{dMan^c}$, where $i : [0, \infty) \hookrightarrow \mathbb{R}$ is the inclusion. Then $\mathbf{X} = (X, \partial X, i_{\mathbf{X}}, \omega_{\mathbf{X}})$ is 'a point with point boundary', of virtual dimension 0, and its boundary $\partial \mathbf{X}$ is an 'obstructed point', a point with obstruction space $\mathbb{R}$, of virtual dimension -1.

The conormal bundle $\mathcal{N}_{\mathbf{X}}$ of ∂X in X is the obstruction space $\mathbb{R}$ of ∂X. In this case, the orientation $\omega_{\mathbf{X}}$ on $\mathcal{N}_{\mathbf{X}}$ cannot be determined from $X, \partial X, i_{\mathbf{X}}$, in fact, there is an automorphism of $X, \partial X, i_{\mathbf{X}}$ which reverses the orientation of $\mathcal{N}_{\mathbf{X}}$. So $\omega_{\mathbf{X}}$ really is extra data. We include $\omega_{\mathbf{X}}$ in the definition of d-manifolds with corners to ensure that orientations of d-manifolds with corners are well-behaved. If we omitted $\omega_{\mathbf{X}}$ from the definition, there would exist oriented d-manifolds with corners $\mathbf{X}$ whose boundaries $\partial \mathbf{X}$ are not orientable.

(ii) The fibre product $[0, \infty) \times_{i,[0,\infty),0} *$ is a point $*$ without boundary. The only difference with **(i)** is that we have replaced the target $\mathbb{R}$ with $[0, \infty)$, adding a boundary. So in a fibre product $\mathbf{W} = \mathbf{X} \times_{\mathbf{Z}} \mathbf{Y}$ in $\mathbf{dMan^c}$, the boundary of $\mathbf{Z}$ affects the boundary of $\mathbf{W}$. This does not happen for fibre products in $\mathbf{Man^c}$.

(iii) Let $\mathbf{X}'$ be the fibre product $[0, \infty) \times_{i,\mathbb{R},i} (-\infty, 0]$ in $\mathbf{dMan^c}$, that is, the derived intersection of submanifolds $[0, \infty), (-\infty, 0]$ in $\mathbb{R}$. Topologically, $\mathbf{X}'$ is just the point $\{0\}$, but as a d-manifold with corners $\mathbf{X}'$ has virtual dimension 1. The boundary $\partial \mathbf{X}'$ is the disjoint union of two copies of $\mathbf{X}$ in **(i)**. The C^∞-scheme $\underline{X}$ in X is the spectrum of the C^∞-ring $C^\infty\big([0, \infty)^2\big)/(x + y)$, which is infinite-dimensional, although its topological space is a point.

Orbifolds are generalizations of manifolds locally modelled on $\mathbb{R}^n / G$ for G a finite group. They are related to manifolds as Deligne–Mumford stacks are related to schemes in algebraic geometry, and form a 2-category $\mathbf{Orb}$. Lerman [23] surveys definitions of orbifolds, and explains why $\mathbf{Orb}$ is a 2-category. As for $\mathbf{Man^b}, \mathbf{Man^c}$ one can also consider 2-categories of *orbifolds with boundary* $\mathbf{Orb^b}$ and *orbifolds with corners* $\mathbf{Orb^c}$, discussed in [20, §8].

In [20, §9 & §11] we define 2-categories of *d-stacks* $\mathbf{dSta}$, *d-stacks with boundary* $\mathbf{dSta^b}$ and *d-stacks with corners* $\mathbf{dSta^c}$, which are orbifold versions of $\mathbf{dSpa}, \mathbf{dSpa^b}, \mathbf{dSpa^c}$. Broadly, to go from d-spaces $X = (\underline{X}, \mathcal{O}'_X, \mathcal{E}_X, \iota_X, \jmath_X)$ to d-stacks we just replace the C^∞-scheme $\underline{X}$ by a *Deligne–Mumford C^∞-stack* $\mathcal{X}$, where the theory of the 2-category of C^∞-stacks $\mathbf{C^\infty Sta}$ is developed in [18, §7–§9] and summarized in [19, §4]. Then in [20, §10 & §12] we define 2-categories of *d-orbifolds* $\mathbf{dOrb}$, *d-orbifolds with boundary* $\mathbf{dOrb^b}$ and *d-orbifolds with corners* $\mathbf{Orb^c}$, which are orbifold versions of $\mathbf{dMan}, \mathbf{dMan^b}, \mathbf{dMan^c}$.

One might expect that combining the 2-categories $\mathbf{Orb}$ and $\mathbf{dMan}$ should result in a 3-category $\mathbf{dOrb}$, but in fact a 2-category is sufficient. For 1-morphisms $f, g : \mathcal{X} \to \mathcal{Y}$ in $\mathbf{dOrb}$, a 2-morphism $\eta : f \Rightarrow g$ in $\mathbf{dOrb}$ is a pair (η, η'), where $\eta : f \Rightarrow g$ is a 2-morphism in $\mathbf{C^\infty Sta}$, and $\eta' : f^*(\mathcal{F}_{\mathcal{Y}}) \to \mathcal{E}_{\mathcal{X}}$ is as for 2-morphisms in $\mathbf{dMan}$. These η, η' do not interact very much.

The generalizations to d-orbifolds are mostly straightforward, with few surprises. Almost all the results of §5.3–§5.4.8, and Theorem 5.68, extend to d-stacks and d-orbifolds with only cosmetic changes. One exception is that the generalizations of Theorems 5.28 and 5.46 to d-stacks and d-orbifolds need extra conditions on the C^∞-stack 2-morphism components η_{ijk} of $\boldsymbol{\eta}_{ijk}$ on quadruple overlaps $\boldsymbol{\mathcal{X}}_i \cap \boldsymbol{\mathcal{X}}_j \cap \boldsymbol{\mathcal{X}}_k \cap \boldsymbol{\mathcal{X}}_l$, as in Remark 5.29. This is because 2-morphisms η_{ijk} in $\mathbf{C}^\infty\mathbf{Sta}$ are discrete, and cannot be glued using partitions of unity.

5.4.10 D-manifold bordism, and virtual cycles

Classical bordism groups $MSO_k(Y)$ were defined by Atiyah [1] for topological spaces Y, using continuous maps $f : X \to Y$ for X a compact oriented manifold. Conner [7, §I] gives a good introduction. We define bordism $B_k(Y)$ only for manifolds Y, using smooth $f : X \to Y$, following Conner's *differential bordism groups* [7, §I.9]. By [7, Th. I.9.1], the natural projection $B_k(Y) \to MSO_k(Y)$ is an isomorphism, so our notion of bordism agrees with the usual definition.

Definition 5.70. Let Y be a manifold without boundary, and $k \in \mathbb{Z}$. Consider pairs (X, f), where X is a compact, oriented manifold without boundary with $\dim X = k$, and $f : X \to Y$ is a smooth map. Define an equivalence relation $\sim$ on such pairs by $(X, f) \sim (X', f')$ if there exists a compact, oriented $(k + 1)$-manifold with boundary W, a smooth map $e : W \to Y$, and a diffeomorphism of oriented manifolds $j : -X \amalg X' \to \partial W$, such that $f \amalg f' = e \circ i_W \circ j$, where $-X$ is X with the opposite orientation.

Write $[X, f]$ for the $\sim$-equivalence class (*bordism class*) of a pair (X, f). For each $k \in \mathbb{Z}$, define the kth *bordism group* $B_k(Y)$ of Y to be the set of all such bordism classes $[X, f]$ with $\dim X = k$. We give $B_k(Y)$ the structure of an abelian group, with zero element $0_Y = [\emptyset, \emptyset]$, and addition given by $[X, f] + [X', f'] = [X \amalg X', f \amalg f']$, and additive inverses $-[X, f] = [-X, f]$.

Define $\Pi_{\mathrm{bo}}^{\mathrm{hom}} : B_k(Y) \to H_k(Y; \mathbb{Z})$ by $\Pi_{\mathrm{bo}}^{\mathrm{hom}} : [X, f] \longmapsto f_*([X])$, where $H_*(-; \mathbb{Z})$ is singular homology, and $[X] \in H_k(X; \mathbb{Z})$ is the fundamental class.

If Y is oriented and of dimension n, there is a biadditive, associative, supercommutative *intersection product* $\bullet : B_k(Y) \times B_l(Y) \to B_{k+l-n}(Y)$, such that if $[X, f], [X', f']$ are classes in $B_*(Y)$, with f, f' transverse, then the fibre product $X \times_{f,Y,f'} X'$ exists as a compact oriented manifold, and

$$[X, f] \bullet [X', f'] = [X \times_{f,Y,f'} X', f \circ \pi_X].$$

As in [7, §I.5], bordism is a generalized homology theory. Results of Thom, Wall and others in [7, §I.2] compute the bordism groups $B_k(*)$ of the point $*$. This partially determines the bordism groups of general manifolds Y, as there is a spectral sequence $H_i\big(Y; B_j(*)\big) \Rightarrow B_{i+j}(Y)$. We define *d-manifold bordism* by replacing manifolds X in $[X, f]$ by d-manifolds X:

Definition 5.71. Let Y be a compact manifold without boundary, and $k \in \mathbb{Z}$. Consider pairs (X, f), where $X \in \mathbf{dMan}$ is a compact, oriented d-manifold without boundary with $\mathrm{vdim}\, X = k$, and $f : X \to Y$ is a 1-morphism in $\mathbf{dMan}$, where $Y = F_{\mathbf{Man}}^{\mathbf{dMan}}(Y)$.

Define an equivalence relation $\sim$ between such pairs by $(X, f) \sim (X', f')$ if there exists a compact, oriented d-manifold with boundary $\mathbf{W}$ with $\mathrm{vdim}\, \mathbf{W} = k + 1$, a 1-morphism $e : \mathbf{W} \to Y$ in $\mathbf{dMan^b}$, an equivalence of oriented d-manifolds $j : -X \amalg X' \to \partial \mathbf{W}$, and a 2-morphism $\eta : f \amalg f' \Rightarrow e \circ i_{\mathbf{W}} \circ j$.

Write $[X, f]$ for the $\sim$-equivalence class (*d-bordism class*) of a pair (X, f). For each $k \in \mathbb{Z}$, define the kth *d-manifold bordism group*, or *d-bordism group*, $dB_k(Y)$ of Y to be the set of all such d-bordism classes $[X, f]$ with $\mathrm{vdim}\, X = k$. As for $B_k(Y)$, we give $dB_k(Y)$ the structure of an abelian group, with zero element $0_Y = [\emptyset, \emptyset]$, addition $[X, f] + [X', f'] = [X \amalg X', f \amalg f']$, and additive inverses $-[X, f] = [-X, f]$.

If Y is oriented and of dimension n, we define a biadditive, associative, supercommutative *intersection product* $\bullet : dB_k(Y) \times dB_l(Y) \to dB_{k+l-n}(Y)$ by

$$[X, f] \bullet [X', f'] = [X \times_{f,Y,f'} X', f \circ \pi_X].$$

Here $X \times_{f,Y,f'} X'$ exists as a d-manifold by Theorem 5.52(a), and is oriented by Theorem 5.66. Note that we do not need to restrict to $[X, f], [X', f']$ with f, f' transverse as in Definition 5.70. Define a morphism $\Pi_{\mathrm{bo}}^{\mathrm{dbo}} : B_k(Y) \to dB_k(Y)$ for $k \geqslant 0$ by $\Pi_{\mathrm{bo}}^{\mathrm{dbo}} : [X, f] \longmapsto \big[F_{\mathbf{Man}}^{\mathbf{dMan}}(X), F_{\mathbf{Man}}^{\mathbf{dMan}}(f)\big]$.

In [20, §13.2] we prove that $B_*(Y)$ and $dB_*(Y)$ are isomorphic. See [30, Th. 2.6] for the analogous result for Spivak's derived manifolds.

Theorem 5.72. *For any manifold Y, we have $dB_k(Y) = 0$ for $k < 0$, and $\Pi_{\mathrm{bo}}^{\mathrm{dbo}} : B_k(Y) \to dB_k(Y)$ is an isomorphism for $k \geqslant 0$. When Y is oriented, $\Pi_{\mathrm{bo}}^{\mathrm{dbo}}$ identifies the intersection products $\bullet$ on $B_*(Y)$ and $dB_*(Y)$.*

Here is the main idea in the proof of Theorem 5.72. Let $[X, f] \in dB_k(Y)$. By Corollary 5.60 there exists an embedding $g : X \to \mathbb{R}^n$ for $n \gg 0$. Then the direct product $(f, g) : X \rightarrow Y \times \mathbb{R}^n$ is also an embedding. Theorem 5.61 shows that there exist an open set $V \subseteq Y \times \mathbb{R}^n$, a vector bundle $E \rightarrow V$ and $s \in C^\infty(E)$ such that $X \simeq S_{V,E,s}$. Let $\tilde{s} \in C^\infty(E)$ be a small, generic perturbation of s. As $\tilde{s}$ is generic, the graph of $\tilde{s}$ in E intersects the zero section transversely. Hence $\tilde{X} = \tilde{s}^{-1}(0)$ is a k-manifold for $k \geqslant 0$, which is compact and oriented for $\tilde{s} - s$ small, and $\tilde{X} = \emptyset$ for $k < 0$. Set $\tilde{f} = \pi_Y|_{\tilde{X}} : \tilde{X} \rightarrow Y$. Then $\Pi^{\text{dbo}}_{\text{bo}}([\tilde{X}, \tilde{f}]) = [X, f]$, so that $\Pi^{\text{dbo}}_{\text{bo}}$ is surjective. A similar argument for $\mathbf{W}, e$ in Definition 5.71 shows that $\Pi^{\text{dbo}}_{\text{bo}}$ is injective.

By Theorem 5.72, we may define a projection $\Pi^{\text{hom}}_{\text{dbo}} : dB_k(Y) \rightarrow H_k(Y; \mathbb{Z})$ for $k \geqslant 0$ by $\Pi^{\text{hom}}_{\text{dbo}} = \Pi^{\text{hom}}_{\text{bo}} \circ (\Pi^{\text{dbo}}_{\text{bo}})^{-1}$. We think of $\Pi^{\text{hom}}_{\text{dbo}}$ as a *virtual class map*. Virtual classes (or virtual cycles, or virtual chains) are used in several areas of geometry to construct enumerative invariants using moduli spaces. In algebraic geometry, Behrend and Fantechi [4] construct virtual classes for schemes with obstruction theories. In symplectic geometry, there are many versions — see for example Fukaya et al. [10, §6], [9, §A1], Hofer et al. [13], and McDuff [26].

The main message we want to draw from this is that *oriented d-manifolds and d-orbifolds admit virtual classes* (or virtual cycles, or virtual chains, as appropriate). Thus, we can use d-manifolds and d-orbifolds as the geometric structure on moduli spaces in enumerative invariants problems such as Gromov–Witten invariants, Lagrangian Floer cohomology, Donaldson–Thomas invariants, ..., as this structure is strong enough to contain all the 'counting' information.

In future work the author intends to define a virtual chain construction for d-manifolds and d-orbifolds, expressed in terms of new (co)homology theories whose (co)chains are built from d-manifolds or d-orbifolds, as for the 'Kuranishi (co)homology' described in [16].

5.4.11 Relation to other classes of spaces in mathematics

In [20, §14] the author studied the relationships between d-manifolds and d-orbifolds and other classes of geometric spaces in the literature. The next theorem summarizes our results:

Theorem 5.73. *We may construct 'truncation functors' from various classes of geometric spaces to d-manifolds and d-orbifolds, as follows:*

(a) *There is a functor $\Pi^{\text{dMan}}_{\text{BManFS}} : \mathbf{BManFS} \rightarrow \text{Ho}(\mathbf{dMan})$, where $\mathbf{BManFS}$ is a category whose objects are triples (V, E, s) of a Banach manifold V, Banach vector bundle $E \rightarrow V$, and smooth section $s : V \rightarrow E$ whose* *linearization* $ds|_x : T_x V \to E|_x$ *is Fredholm with index* $n \in \mathbb{Z}$ *for each* $s \in V$ *with* $s|_x = 0$, *and* Ho(**dMan**) *is the homotopy category of the 2-category of d-manifolds* **dMan**.

There is also an orbifold version $\Pi^{\text{dOrb}}_{\text{BOrbFS}} : \text{Ho}(\mathbf{BOrbFS}) \to \text{Ho}(\mathbf{dOrb})$ *of this using Banach orbifolds* V, *and 'corners' versions of both.*

(b) *There is a functor* $\Pi^{\text{dMan}}_{\text{MPolFS}} : \mathbf{MPolFS} \to \text{Ho}(\mathbf{dMan})$, *where* **MPolFS** *is a category whose objects are triples* (V, E, s) *of an **M-polyfold** without boundary* V *as in Hofer, Wysocki and Zehnder* [12, §3.3], *a fillable strong M-polyfold bundle* E *over* V [12, §4.3], *and an sc-smooth Fredholm section* s *of* E [12, §4.4] *whose linearization* $ds|_x : T_x V \to E|_x$ [12, §4.4] *has Fredholm index* $n \in \mathbb{Z}$ *for all* $x \in V$ *with* $s|_x = 0$.

There is also an orbifold version $\Pi^{\text{dOrb}}_{\text{PolFS}} : \text{Ho}(\mathbf{PolFS}) \to \text{Ho}(\mathbf{dOrb})$ *of this using **polyfolds*** V, *and 'corners' versions of both.*

(c) *Given a d-orbifold with corners* $\mathcal{X}$, *we can construct a **Kuranishi space** (X, κ) in the sense of Fukaya, Oh, Ohta and Ono* [9, §A], *with the same underlying topological space* X. *Conversely, given a Kuranishi space* (X, κ), *we can construct a d-orbifold with corners* $\mathcal{X}'$. *Composing the two constructions,* $\mathcal{X}$ *and* $\mathcal{X}'$ *are equivalent in* $\mathbf{dOrb^c}$.

Very roughly speaking, this means that the 'categories' of d-orbifolds with corners, and Kuranishi spaces, are equivalent. However, Fukaya et al. [9] *do not define morphisms of Kuranishi spaces, nor even when two Kuranishi spaces are 'the same', so we have no category of Kuranishi spaces.*

(d) *There is a functor* $\Pi^{\text{dMan}}_{\text{SchObs}} : \mathbf{Sch_{\mathbb{C}}Obs} \to \text{Ho}(\mathbf{dMan})$, *where* $\mathbf{Sch_{\mathbb{C}}Obs}$ *is a category whose objects are triples* $(X, E^{\bullet}, \phi)$, *where* X *is a separated, second countable* $\mathbb{C}$-scheme *and* $\phi : E^{\bullet} \to \tau_{\geqslant -1}(L_X)$ *a perfect obstruction theory on* X *with constant virtual dimension. We may define a natural orientation on* $\Pi^{\text{dMan}}_{\text{SchObs}}(X, E^{\bullet}, \phi)$ *for each* $(X, E^{\bullet}, \phi)$.

There is also an orbifold version $\Pi^{\text{dOrb}}_{\text{StaObs}} : \text{Ho}(\mathbf{Sta_{\mathbb{C}}Obs}) \to \text{Ho}(\mathbf{dOrb})$, *taking* X *to be a Deligne–Mumford* $\mathbb{C}$-stack.

(e) *There is a functor* $\Pi^{\text{dMan}}_{\text{QsDSch}} : \text{Ho}(\mathbf{QsDSch_{\mathbb{C}}}) \longrightarrow \text{Ho}(\mathbf{dMan})$, *where* $\mathbf{QsDSch_{\mathbb{C}}}$ *is the* ∞-category *of separated, second countable, quasi-smooth derived* $\mathbb{C}$-schemes X *of constant dimension, as in Toën and Vezzosi* [32, 33]. *We may define a natural orientation on* $\Pi^{\text{dMan}}_{\text{QsDSch}}(X)$ *for each* X.

There is also an orbifold version $\Pi^{\text{dOrb}}_{\text{QsDSta}} : \text{Ho}(\mathbf{QsDSta_{\mathbb{C}}}) \to \text{Ho}(\mathbf{dOrb})$, *taking* X *to be a derived Deligne–Mumford* $\mathbb{C}$-stack.

(f) *There is a functor* $\Pi_{\mathbf{DerMan}}^{\mathbf{dMan}} : \mathbf{DerMan} \longrightarrow \mathbf{dMan}$, *where* $\mathbf{DerMan}$ *is the* ∞-*category of **derived manifolds** of constant dimension, in the sense of Spivak* [30].

One moral of Theorem 5.73 is that essentially every geometric structure on moduli spaces which is used to define enumerative invariants, either in differential geometry, or in algebraic geometry over $\mathbb{C}$, has a truncation functor to d-manifolds or d-orbifolds. Combining Theorem 5.73 with proofs from the literature of the existence on moduli spaces of the geometric structures listed in Theorem 5.73, in [20, §14] we deduce:

Theorem 5.74. (i) *Any solution set of a smooth nonlinear elliptic equation with fixed topological invariants on a compact manifold naturally has the structure of a d-manifold, uniquely up to equivalence in* **dMan**.

For example, let (M, g), (N, h) *be Riemannian manifolds, with* M *compact. Then the family of **harmonic maps*** $f : M \to N$ *is a d-manifold* $\mathcal{H}_{M,N}$ *with* $\mathrm{vdim}\,\mathcal{H}_{M,N} = 0$. *If* $M = \mathcal{S}^1$, *then* $\mathcal{H}_{M,N}$ *is the moduli space of **parametrized closed geodesics** in* (N, h).

(ii) *Let* (M, ω) *be a compact symplectic manifold of dimension* $2n$, *and* J *an almost complex structure on* M *compatible with* ω. *For* $\beta \in H_2(M, \mathbb{Z})$ *and* $g, m \geqslant 0$, *write* $\overline{\mathcal{M}}_{g,m}(M, J, \beta)$ *for the moduli space of stable triples* $(\Sigma, \vec{z}, u)$ *for* Σ *a genus* g *prestable Riemann surface with* m *marked points* $\vec{z} = (z_1, \ldots, z_m)$ *and* $u : \Sigma \to M$ *a* J-*holomorphic map with* $[u(\Sigma)] = \beta$ *in* $H_2(M, \mathbb{Z})$. *Using results of Hofer, Wysocki and Zehnder* [14] *involving their theory of polyfolds, we can make* $\overline{\mathcal{M}}_{g,m}(M, J, \beta)$ *into a compact, oriented d-orbifold* $\overline{\boldsymbol{\mathcal{M}}}_{g,m}(M, J, \beta)$.

(iii) *Let* (M, ω) *be a compact symplectic manifold,* J *an almost complex structure on* M *compatible with* ω, *and* L *a compact, embedded Lagrangian submanifold in* M. *For* $\beta \in H_2(M, L; \mathbb{Z})$ *and* $k \geqslant 0$, *write* $\overline{\mathcal{M}}_k(M, L, J, \beta)$ *for the moduli space of* J-***holomorphic stable maps*** $(\Sigma, \vec{z}, u)$ *to* M *from a prestable holomorphic disc* Σ *with* k *boundary marked points* $\vec{z} = (z_1, \ldots, z_k)$, *with* $u(\partial\Sigma) \subseteq L$ *and* $[u(\Sigma)] = \beta$ *in* $H_2(M, L; \mathbb{Z})$. *Using results of Fukaya, Oh, Ohta and Ono* [9, §7–§8] *involving their theory of Kuranishi spaces, we can make* $\overline{\mathcal{M}}_k(M, L, J, \beta)$ *into a compact d-orbifold with corners* $\overline{\boldsymbol{\mathcal{M}}}_k(M, L, J, \beta)$. *Given a relative spin structure for* (M, L), *we may define an orientation on* $\overline{\boldsymbol{\mathcal{M}}}_k(M, L, J, \beta)$.

(iv) *Let* X *be a complex projective manifold, and* $\overline{\mathcal{M}}_{g,m}(X, \beta)$ *the Deligne–Mumford moduli* $\mathbb{C}$-*stack of stable triples* $(\Sigma, \vec{z}, u)$ *for* Σ *a genus* g *prestable Riemann surface with* m *marked points* $\vec{z} = (z_1, \ldots, z_m)$ *and* $u : \Sigma \to X$ *a morphism with* $u_*([\Sigma]) = \beta \in H_2(X; \mathbb{Z})$. *Then Behrend* [2] *defines a* *perfect obstruction theory on* $\overline{\mathcal{M}}_{g,m}(X, \beta)$, *so we can make* $\overline{\mathcal{M}}_{g,m}(X, \beta)$ *into a compact, oriented d-orbifold* $\overline{\boldsymbol{\mathcal{M}}}_{g,m}(X, \beta)$.

(v) *Let X be a complex algebraic surface, and* $\mathcal{M}$ *a stable moduli* $\mathbb{C}$*-scheme of vector bundles or coherent sheaves E on X with fixed Chern character. Then Mochizuki [27] defines a perfect obstruction theory on* $\mathcal{M}$*, so we can make* $\mathcal{M}$ *into an oriented d-manifold* $\boldsymbol{\mathcal{M}}$.

(vi) *Let X be a complex Calabi–Yau 3-fold or smooth Fano 3-fold, and* $\mathcal{M}$ *a stable moduli* $\mathbb{C}$*-scheme of coherent sheaves E on X with fixed Hilbert polynomial. Then Thomas [31] defines a perfect obstruction theory on* $\mathcal{M}$*, so we can make* $\mathcal{M}$ *into an oriented d-manifold* $\boldsymbol{\mathcal{M}}$.

(vii) *Let X be a smooth complex projective 3-fold, and* $\mathcal{M}$ *a moduli* $\mathbb{C}$*-scheme of 'stable PT pairs'* (C, D) *in X, where* $C \subset X$ *is a curve and* $D \subset C$ *is a divisor. Then Pandharipande and Thomas [29] define a perfect obstruction theory on* $\mathcal{M}$*, so we can make* $\mathcal{M}$ *into a compact, oriented d-manifold* $\boldsymbol{\mathcal{M}}$.

(ix) *Let X be a complex Calabi–Yau 3-fold, and* $\mathcal{M}$ *a separated moduli* $\mathbb{C}$*-scheme of simple perfect complexes in the derived category* $D^b \operatorname{coh}(X)$*. Then Huybrechts and Thomas [15] define a perfect obstruction theory on* $\mathcal{M}$*, so we can make* $\mathcal{M}$ *into an oriented d-manifold* $\boldsymbol{\mathcal{M}}$.

We can use d-manifolds and d-orbifolds to construct *virtual classes* or *virtual chains* for all these moduli spaces.

Remark 5.75. D-manifolds should not be confused with *differential graded manifolds*, or *dg-manifolds*. This term is used in two senses, in algebraic geometry to mean a special kind of dg-scheme, as in Ciocan-Fontanine and Kapranov [6, Def. 2.5.1], and in differential geometry to mean a supermanifold with extra structure, as in Cattaneo and Schätz [5, Def. 3.6]. In both cases, a dg-manifold $\mathfrak{E}$ is roughly the total space of a graded vector bundle $E^{\bullet}$ over a manifold V, with a vector field Q of degree 1 satisfying $[Q, Q] = 0$.

For example, if E is a vector bundle over V and $s \in C^{\infty}(E)$, we can make E into a dg-manifold $\mathfrak{E}$ by giving E the grading -1, and taking Q to be the vector field on E corresponding to s. To this $\mathfrak{E}$ we can associate the d-manifold $S_{V,E,s}$ from Example 5.33. Note that $S_{V,E,s}$ only knows about an infinitesimal neighbourhood of $s^{-1}(0)$ in V, but $\mathfrak{E}$ remembers all of V, E, s.

A Basics of 2-categories

Finally we discuss 2-categories. A good reference is Behrend et al. [3, App. B].

Definition 5.76. A (*strict*) 2-*category* $\mathfrak{C}$ consists of a proper class of *objects* $\mathrm{Obj}(\mathfrak{C})$, for all $X, Y \in \mathrm{Obj}(\mathfrak{C})$ a category $\mathrm{Hom}(X, Y)$, for all X in $\mathrm{Obj}(\mathfrak{C})$ an object id_X in $\mathrm{Hom}(X, X)$ called the *identity* 1-*morphism*, and for all X, Y, Z in $\mathrm{Obj}(\mathfrak{C})$ a functor $\mu_{X,Y,Z} : \mathrm{Hom}(X, Y) \times \mathrm{Hom}(Y, Z) \to \mathrm{Hom}(X, Z)$. These must satisfy the *identity property*, that

$$\mu_{X,X,Y}(\mathrm{id}_X, -) = \mu_{X,Y,Y}(-, \mathrm{id}_Y) = \mathrm{id}_{\mathrm{Hom}(X,Y)}$$

as functors $\mathrm{Hom}(X, Y) \to \mathrm{Hom}(X, Y)$, and the *associativity property*, that

$$\mu_{W,Y,Z} \circ (\mu_{W,X,Y} \times \mathrm{id}_{\mathrm{Hom}(Y,Z)}) = \mu_{W,X,Z} \circ (\mathrm{id}_{\mathrm{Hom}(W,X)} \times \mu_{X,Y,Z})$$

as functors $\mathrm{Hom}(W, X) \times \mathrm{Hom}(X, Y) \times \mathrm{Hom}(Y, Z) \to \mathrm{Hom}(W, X)$.

Objects f of $\mathrm{Hom}(X, Y)$ are called 1-*morphisms*, written $f : X \to Y$. For 1-morphisms $f, g : X \to Y$, morphisms $\eta \in \mathrm{Hom}_{\mathrm{Hom}(X,Y)}(f, g)$ are called 2-*morphisms*, written $\eta : f \Rightarrow g$. Thus, a 2-category has objects X, and two kinds of morphisms, 1-morphisms $f : X \to Y$ between objects, and 2-morphisms $\eta : f \Rightarrow g$ between 1-morphisms.

There are three kinds of composition in a 2-category, satisfying various associativity relations. If $f : X \to Y$ and $g : Y \to Z$ are 1-morphisms then $\mu_{X,Y,Z}(f, g)$ is the *horizontal composition of* 1-*morphisms*, written $g \circ f : X \to Z$. If $f, g, h : X \to Y$ are 1-morphisms and $\eta : f \Rightarrow g$, $\zeta : g \Rightarrow h$ are 2-morphisms then composition of η, ζ in $\mathrm{Hom}(X, Y)$ gives the *vertical composition of* 2-*morphisms* of η, ζ, written $\zeta \odot \eta : f \Rightarrow h$, as a diagram

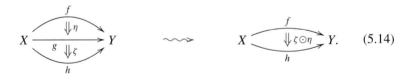

$$\tag{5.14}$$

And if $f, \widetilde{f} : X \to Y$ and $g, \widetilde{g} : Y \to Z$ are 1-morphisms and $\eta : f \Rightarrow \widetilde{f}$, $\zeta : g \Rightarrow \widetilde{g}$ are 2-morphisms then $\mu_{X,Y,Z}(\eta, \zeta)$ is the *horizontal composition of* 2-*morphisms*, written $\zeta * \eta : g \circ f \Rightarrow \widetilde{g} \circ \widetilde{f}$, as a diagram

$$X \underset{\widetilde{f}}{\overset{f}{\rightrightarrows}} Y \underset{\widetilde{g}}{\overset{g}{\rightrightarrows}} Z \quad \rightsquigarrow \quad X \underset{\widetilde{g} \circ \widetilde{f}}{\overset{g \circ f}{\rightrightarrows}} Z. \tag{5.15}$$

There are also two kinds of identity: *identity* 1-*morphisms* $\mathrm{id}_X : X \to X$ and *identity* 2-*morphisms* $\mathrm{id}_f : f \Rightarrow f$.

A basic example is the 2-*category of categories* $\mathfrak{Cat}$, with objects categories $\mathcal{C}$, 1-morphisms functors $F : \mathcal{C} \to \mathcal{D}$, and 2-morphisms natural transformations $\eta : F \Rightarrow G$ for functors $F, G : \mathcal{C} \rightarrow \mathcal{D}$. Orbifolds naturally form a 2-category, as do stacks in algebraic geometry.

In a 2-category $\mathfrak{C}$, there are three notions of when objects X, Y in $\mathfrak{C}$ are 'the same': *equality* $X = Y$, and *isomorphism*, that is we have 1-morphisms $f : X \rightarrow Y, g : Y \rightarrow X$ with $g \circ f = \mathrm{id}_X$ and $f \circ g = \mathrm{id}_Y$, and *equivalence*, that is we have 1-morphisms $f : X \rightarrow Y, g : Y \rightarrow X$ and 2-isomorphisms $\eta : g \circ f \Rightarrow \mathrm{id}_X$ and $\zeta : f \circ g \Rightarrow \mathrm{id}_Y$. Usually equivalence is the correct notion.

Commutative diagrams in 2-categories should in general only commute *up to (specified) 2-isomorphisms*, rather than strictly. A simple example of a commutative diagram in a 2-category $\mathfrak{C}$ is

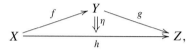

which means that X, Y, Z are objects of $\mathfrak{C}$, $f : X \rightarrow Y, g : Y \rightarrow Z$ and $h : X \rightarrow Z$ are 1-morphisms in $\mathfrak{C}$, and $\eta : g \circ f \Rightarrow h$ is a 2-isomorphism.

We define fibre products in 2-categories, following [3, Def. B.13].

Definition 5.77. Let $\mathfrak{C}$ be a 2-category and $g : X \rightarrow Z, h : Y \rightarrow Z$ be 1-morphisms in $\mathfrak{C}$. A *fibre product* $X \times_Z Y$ in $\mathfrak{C}$ consists of an object W, 1-morphisms $\pi_X : W \rightarrow X$ and $\pi_Y : W \rightarrow Y$ (we usually write $e = \pi_X$ and $f = \pi_Y$) and a 2-isomorphism $\eta : g \circ \pi_X \Rightarrow h \circ \pi_Y$ in $\mathfrak{C}$ with the following universal property: suppose $\pi_X' : W' \rightarrow X$ and $\pi_Y' : W' \rightarrow Y$ are 1-morphisms and $\eta' : g \circ \pi_X' \Rightarrow h \circ \pi_Y'$ is a 2-isomorphism in $\mathfrak{C}$. Then there should exist a 1-morphism $b : W' \rightarrow W$ and 2-isomorphisms $\zeta_X : \pi_X \circ b \Rightarrow \pi_X'$, $\zeta_Y : \pi_Y \circ b \Rightarrow \pi_Y'$ such that the following diagram of 2-isomorphisms commutes:

$$
\begin{array}{ccc}
g \circ \pi_X \circ b & \xrightarrow{\;\;\eta * \mathrm{id}_b\;\;} & h \circ \pi_Y \circ b \\
{\scriptstyle \mathrm{id}_g * \zeta_X} \big\Downarrow & & \big\Downarrow {\scriptstyle \mathrm{id}_h * \zeta_Y} \\
g \circ \pi_X' & \xrightarrow[\;\;\eta'\;\;]{} & h \circ \pi_Y'.
\end{array}
$$

Furthermore, if $\widetilde{b}, \widetilde{\zeta}_X, \widetilde{\zeta}_Y$ are alternative choices of b, ζ_X, ζ_Y then there should exist a unique 2-isomorphism $\theta : \widetilde{b} \Rightarrow b$ with

$$\widetilde{\zeta}_X = \zeta_X \odot (\mathrm{id}_{\pi_X} * \theta) \quad \text{and} \quad \widetilde{\zeta}_Y = \zeta_Y \odot (\mathrm{id}_{\pi_Y} * \theta).$$

If a fibre product $X \times_Z Y$ in $\mathfrak{C}$ exists then it is unique up to equivalence.

Orbifolds, and stacks in algebraic geometry, form 2-categories, and Definition 5.77 is the right way to define fibre products of orbifolds or stacks.

References

[1] M.F. Atiyah, *Bordism and cobordism*, Proc. Camb. Phil. Soc. 57 (1961), 200–208.

[2] K. Behrend, *Gromov-Witten invariants in algebraic geometry*, Invent. Math. 127 (1997), 601–617. alg-geom/9601011.

[3] K. Behrend, D. Edidin, B. Fantechi, W. Fulton, L. Göttsche and A. Kresch, *Introduction to stacks*, book in preparation, 2010.

[4] K. Behrend and B. Fantechi, *The intrinsic normal cone*, Invent. Math. 128 (1997), 45–88. alg-geom/9601010.

[5] A.S. Cattaneo and F. Schätz, *Introduction to supergeometry*, Rev. Math. Phys. 23 (2011), 669–690. arXiv:1011.3401.

[6] I. Ciocan-Fontanine and M. Kapranov, *Derived Quot schemes*, Ann. Sci. Ec. Norm. Sup. 34 (2001), 403–440. math.AG/9905174.

[7] P.E. Conner, *Differentiable Periodic Maps*, second edition, Springer Lecture Notes in Mathematics 738, Springer-Verlag, Berlin, 1979.

[8] E.J. Dubuc, C^∞-schemes, Amer. J. Math. 103 (1981), 683–690.

[9] K. Fukaya, Y.-G. Oh, H. Ohta and K. Ono, *Lagrangian intersection Floer theory – anomaly and obstruction*, Parts I & II. AMS/IP Studies in Advanced Mathematics, 46.1 & 46.2, A.M.S./International Press, 2009.

[10] K. Fukaya and K. Ono, *Arnold Conjecture and Gromov–Witten invariant*, Topology 38 (1999), 933–1048.

[11] R. Hartshorne, *Algebraic Geometry*, Graduate Texts in Math. 52, Springer, New York, 1977.

[12] H. Hofer, K. Wysocki and E. Zehnder, *A general Fredholm theory I: A splicing-based differential geometry*, J. Eur. Math. Soc. 9 (2007), 841–876. math.FA/0612604.

[13] H. Hofer, K. Wysocki and E. Zehnder, *Integration theory for zero sets of polyfold Fredholm sections*, arXiv:0711.0781, 2007.

[14] H. Hofer, K. Wysocki and E. Zehnder, *Applications of polyfold theory I: the polyfolds of Gromov–Witten theory*, arXiv:1107.2097, 2011.

[15] D. Huybrechts and R.P. Thomas, *Deformation-obstruction theory for complexes via Atiyah and Kodaira–Spencer classes*, Math. Ann. 346 (2010), 545–569. arXiv:0805.3527.

[16] D. Joyce, *Kuranishi homology and Kuranishi cohomology: a User's Guide*, math.SG/0710.5634, 2007.

[17] D. Joyce, *On manifolds with corners*, arXiv:0910.3518, 2009.

[18] D. Joyce, *Algebraic Geometry over C^∞-rings*, arXiv:1001.0023, 2010.

[19] D. Joyce, *An introduction to C^∞-schemes and C^∞-algebraic geometry*, pages 299–325 in H.-D. Cao and S.-T. Yau, editors, *In memory of C.C. Hsiung: Lectures given at the JDG symposium, Lehigh University, June 2010*, Surveys in Differential Geometry 17, 2012. arXiv:1104.4951.

[20] D. Joyce, *D-manifolds and d-orbifolds: a theory of derived differential geometry*, book in preparation, 2012. Preliminary version available at http://people.maths.ox.ac.uk/~joyce/dmanifolds.html.

[21] D. Joyce, *D-manifolds, d-orbifolds and derived differential geometry: a detailed summary*, arXiv: 1208.4948, 2012.

[22] M. Kontsevich, *Enumeration of rational curves via torus actions*, pages 335–368 in R. Dijkgraaf, C. Faber and G. van der Geer, editors, *The moduli space of curves*, Progr. Math. 129, Birkhäuser, 1995. hep-th/9405035.

[23] E. Lerman, *Orbifolds as stacks?*, arXiv:0806.4160, 2008.

[24] J. Lurie, *Higher Topos Theory*, Annals of Math. Studies, 170, Princeton University Press, Princeton, NJ, 2009. math.CT/0608040.

[25] J. Lurie, *Derived Algebraic Geometry V: Structured spaces*, arXiv:0905.0459, 2009.

[26] D. McDuff, *The virtual moduli cycle*, pages 73–102 in Y. Eliashberg, D. Fuchs, T. Ratiu and A. Weinstein, editors, *Northern California Symplectic Geometry Seminar*, A.M.S. Translations 196, A.M.S., Providence, RI, 1999.

[27] T. Mochizuki, *Donaldson type invariants for algebraic surfaces*, Lecture Notes in Math. 1972, Spinger, 2009.

[28] I. Moerdijk and G.E. Reyes, *Models for smooth infinitesimal analysis*, Springer-Verlag, New York, 1991.

[29] R. Pandharipande and R.P. Thomas, *Curve counting via stable pairs in the derived category*, Invent. math. 178 (2009) 407–447. arXiv:0707.2348.

[30] D.I. Spivak, *Derived smooth manifolds*, Duke Math. J. 153 (2010), 55–128. arXiv:0810.5174.

[31] R.P. Thomas, *A holomorphic Casson invariant for Calabi–Yau 3-folds, and bundles on $K3$ fibrations*, J. Diff. Geom. 54 (2000), 367–438. math.AG/9806111.

[32] B. Toën, *Higher and derived stacks: a global overview*, pages 435–487 in Proc. Symp. Pure Math. vol 80, part 1, A.M.S., 2009. math.AG/0604504.

[33] B. Toën and G. Vezzosi, *Homotopical Algebraic Geometry II: Geometric Stacks and Applications*, Memoirs of the A.M.S. vol. 193, no. 902, 2008. math.AG/0404373.

[34] H. Whitney, *Differentiable manifolds*, Ann. Math. 37 (1936), 645–680.

THE MATHEMATICAL INSTITUTE, ANDREW WILES BUILDING, RADCLIFFE OBSERVATORY QUARTER, WOODSTOCK ROAD, OXFORD, OX2 6GG, U.K. E-MAIL: joyce@maths.ox.ac.uk

6

13/2 ways of counting curves

R. Pandharipande and R. P. Thomas

I was of three minds,
Like a tree
In which there are three blackbirds.
(Wallace Stevens)

Abstract

In the past 20 years, compactifications of the families of curves in algebraic varieties X have been studied via stable maps, Hilbert schemes, stable pairs, unramified maps, and stable quotients. Each path leads to a different enumeration of curves. A common thread is the use of a 2-term deformation/obstruction theory to define a virtual fundamental class. The richest geometry occurs when X is a nonsingular projective variety of dimension 3.

We survey here the 13/2 principal ways to count curves with special attention to the 3-fold case. The different theories are linked by a web of conjectural relationships which we highlight. Our goal is to provide a guide for graduate students looking for an elementary route into the subject.

Contents

Moduli Spaces, eds. L. Brambila-Paz, O. García-Prada, P. Newstead and R. Thomas. Published by Cambridge University Press. © Cambridge University Press 2014.

0 Introduction

Counting

Let X be a nonsingular projective variety (over $\mathbb{C}$), and let $\beta \in H_2(X, \mathbb{Z})$ be a homology class. We are interested here in counting the algebraic curves of X in class β. For example, how many twisted cubics in $\mathbb{P}^3$ meet 12 given lines? Mathematicians such as Hurwitz, Schubert, and Zeuthen have considered such questions since the nineteenth century. Towards the end of the twentieth century and continuing to the present, the subject has been greatly enriched by new insights from symplectic geometry and topological string theory.

Under appropriate genericity conditions, counting solutions in algebraic geometry often yields deformation invariant answers. A simple example is provided by Bezout's Theorem concerning the intersections of plane curves. Two generic algebraic curves in $\mathbb{C}^2$ of degrees d_1 and d_2 intersect transversally in finitely many points. Counting these points yields the topological intersection number $d_1 d_2$. But in nongeneric situations, we can find fewer solutions or an infinite number. The curves may intersect with tangencies in a smaller number of points (remedied by counting intersection points with multiplicities). If the curves intersect "at infinity", we will again find fewer intersection points in $\mathbb{C}^2$ whose total we do not consider to be a "sensible" answer. Instead, we compactify $\mathbb{C}^2$ by $\mathbb{P}^2$ and count there. Finally, the curves may intersect in an entire component. The technique of excess intersection theory is required then to obtain the correct answer. Compactification and transversality already play a important role in the geometry of Bezout's Theorem.

Having deformation invariant answers for the enumerative geometry of curves in X is desirable for several reasons. The most basic is the possibility of deforming X to a more convenient space. To achieve deformation invariance, two main issues must be considered:

(i) compactification of the moduli space $\mathcal{M}(X, \beta)$ of curves $C \subset X$ of class β,

(ii) transversality of the solutions.

What we mean by the moduli space $\mathcal{M}(X, \beta)$ is to be explained and will differ in each of the sections below. Transversality concerns both the possible excess dimension of $\mathcal{M}(X, \beta)$ and the transversality of the constraints.

Compactness

For Bezout's Theorem, we compactify the geometry so intersection points running to infinity do not escape our counting. The result is a deformation invariant answer.

A compact space $\mathcal{M}(X, \beta)$ which parameterises all nonsingular embedded curves in class β will usually have to contain singular curves of some sort. Strictly speaking, the compact moduli spaces $\mathcal{M}(X, \beta)$ will often not be compactifications of the spaces of nonsingular embedded curves – the latter need not be dense in $\mathcal{M}(X, \beta)$. For instance $\mathcal{M}(X, \beta)$ might be nonempty when there are *no* nonsingular embedded curves. The singular strata are important for deformation invariance. As we deform X, curves can "wander off to infinity" in $\mathcal{M}(X, \beta)$ by becoming singular.

Transversality

A simple question to consider is the number of elliptic cubics in $\mathbb{P}^2$ passing through nine points $p_1, \ldots, p_9 \in \mathbb{P}^2$. The linear system

$$\mathbb{P}(H^0(\mathbb{P}^2, \mathcal{O}_{\mathbb{P}^2}(3))) \cong \mathbb{P}^9$$

provides a natural compactification of the moduli space. Each p_i imposes a single linear condition which determines a hyperplane

$$\mathbb{P}_i^8 \subset \mathbb{P}^9,$$

of curves passing through p_i. For general p_i, these nine hyperplanes are transverse and intersect in a single point. Hence, we expect our count to be 1. But if the p_i are the nine intersection points of two cubics, then we obtain an entire pencil of solutions given by the linear combinations of the two cubics.

An alternative way of looking at the same enumerative question is the following. Let

$$\epsilon : S \to \mathbb{P}^2$$

be the blow-up of $\mathbb{P}^2$ at nine points p_i and consider curves in the class

$$\beta = 3H - E_1 - E_2 - \ldots - E_9$$

where H is the ϵ pull-back of the hyperplane class and the E_i are the exceptional divisors. In general there will be a unique elliptic curve embedded in class β. But if the nine points are the intersection of two cubics, then S is a rational elliptic surface via the pencil

$$\pi : S \to \mathbb{P}^1.$$

How to sensibly "count" the pencil of elliptic fibres on S is not obvious.

A temptation based on the above discussion is to define the enumeration of curves by counting after taking a generic perturbation of the geometry. Unfortunately, we often do not have enough perturbations to make the situation fully transverse. A basic rigid example is given by counting the intersection points of a (-1)-curve with itself on a surface. Though we cannot algebraically move the curve to be transverse to itself, we know another way to get the "sensible" answer of topology: take the Euler number -1 of the normal bundle. In curve counting, there is a similar excess intersection theory approach to getting a sensible, deformation invariant answer using *virtual fundamental classes*.

For the rational elliptic surface S, the base $\mathbb{P}^1$ is a natural compact moduli space parameterising the elliptic curves in the pencil. The count of elliptic fibres is the Euler class of the *obstruction bundle* over the pencil $\mathbb{P}^1$. Calculating the obstruction bundle to be $\mathcal{O}_{\mathbb{P}^1}(1)$, we recover the answer 1 expected from deformation invariance.

Why is the obstruction bundle $\mathcal{O}_{\mathbb{P}^1}(1)$? In Section $1\frac{1}{2}$, a short introduction to the deformation theory of maps is presented. Let $E \subset S$ be the fibre of π over $[E] \in \mathbb{P}^1$. Let ν_E be the normal bundle of E in S. The obstruction space at $[E] \in \mathbb{P}^1$ is

$$ H^1(E, \nu_E) = H^1(E, \mathcal{O}_E) \otimes \mathcal{O}_{\mathbb{P}^1}(2)|_{[E]} \, . $$

The term $H^1(E, \mathcal{O}_E)$ yields the dual of the Hodge bundle as E varies and is isomorphic to $\mathcal{O}_{\mathbb{P}^1}(-1)$. Hence, we find the obstruction bundle to be $\mathcal{O}_{\mathbb{P}^1}(1)$.

We will discuss virtual classes in the Appendix. We should think loosely of $\mathcal{M}(X, \beta)$ as being cut out of a nonsingular ambient space by a set of equations. The expected, or *virtual*, dimension of $\mathcal{M}(X, \beta)$ is the dimension of the ambient space minus the number of equations. If the derivatives of the equations are not linearly independent along the common zero locus $\mathcal{M}(X, \beta)$, then $\mathcal{M}(X, \beta)$ will be singular or have dimension higher than expected. In practice, $\mathcal{M}(X, \beta)$ is very rarely nonsingular of the expected dimension. We should think of the virtual class as representing the fundamental cycle of the "correct" moduli space (of dimension *equal* to the virtual dimension) inside the actual moduli space. The virtual class may be considered to give the result of perturbing the setup to a transverse geometry, even when such perturbations do not actually exist.

Overview

A nonsingular embedded curve $C \subset X$ can be described in two fundamentally different ways:

 (i) as an algebraic map $C \to X$

 (ii) as the zero locus of an ideal of algebraic functions on X.

In other words, C can be seen as a *parameterised* curve with a map or an *unparameterised* curve with an embedding. Both realisations arise naturally in physics — the first as the worldsheet of a string moving in X, the second as a D-brane or boundary condition embedded in X.

Associated to the two basic ways of thinking of curves, there are two natural paths for compactifications. The first allows the map f to degenerate badly while keeping the domain curve as nice as possible. The second keeps the map as an embedding but allows the curve to degenerate arbitrarily.

We describe here $6\frac{1}{2}$ methods for defining curve counts in algebraic geometry. We start in Section $\frac{1}{2}$ with a discussion of the successes and limitations of the naive counts pursued by the nineteenth-century geometers (and followed for more than 100 years). Since such counting is not always well-defined and has many drawbacks, we view the naive approach as only $\frac{1}{2}$ a method.

In Sections $1\frac{1}{2} - 6\frac{1}{2}$, six approaches to deformation invariant curve counting are presented. Two (stable maps and unramified maps) fall in class (i), three (BPS invariants, ideal sheaves, stable pairs) in class (ii), and one (stable quotients) straddles both classes (i-ii). The compactifications and virtual class constructions are dealt with differently in the six cases. Of course, each of the six has advantages and drawbacks.

There are several excellent references covering different aspects of the material surveyed here in much greater depth, see for instance [23, 43, 64, 78, 100]. Also, there are many beautiful directions which we do not cover at all. For example, mirror symmetry, integrable hierarchies, descendent invariants, 3-dimensional partitions, and holomorphic symplectic geometry all play significant roles in the subject. Though orbifold and relative geometries have been very important for the development of the ideas presented here, we have chosen to omit a discussion. Our goal is to describe the $6\frac{1}{2}$ counting theories as simply as possible and to present the web of relationships amongst them.

Acknowledgments

We thank our students, collaborators, and colleagues for their contributions to our understanding of the subjects presented here. Special thanks are due to L. Brambila-Paz, J. Bryan, Y. Cooper, S. Katz, A. Kresch, A. MacPherson, P. Newstead and H.-H. Tseng for their specific comments and suggestions about the paper.

R.P. was partially supported by NSF grant DMS-1001154, a Marie Curie fellowship at IST Lisbon, and a grant from the Gulbenkian foundation. R.T.

was partially supported by an EPSRC programme grant. We would both like to thank the Isaac Newton Institute, Cambridge for support and a great research environment.

$\frac{1}{2}$ Naive counting of curves

Let X be a nonsingular projective variety, and let $\beta \in H_2(X, \mathbb{Z})$ be a homology class. Let $C \subset X$ be a nonsingular embedded (or immersed) curve of genus g and class β. The expected dimension of the family of genus g and class β curves containing C is

$$3g - 3 + \chi(T_X|_C) = \int_C c_1(X) + (\dim_{\mathbb{C}} X - 3)(1 - g). \qquad (6.1)$$

The first term on the left comes from the complex moduli of the genus g curve,

$$\dim_{\mathbb{C}} \overline{\mathcal{M}}_g = 3g - 3.$$

The second term arises from infinitesimal deformations of C which do not change the complex structure of C. More precisely,

$$\chi(T_X|_C) = h^0(C, T_X|_C) - h^1(C, T_X|_C)$$

where $H^0(C, T_X|_C)$ is the space of such deformations (at least when C has no continuous families of automorphisms). The "expectation" amounts to the vanishing of $H^1(C, T_X|_C)$. Indeed if $H^1(C, T_X|_C)$ vanishes, the family of curves is nonsingular of expected dimension at C, see [54]. We will return to this deformation theory in Section $1\frac{1}{2}$.

If the open family of embedded (or immersed) curves of genus g and class β is of pure expected dimension (6.1), then naive classical curve counting is sensible to undertake. We can attempt to count the actual numbers of embedded (or immersed) curves of genus g and class β in X subject to incidence conditions.

The main classical[1] examples where naive curve counting with simple incidence is reasonable to consider constitute a rather short list:

  (i) counting Hurwitz coverings of $\mathbb{P}^1$ and curves of higher genus,
 (ii) Severi degrees in $\mathbb{P}^2$ and $\mathbb{P}^1 \times \mathbb{P}^1$ in all genera,
(iii) counting genus 0 curves in general blow-ups of $\mathbb{P}^2$,
 (iv) counting genus 0 curves in homogeneous spaces such as $\mathbb{P}^n$, Grassmannians, and flag varieties,
  (v) counting lines on complete intersections in $\mathbb{P}^n$,
 (vi) counting curves of genus 1 and 2 in $\mathbb{P}^3$.

---

[1] We do not attempt here to give a complete classical bibliography. Rather, the references we list, for the most part, are modern treatments.

The Hurwitz covers of $\mathbb{P}^1$ (or higher genus curves),

$$C \to \mathbb{P}^1,$$

are neither embeddings nor immersions, but rather are counts of ramified maps, see [74] for an introduction. Nevertheless (i) fits naturally in the list of classical examples. The Severi degrees (ii) are the numbers of immersed curves of genus g and class β passing through the expected number of points on a surface. Particularly for the case of $\mathbb{P}^2$, the study of Severi degrees has a long history [16, 37, 87]. Counting genus 0 curves on blow-ups (iii) is equivalent to imposing multiple point singularities for plane curves, see [38] for a treatment. Genus 0 curves behave very well in homogeneous spaces, so the questions (iv) have been considered since Schubert and Zeuthen [90, 107]. Examples of (v) include the famous 27 lines on a cubic surface and the 2875 lines on a quintic 3-fold, see [23]. The genus 1 and 2 enumerative geometry of space curves was much less studied by the classical geometers, but still can be viewed in terms of naive counting.

For particular genera and classes on other varieties X, the families of curves might be pure of expected dimension. The above list addresses the cases of more uniform behavior. Until new ideas from symplectic geometry and topological string theory were introduced in the 1980s and 90s, the classical cases (i-vi) were the main topics of study in enumerative geometry. The subject was an important area, especially for the development of intersection theory in algebraic geometry. See [30] for a historical survey. However, because of the restrictions, we treat naive counting as only $\frac{1}{2}$ of an enumerative theory here.

New approaches to enumerative geometry by tropical methods have been developed extensively in recent years [46, 73]. However, the lack of a virtual fundamental class in tropical geometry restricts the direct[2] applications at the moment to the classical cases.

The counting of rational curves on algebraic $K3$ surfaces is almost a classical question. A $K3$ surface with Picard number 1 has finitely many rational curves in the primitive class (even though the expected dimension of the family of rational curves is -1 by (6.1)). As proved in [18], for a general $K3$ of Picard number 1, all the primitive rational curves are nodal. A proposal for the count was made by Yau and Zaslow [106] in terms of modular forms. The proofs by Beauville [2] and Bryan–Leung [11] certainly use modern methods. The counting of rational curves in all (including imprimitive) classes on $K3$ surfaces shows the fully non-classical nature of the question [52]. A conjectural extension of the Yau–Zaslow formula to all genera is given in [50].

---

[2] Tropical methods do interact in an intricate way with virtual curve counts on Calabi–Yau 3-folds in the program of Gross and Siebert [40, 41, 42] to study mirror symmetry.

$1\frac{1}{2}$ Gromov–Witten theory

Moduli

Gromov–Witten theory provided the first modern approach to curve counting which dealt successfully with the issues of compactification and transversality. The subject has origins in Gromov's work on pseudo-holomorphic curves in symplectic geometry [39] and papers of Witten on topological strings [104]. Contributions by Kontsevich, Manin, Ruan, and Tian [56, 57, 88, 89] played an important role in the early development.

In Gromov–Witten theory, curves are viewed as parameterised with an algebraic map

$$C \to X.$$

The compactification strategy is to admit only nodal singularities in the domain while allowing the map to become rather degenerate. More precisely, define $\overline{\mathcal{M}}_g(X, \beta)$ to be the moduli space of *stable maps*:

$$\left\{ f : C \to X \ \middle| \ \begin{array}{l} C \text{ a nodal curve of arithmetic genus } g, \\ \quad f_*[C] = \beta, \text{ and Aut}(f) \text{ finite} \end{array} \right\}.$$

The map f is invariant under an automorphism ϕ of the domain C if

$$f = f \circ \phi.$$

By definition, $\text{Aut}(f) \subset \text{Aut}(C)$ is the subgroup of elements for which f is invariant. The finite automorphism condition for a stable map implies the moduli space $\overline{\mathcal{M}}_g(X, \beta)$ is naturally a Deligne–Mumford stack.

The compactness of $\overline{\mathcal{M}}_g(X, \beta)$ is not immediate. A proof can be found in [32] using standard properties of semistable reduction for curves. In Section $3\frac{1}{2}$ below, we will discuss nontrivial limits in the space of stable maps, see for instance (6.12) and (6.14).

Deformation theory

We return now to the deformation theory for embedded curves briefly discussed in Section $\frac{1}{2}$. The deformation theory for arbitrary stable maps is very similar.

Let $C \subset X$ be a nonsingular embedded curve with normal bundle ν_C. The Zariski tangent space to the moduli space $\overline{\mathcal{M}}_g(X, \beta)$ at the point $[C \to X]$ is given by $H^0(C, \nu_C)$. Locally, we can lift a section of ν_C to a section of $T_X|_C$ and deform C along the lift to first order. Since globally ν_C is not usually a summand of $T_X|_C$ but only a quotient, the lifts will differ over overlaps by vector fields along C. The deformed curve will have a complex structure whose transition functions differ by these vector fields. In other words, from

$$0 \to T_C \to T_X|_C \to v_C \to 0,$$

we obtain the sequence

$$0 \to H^0(C, T_C) \to H^0(C, T_X|_C) \to H^0(C, v_C) \to H^1(C, T_C) \quad (6.2)$$

which expresses how deformations in $H^0(C, v_C)$ change the complex structure on C through the boundary map to $H^1(C, T_C)$. The kernel

$$H^0(C, T_X|_C)/H^0(C, T_C)$$

consists of the deformations given by moving C along vector fields in X, thus preserving the complex structure of C, modulo infinitesimal automorphisms of C. Similarly, obstructions to deformations lie in $H^1(C, v_C)$.

The expected dimension $\chi(v_C) = h^0(v_C) - h^1(v_C)$ of the moduli space is given by the calculation

$$\chi(v_C) = \int_C c_1(X) + (\dim_{\mathbb{C}} X - 3)(1 - g) \quad (6.3)$$

obtained from sequence (6.2). If $H^1(C, T_X|_C)$ vanishes, so does the obstruction space $H^1(C, v_C)$. Formula (6.3) then computes the actual dimension of the Zariski tangent space.

For arbitrary stable maps $f : C \to X$, we replace the dual of v_C by the complex

$$\{f^*\Omega_X \to \Omega_C\} \quad (6.4)$$

on C. If C is nonsingular and f is an embedding, the complex (6.4) is quasi-isomorphic to its kernel v_C^*. The deformations/obstructions of f are governed by

$$\mathrm{Ext}^i\left(\{f^*\Omega_X \to \Omega_C\}, \mathcal{O}_C\right) \quad (6.5)$$

for $i = 0, 1$. Similarly the deformations/obstructions of f with the curve C fixed are governed by $\mathrm{Ext}^i(f^*\Omega_X, \mathcal{O}_C) = H^i(f^*T_X)$.

Since the Ext groups (6.5) vanish for $i \neq 0, 1$, the deformation/obstruction theory is 2-term. The moduli space admits a virtual fundamental class[3]

$$[\overline{\mathcal{M}}_g(X, \beta)]^{vir} \in H_*(\overline{\mathcal{M}}_g(X, \beta), \mathbb{Q})$$

---

[3] The virtual fundamental class is algebraic, so should be more naturally considered in the Chow group $A_*(\overline{\mathcal{M}}_g(X, \beta), \mathbb{Q})$.

of complex dimension equal to the virtual dimension

$$\mathrm{ext}^0 - \mathrm{ext}^1 = \int_\beta c_1(X) + (\dim_{\mathbb{C}} X - 3)(1 - g). \qquad (6.6)$$

An introduction to the virtual fundamental class is provided in the Appendix.

Invariants

To obtain numerical invariants, we must cut the virtual class from dimension (6.6) to zero. The simplest way is by imposing incidence conditions: we count only those curves which pass though fixed cycles in X. Let

$$\mathcal{C} \to X \times \overline{\mathcal{M}}_g(X, \beta)$$

be the universal curve. We would like to intersect $\mathcal{C}$ with a cycle α pulled back from X. Transversality issues again arise here, so we use Poincaré dual cocyles.[4] Let

$$f : \mathcal{C} \to X \quad \text{and} \quad \pi : \mathcal{C} \to \overline{\mathcal{M}}_g(X, \beta)$$

be the universal map and the projection to $\overline{\mathcal{M}}_g(X, \beta)$ respectively. Let

$$\widetilde{\alpha} = \pi_* \big(f^* \mathrm{PD}(\alpha) \big) \in H^*(\overline{\mathcal{M}}_g(X, \beta)).$$

If α is a cycle of real codimension a, then $\widetilde{\alpha}$ is a cohomology class[5] in degree $a - 2$. When transversality is satisfied, $\widetilde{\alpha}$ is Poincaré dual to the locus of curves in $\overline{\mathcal{M}}_g(X, \beta)$ which intersect α. After imposing sufficiently many incidence conditions to cut the virtual dimension to zero, we define the *Gromov–Witten invariant*

$$N_{g,\beta}^{\mathrm{GW}}(\alpha_1, \dots, \alpha_k) = \int_{[\overline{\mathcal{M}}_g(X,\beta)]^{vir}} \widetilde{\alpha}_1 \wedge \dots \wedge \widetilde{\alpha}_k \in \mathbb{Q}.$$

We view the Gromov–Witten invariant[6] $N_{g,\beta}$ as counting the curves in X which pass through the cycles α_i. The deformation invariance of $N_{g,\beta}$ follows from the construction of the virtual class. We are free to deform X and the cycles α_i in order to compute $N_{g,\beta}$.

The projective variety X may be viewed as a symplectic manifold with symplectic form obtained from the projective embedding. In fact, $N_{g,\beta}$ can be defined on any symplectic manifold X by picking a compatible almost complex

---

[4] Even if two submanifolds do not intersect transversally, the integral of the Poincaré dual cohomology class of one over the other still gives the correct topological intersection.

[5] The cohomological push-forward here uses the fact that π is an lci morphism. Alternatively flatness can be used [28].

[6] We drop the superscript GW when clear from context.

structure and using pseudo-holomorphic maps of curves. The resulting invariants do not depend on the choice of compatible almost complex structure, so define invariants of the symplectic structure.[7]

We can try to perturb the almost complex structures to make the moduli space transverse of the correct dimension. But even when embedded pseudo-holomorphic curves in X are well-behaved, their multiple covers invariably are not. Even within symplectic geometry, the correct treatment of Gromov–Witten theory currently involves virtual classes.

Advantages

Gromov–Witten theory is defined for spaces X of all dimensions and has been proved to be a symplectic invariant (unlike most of the theories we will describe below). As the first deformation invariant theory constructed, Gromov–Witten theory has been intensively studied for more than 20 years – by now there are many exact calculations and significant structural results related to integrable hierarchies and mirror symmetry.

Since the moduli space of stable maps $\overline{\mathcal{M}}_g(X, \beta)$ lies over the moduli space $\overline{\mathcal{M}}_g$ of stable curves, Gromov–Witten theory is intertwined with the geometry of $\overline{\mathcal{M}}_g$. For a famous early application, see [55]. Relations in the cohomology of $\overline{\mathcal{M}}_{g,n}$ yield universal differential equations for the generating functions of Gromov–Witten invariants. The most famous case is the WDVV equation [26, 103] obtained by the linear equivalence of the boundary strata of $\overline{\mathcal{M}}_{0,4}$. The WDVV equation implies the associativity of the quantum cohomology ring of X defined via the genus 0 Gromov–Witten invariants. For example, associativity for $\mathbb{P}^3$ implies 80160 twisted cubics meet 12 general lines [25, 32]. Higher genus relations such as Getzler's [33] in genus 1 and the BP equation [8] in genus 2 also exist.

Gromov–Witten theory has links in many directions. When X is a curve, Gromov–Witten theory is related to counts of Hurwitz covers [75]. For the Severi degrees of curves in $\mathbb{P}^2$ and $\mathbb{P}^1 \times \mathbb{P}^1$, Gromov–Witten theory agrees with naive counts (when the latter are sensible). For surfaces of general type, Gromov–Witten theory links beautifully with Seiberg–Witten theory [95]. For 3-folds, there is a subtle and surprising relationship between Gromov–Witten theory and the sheaf counting theories discussed here in later sections. The relation with mirror symmetry [15, 34, 66] is a high point of the subject.

---

[7] The role of the symplectic structure in the definition of the invariants is well hidden. Via Gromov's results, the symplectic structure is crucial for the compactness of the moduli space of stable maps.

Drawbacks

The theory is extremely hard to compute: even the Gromov–Witten theories of varieties of dimensions 0 and 1 are very complicated. The theory of a point is related to the KdV hierarchy [104], and the theory of $\mathbb{P}^1$ is related to the Toda hierarchy [75]. While such connections are beautiful, using Gromov–Witten theory to actually count curves is difficult, essentially due to the nonlinearity of maps from curves to varieties. The sheaf theories considered in the next sections concern more linear objects.

Because of the finite automorphisms of stable maps, Gromov–Witten invariants are typically rational numbers. An old idea in Gromov–Witten theory is that underlying the rational Gromov–Witten invariants should be integer-valued curve counts. For instance, consider a stable map $f \in \overline{\mathcal{M}}_g(X, \beta)$ double covering an image curve $C \subset X$ in class $\beta/2$. Suppose, for simplicity, f and C are rigid and unobstructed. Then, f counts $1/2$ towards the Gromov–Witten invariant $N_{g,\beta}(X)$ because of its $\mathbb{Z}/2$-stabiliser. Underlying this rational number is an integer 1 counting the embedded curve C in class $\beta/2$.

Serious difficulties

For the case of 3-folds, Gromov–Witten theory is *not* enumerative in the naive sense in genus $g > 0$ due to degenerate contributions. The departure from naive counting happens already in positive genus for $\mathbb{P}^3$.

Let X be a 3-fold. The formula for the expected dimension of the moduli space of stable maps (6.6) is not genus dependent. Consider a nonsingular embedded rigid rational curve

$$\mathbb{P}^1 \subset X \tag{6.7}$$

in homology class β. The curve not only contributes 1 to $N_{0,\beta}$, but also contributes in a complicated way to $N_{g \geq 1,\beta}$. By attaching to the $\mathbb{P}^1$ any stable curve C at a nonsingular point, we obtain a stable map in the same class β which collapses C to a point. The contribution of (6.7) to the Gromov–Witten invariants $N_{g \geq 1,\beta}$ of X must be computed via integrals over the moduli spaces of stable curves. The latter integrals are hard to motivate from the point of view of curve counting.

A rather detailed study of the *Hodge integrals* over the moduli spaces of curves which arise in such degenerate contributions in Gromov–Witten theory has been pursued [29, 76]. A main outcome has been an understanding of the relationship of Gromov–Witten theory to naive curve counting on 3-folds in the Calabi–Yau and Fano cases. The conclusion is a precise conjecture expressing integer counts in terms of Gromov–Witten invariants (see the BPS conjecture in the next section). The sheaf counting theories developed later are now viewed as a more direct path to the integers underlying Gromov–Witten theory in dimension 3.

What happens in higher dimensions? Results of [53, 85] for spaces X of dimensions 4 and 5 show a similar underlying integer structure for Gromov–Witten theory. However, a direct interpretation of the integer counts (in terms of sheaves or other structures) in dimensions higher than 3 awaits discovery.

$2\frac{1}{2}$ Gopakumar–Vafa / BPS invariants

Invariants

BPS invariants were introduced for Calabi–Yau 3-folds by Gopakumar–Vafa in [35, 36] using an M-theoretic construction. The multiple cover calculations [29, 76] in Gromov–Witten theory provided basic motivation. The definitions and conjectures related to BPS states were generalised to arbitrary 3-folds in [76, 77]. While the original approach to the subject is not yet on a rigorous footing, the hope is to define curve counting invariants which avoid the multiple cover and degenerate contributions of Gromov–Witten theory. The BPS counts should be the integers underlying the rational Gromov–Witten invariants of 3-folds.

To simplify the discussion here, let X be a Calabi–Yau 3-fold. Gopakumar and Vafa consider a moduli space $\mathcal{M}$ of D-branes supported on curves in class β. While the precise mathematical definition is not clear, for a nonsingular embedded curve $C \subset X$ of genus g and class

$$[C] = \beta \in H_2(X, \mathbb{Z}),$$

the D-branes are believed to be (the pushforward to X of) line bundles on C of a fixed degree, with moduli space a Jacobian torus diffeomorphic to T^{2g}. For singular curves, the D-brane moduli space should be a type of relative compactified Jacobian over the "space" of curves of class β.

Mathematicians have tended to interpret $\mathcal{M}$ as a moduli space of stable sheaves with 1-dimensional support in class β and holomorphic Euler characteristic $\chi = 1$. The latter condition is a technical device to rule out strictly semistable sheaves. Over nonsingular curves C, the moduli space is simply $\mathrm{Pic}_g(C)$. For singular curves, more exotic sheaves in the compactified Picard scheme arise. For nonreduced curves, we can find higher rank sheaves supported on the underlying reduced curve. The support map $\mathcal{M} \to B$, taking such a sheaf to the underlying support curve, is also required for the geometric path to the BPS invariants. Here, B is an appropriate (unspecified) parameter space of curves in X. For instance, there is certainly such a support map to the Chow variety of 1-cycles in X.

Let us now imagine that we are in the ideal situation where the parameter space $B = \coprod_i B_i$ is a disjoint union of connected components over which the map $\mathcal{M} \to B$ is a product,

$$\mathcal{M} = \coprod \mathcal{M}_i \quad \text{and} \quad \mathcal{M}_i = B_i \times F_i$$

with fibres F_i. The supposition is not ridiculous: the virtual dimension of curves in a Calabi–Yau 3-fold is 0, so we might hope that B is a finite set of points. Then,

$$H^*(\mathcal{M}) = \bigoplus_i H^*(B_i) \otimes H^*(F_i). \tag{6.8}$$

When each B_i parameterises nonsingular curves of genus g_i only,

$$H^*(F_i) = H^*(T^{2g_i}) = (H^*(S^1))^{\otimes 2g_i}$$

has normalised Poincaré polynomial

$$P_y(F_i) = y^{-g_i}(1+y)^{2g_i}.$$

Here, we normalise by shifting cohomological degrees by $-\dim_{\mathbb{C}}(F_i)$ to make $P_y(F_i)$ symmetric about degree 0. Then, $P_y(F_i)$ is a palindromic Laurent polynomial invariant under $y \leftrightarrow y^{-1}$ by Poincaré duality.

For more general F_i, the normalised Poincaré polynomial $P_y(F_i)$ is again invariant under $y \leftrightarrow y^{-1}$ if $H^*(F_i)$ satisfies even dimensional Poincaré duality. Therefore, $P_y(F_i)$ may be written as a finite integral combination of terms $y^{-r}(1+y)^{2r}$, since the latter form a basis for the palindromic Laurent polynomials. Thus we can express $H^*(F_i)$ as a virtual combination of cohomologies of even dimensional tori. For instance, a cuspidal elliptic curve is topologically S^2 with

$$P_y = y^{-1}(1+y^2) = (y^{-1} + 2 + y) - 2 = P_y(T^2) - 2P_y(T^0).$$

Cohomologically, we interpret the cuspidal elliptic curve as 1 Jacobian of a genus 1 curve minus 2 Jacobians of genus 0 curves.

To tease the "number of genus r curves" in class β from (6.8), Gopakumar and Vafa write

$$\sum_i (-1)^{\dim B_i} e(B_i) P_y(F_i) \quad \text{as} \quad \sum_r n_r(\beta) y^{-r}(1+y)^{2r} \tag{6.9}$$

and define the integers $n_r(\beta)$ to be the BPS invariants counting genus r curves in class β. In Section $3\frac{1}{2}$, we will see that when B is nonsingular and can be broken up into a finite number of points by a generic deformation, that number of points is $(-1)^{\dim B}e(B)$, see for instance (6.19). In other words, the virtual class of B consists of $(-1)^{\dim B}e(B)$ points, explaining the first term in (6.9).

The Künneth decomposition (6.8) does *not* hold for general $\mathcal{M} \to B$, but can be replaced by the associated Leray spectral sequence. According to [44], the *perverse* Leray spectral sequence on intersection cohomology is preferable since it collapses and its terms satisfy the Hard Lefschetz theorem (which replaces the Poincaré duality used above). At least when B is nonsingular, $\mathcal{M}$ is reduced with sufficiently mild singularities, and

$$\pi : \mathcal{M} \to B$$

is equidimensional of fibre dimension $f = \dim \mathcal{M} - \dim B$, we can take

$$y^{-f} \sum_j (-1)^{\dim B} e({}^p R^j \pi_* \mathcal{IC}(\mathbb{C})) y^j = \sum_r n_r(\beta) y^{-r}(1+y)^{2r} \qquad (6.10)$$

as the Hosono–Saito–Takahashi definition[8] of the BPS invariants $n_r(\beta)$.

The entire preceding discussion of BPS invariants is only motivational. We have not been precise about the definition of the moduli space B. Moreover, the hypotheses imposed in the above constructions are rarely met (and when the hypotheses fail, the constructions are usually unreasonable or just wrong). Nevertheless, there should exist BPS invariants $n_{g,\beta} \in \mathbb{Z}$ "counting" curves of genus g and class β in X.

In addition to the M-theoretic construction, Gopakumar and Vafa have made a beautiful prediction of the relationship of the BPS counts to Gromov–Witten theory. For Calabi–Yau 3-folds, the conjectural formula is

$$\sum_{g\geq 0,\ \beta\neq 0} N_{g,\beta}^{\mathrm{GW}} u^{2g-2} v^\beta =$$

$$\sum_{g\geq 0,\ \beta\neq 0} n_{g,\beta} u^{2g-2} \sum_{d>0} \frac{1}{d}\left(\frac{\sin(du/2)}{u/2}\right)^{2g-2} v^{d\beta}. \qquad (6.11)$$

The trigonometric terms on the right are motivated by multiple cover formulas in Gromov–Witten theory [29, 76]. The entire geometric discussion can be bypassed by *defining* the BPS invariants via Gromov–Witten theory by equation (6.11). A precise conjecture [12] then arises.

---

[8]   The original sources [35, 36, 22] make a great deal of use of $\mathfrak{sl}_2 \times \mathfrak{sl}_2$-actions on the cohomology of $\mathcal{M}$, but the end result is equivalent to the above intuitive description: decompose the fibrewise cohomology of $\mathcal{M}$ into the cohomologies of Jacobian tori, then take signed Euler characteristics in the base direction.

BPS conjecture I. *For the $n_{g,\beta}$ defined via Gromov–Witten theory and formula* (6.11), *the following properties hold:*

(i) $n_{g,\beta} \in \mathbb{Z}$,
(ii) for fixed β, the $n_{g,\beta}$ vanish except for finitely many $g \geq 0$.

For other 3-folds X, when the virtual dimension is positive

$$\int_{\beta} c_1(X) > 0 \,,$$

incidence conditions to cut down the virtual dimension to 0 must be included. This case will be discussed in Section $5\frac{1}{2}$. The conjectural formula for the BPS counts is similar, see (6.34).

Advantages

For 3-folds, BPS invariants should be the ideal curve counts. The BPS invariants are integer valued and coincide with naive counts in many cases where the latter make sense. For example, the BPS counts (defined via Gromov–Witten theory) agree with naive curve counting in $\mathbb{P}^3$ in genus 0, 1, and 2. The definition via Gromov–Witten theory shows $n_{g,\beta}$ is a symplectic invariant.

For Calabi–Yau 3-folds X, the BPS counts do not always agree with naive counting. A trivial example is the slightly different treatment of an embedded super-rigid elliptic curve $E \subset X$, see [76]. Such an E contributes a single BPS count to each multiple degree $n[E]$. A much more subtle BPS contribution is given by a super-rigid genus 2 curve C in class 2[C] [13, 14]. We view BPS counting now as more fundamental than naive curve counting (and equivalent to, but *not* always equal to, naive counting).

Drawbacks

The main drawback is the murky foundation of the geometric construction of the BPS invariants. For nonreduced curves, the contributions of the higher rank moduli spaces of sheaves on the underlying support curves remain mysterious. The real strength of the theory will only be realised after the foundations are clarified. For example, properties (i) and (ii) of the BPS conjecture should be immediate from a geometric construction. The definition via Gromov–Witten theory is far from adequate.

A significant limitation of the BPS counts is the restriction to 3-folds. However, calculations [53, 85] show some hope of parallel structures in higher dimensions, see also [47].

Serious difficulties

The geometric foundations appear very hard to establish. There is no likely path in sight (except in genus 0 where Katz has made a rigorous proposal [49], see Section $4\frac{1}{2}$). The Hosono–Saito–Takahashi approach does not incorporate the virtual class (the term $(-1)^{\dim B} e(B)$ is a crude approximation for the virtual class of the base B) and fails in general.

Developments concerning motivic invariants [48, 26] and the categorification of invariants with cohomology theories instead of Euler characteristics appear somewhat closer to the methods required in the Calabi–Yau 3-fold case. For instance, Behrend has been working to categorify his constructible function [3] to give a perverse sheaf that could replace $\mathcal{IC}(\mathbb{C})$ in the HST definition, perhaps yielding a deformation invariant theory. Even then, why formula (6.11) should hold is a mystery.

An approach to BPS invariants via stable pairs (instead of Gromov–Witten theory) will be discussed in Section $4\frac{1}{2}$ below. The BPS invariants $n_{g,\beta}$ are there again defined by a formula similar to (6.11). The stable pairs perspective is better than the Gromov–Witten approach and has led to substantial recent progress [19, 71, 72, 92, 97]. Nevertheless, the hole in the subject left by the lack of a direct geometric construction is not yet filled.

$3\frac{1}{2}$ Donaldson–Thomas theory

Moduli

Instead of considering maps of curves into X, we can instead study embedded curves. Let a *subcurve* $Z \subset X$ be a subscheme of dimension 1. The *Hilbert scheme* compactifies embedded curves by allowing them to degenerate to arbitrary subschemes. Let $I_n(X, \beta)$ be the Hilbert scheme parameterising subcurves $Z \subset X$ with

$$\chi(\mathcal{O}_Z) = n \in \mathbb{Z} \quad \text{and} \quad [Z] = \beta \in H_2(X).$$

Here, χ denotes the holomorphic Euler characteristic and $[Z]$ denotes the class of the subcurve (involving only the 1-dimensional components). By the above conditions, $I_n(X, \beta)$ parameterises subschemes which are unions of possibly nonreduced curves and points in X.

We give a few examples to show how the Hilbert scheme differs from the space of stable maps. First, consider a family of nonsingular conics

$$C_{t \neq 0} = \{x^2 + ty = 0\} \subset \mathbb{C}^2 \tag{6.12}$$

as a local model which can, of course, be further embedded in any higher dimension. The natural limit as $t \to 0$,

$$C_0 = \{x^2 = 0\} \subset \mathbb{C}^2, \tag{6.13}$$

is indeed the limit in the Hilbert scheme.

The limit (6.13) is the y-axis with multiplicity two thickened in the x-direction.

In the stable map case, the limit of the family (6.12) is very different. There we take the limit of the associated map from $\mathbb{C}$ to C_t given by[9]

$$\xi \longmapsto (-t^{1/2}\xi, \xi^2).$$

The result is the *double cover* $\xi \longmapsto (0, \xi^2)$ of the y-axis. So the thickened scheme in the Hilbert scheme is replaced by the double cover. The latter is an orbifold point in the space of stable maps with $\mathbb{Z}/2$-stabiliser given by $\xi \longmapsto -\xi$.

In the next example, we illustrate the phenomenon of genus change which occurs only in dimension at least 3. A global model is given by a twisted rational cubic in $\mathbb{P}^3$ degenerating to a plane cubic of genus 1 [86]. An easier

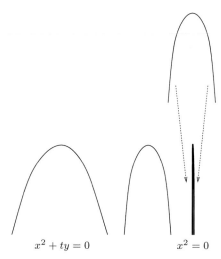

Figure 6.1. The degeneration (6.12) with the limiting stable map double covering $x = 0$.

$x^2 + ty = 0$ $x^2 = 0$

---

[9] The formula gives a well-defined map only modulo automorphisms of the curve – specifically the automorphism $\xi \longmapsto -\xi$.

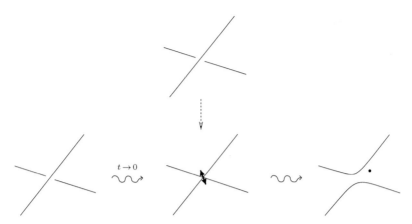

Figure 6.2. The family (6.14) with the subscheme limit below and the stable map limit above. On the right is a deformation of the limit subscheme with a free point breaking off.

local model $C_t \subset \mathbb{C}^3$ has two components: the x-axis in the plane $z = 0$, and the y-axis moved up into the plane $z = t$,

$$C_t = \{x = 0 = z\} \sqcup \{y = 0 = z - t\} \subset \mathbb{C}^3, \qquad (6.14)$$

see Figure 6.2. As a stable map, we take the associated inclusion of two copies of the affine line $\mathbb{C}$. The stable map limit at $t = 0$ takes the same domain $\mathbb{C} \sqcup \mathbb{C}$ onto the x- and y- axes, an embedding away from the origin where the map is 2:1. In other words, the limit stable map[10] is the normalisation of the image

$$\{xy = 0 = z\} \subset \mathbb{C}^3. \qquad (6.15)$$

In the Hilbert scheme, the limit of the family (6.14) is rather worse. The ideal of C_t is

$$(x, z) \cdot (y, z - t) = \big(xy, x(z - t), yz, z(z - t)\big).$$

We take the limit as $t \to 0$. The flat limit here happens to be the ideal generated by the limit of the above generators. The limit ideal does not contain z:

$$(xy, xz, yz, z^2) \subsetneq (xy, z). \qquad (6.16)$$

However, after multiplying z by any element of the maximal ideal (x, y, z) of the origin, we land inside the limit ideal. Therefore, the limit curve is given by $\{xy = 0 = z\}$ with a scheme-theoretic embedded point added at the origin

[10] There is another stable map given by the embedding of the image (6.15). In a compact global model, the latter would be a map from a nodal stable curve of genus one larger so would *not* feature in the compactification of the family we are considering.

pointing along the z-axis — in the direction along which the two components came together. The embedded point "makes up for" the point lost in the intersection and ensures that the family of curves is flat over $t = 0$.

In a further flat family, the embedded point can break off, and the curve can be smoothed $\{xy = \epsilon, z = 0\}$ to a curve of higher genus. In the Hilbert scheme, we have all 1-dimensional subschemes made up of curves *and* points, with curves of different genus balanced by extra free points. The constant n in $I_n(X, \beta)$ is $1 - g + k$ for a reduced curve of arithmetic genus g with k free and embedded points added, so we can increase g at the expense of increasing k by the same amount.

Deformation theory

Hilbert schemes of curves can have arbitrary dimensional components and terrible singularities. Worse still, the natural deformation/obstruction theory of the Hilbert scheme does not lead to a virtual class. However, if we restrict attention to 3-folds X and view $I_n(X, \beta)$ as a moduli space of *sheaves*, then we obtain a different obstruction theory which does admit a virtual class. The latter observation is the starting point of Donaldson–Thomas theory.

Given a 1-dimensional subscheme $Z \subset X$, the associated ideal sheaf $\mathscr{I}_Z$ is a stable sheaf with Chern character

$$(1, 0, -\beta, -n) \in H^0 \oplus H^2 \oplus H^4 \oplus H^6$$

and trivial determinant. Conversely, all such stable sheaves with trivial determinant can be shown to embed in their double duals $\mathcal{O}_X$ and thus are all ideal sheaves. Hence, $I_n(X, \beta)$ is a moduli space of sheaves, at least set theoretically. With more work, an isomorphism of schemes can be established, see [82, Theorem 2.7].

The moduli space of sheaves $I_n(X, \beta)$ also admits a virtual class [96, 68]. The main point is that deformations and obstructions are governed by

$$\mathrm{Ext}^1(\mathscr{I}_Z, \mathscr{I}_Z)_0 \quad \text{and} \quad \mathrm{Ext}^2(\mathscr{I}_Z, \mathscr{I}_Z)_0 \qquad (6.17)$$

respectively, where the subscript 0 denotes the trace-free part governing deformations with *fixed determinant*. Since $\mathrm{Hom}(\mathscr{I}_Z, \mathscr{I}_Z) = \mathbb{C}$ consists of only the scalars, the trace-free part vanishes. By Serre duality,

$$\mathrm{Ext}^3(\mathscr{I}_Z, \mathscr{I}_Z) \cong \mathrm{Hom}(\mathscr{I}_Z, \mathscr{I}_Z \otimes K_X)^* \cong H^0(K_X)^* \cong H^3(\mathcal{O}_X).$$

The last group $H^3(\mathcal{O}_X)$ is removed when taking trace-free parts. Hence, the terms (6.17) are the only nonvanishing trace-free Exts, and there are *no* higher

obstruction spaces. The Exts (6.17) govern a perfect obstruction theory of
virtual dimension equal to

$$\text{ext}^1(\mathscr{I}_Z, \mathscr{I}_Z)_0 - \text{ext}^2(\mathscr{I}_Z, \mathscr{I}_Z)_0 = \int_\beta c_1(X),$$

compare (6.6). If the virtual dimension is positive, insertions are needed to
produce invariants [68].

On Calabi–Yau 3-folds, moduli of sheaves admit a particularly nice
deformation-obstruction theory [27, 96]. The deformation and obstruction
spaces (6.17) are *dual* to each other,

$$\text{Ext}^2(\mathscr{I}_Z, \mathscr{I}_Z)_0 \cong \text{Ext}^1(\mathscr{I}_Z, \mathscr{I}_Z)_0^*, \tag{6.18}$$

by Serre duality. Any moduli space of sheaves on a Calabi–Yau 3-fold can be
realized as the critical locus of a holomorphic function on an ambient nonsingular space: the holomorphic Chern–Simons functional in infinite dimensions
[105, 27] or locally on an appropriate finite dimensional slice [48]. Since the
moduli space is the zero locus of a closed 1-form, the obstruction space is
the cotangent space at any point of moduli space. More generally, Behrend [3]
calls obstruction theories satisfying the global version of (6.18) *symmetric*. The
condition is equivalent to asking for the moduli space to be locally the zeros of
an *almost closed* 1-form on a smooth ambient space – a 1-form with exterior
derivative vanishing scheme theoretically on the moduli space.[11]

If the moduli space of sheaves is nonsingular (but of too high dimension),
then the symmetric obstruction theory forces the obstruction bundle to be
globally isomorphic to the cotangent bundle. The virtual class, here the top
Chern class of the obstruction bundle, is then just the signed topological Euler
characteristic of the moduli space

$$(-1)^{\dim I_n(X,\beta)} e(I_n(X, \beta)). \tag{6.19}$$

Remarkably, Behrend shows that for any moduli space $\mathcal{M}$ with a symmetric
obstruction theory there is a constructible function

$$\chi^B : \mathcal{M} \to \mathbb{Z}$$

with respect to which the weighted Euler characteristic gives the integral of
the virtual class [3]. Therefore, each point of the moduli space contributes
in a local way to the global invariant, by $(-1)^{\dim \mathcal{M}}$ for a nonsingular point
and by a complicated number taking multiplicities into account for singular points. When $\mathcal{M}$ is locally the critical locus of a function, the number is

---

[11] By [84], the condition is strictly weaker than asking for the moduli space to be locally the
zeros of a closed 1-form.

$(-1)^{\dim \mathcal{M}}(1 - e(F))$ where F is the Milnor fibre of our point. Unfortunately, how to find a parallel approach to the virtual class when X is not Calabi–Yau is not currently known.

Integration against the virtual class of $I_n(X, \beta)$ yields the Donaldson–Thomas invariants. In the Calabi–Yau case, no insertions are required:

$$I_{n,\beta} = \int_{[I_n(X,\beta)]^{vir}} 1 = e\big(I_n(X, \beta), \chi^B\big).$$

Since $I_n(X, \beta)$ is a scheme (ideal sheaves have no automorphisms) and $[I_n(X, \beta)]^{vir}$ is a cycle class with $\mathbb{Z}$-coefficients, the invariants $I_{n,\beta}$ are *integers*. Deformation invariance of $I_{n,\beta}$ follows from properties of the virtual class.

MNOP conjectures

A series of conjectures linking the Donaldson–Thomas theory of 3-folds to Gromov–Witten theory were advanced in [68, 69]. For simplicity, we restrict ourselves here to the Calabi–Yau case.

For fixed curve class $\beta \in H_2(X, \mathbb{Z})$, the Donaldson–Thomas *partition function* is

$$Z_\beta^{\mathrm{DT}}(q) = \sum_n I_{n,\beta} q^n .$$

Since $I_n(X, \beta)$ is easily seen to be empty for n sufficiently negative, the partition function is a Laurent series in q. To count just curves, and not points and curves, MNOP form the *reduced* generating function [68] by dividing by the contribution of just points:

$$Z_\beta^{\mathrm{red}}(q) = \frac{Z_\beta^{\mathrm{DT}}(q)}{Z_0^{\mathrm{DT}}(q)} . \tag{6.20}$$

MNOP first conjectured the degree $\beta = 0$ contribution can be calculated as

$$Z_0^{\mathrm{DT}}(q) = M(-q)^{e(X)} ,$$

where M is the MacMahon function,

$$M(q) = \prod_{n \geq 1} (1 - q^n)^{-n} ,$$

the generating function for 3d partitions. Proofs can now be found in [7, 63, 60]. Second, MNOP conjectured $Z_\beta^{\mathrm{red}}(q)$ is the Laurent expansion of a *rational function* in q, invariant[12] under $q \leftrightarrow q^{-1}$. Therefore, we can substitute

---

[12] The Laurent series itself need not be $q \leftrightarrow q^{-1}$ invariant. For instance the rational function $q(1 + q)^{-2}$ is invariant, but the associated Laurent series $q - 2q^2 + 3q^3 - \dots$ is not.

$q = -e^{iu}$ and obtain a real-valued function of u. The main conjecture of MNOP in the Calabi–Yau case is the following.

GW/DT Conjecture: $Z_\beta^{GW}(u) = Z_\beta^{red}(-e^{iu})$.

The conjecture asserts a precise equivalence relating Gromov–Witten to Donaldson–Thomas theory. Here,

$$Z_\beta^{GW}(u) = \sum_{g \geq 0} N_{g,\beta}^\bullet \, u^{2g-2}$$

is the generating function of *disconnected* Gromov–Witten invariants $N_{g,\beta}^\bullet$ defined just as in Section $1\frac{1}{2}$ by relaxing the condition that the curves be connected, but excluding maps which contract connected components to points. Equivalently, $Z_\beta^{GW}(u)$ is the exponential of the generating function of connected Gromov–Witten invariants $N_{g,\beta}$,

$$\sum_{\beta \neq 0} Z_\beta^{GW}(u) v^\beta = \sum_{\beta \neq 0, \, g \geq 0} N_{g,\beta}^\bullet \, u^{2g-2} v^\beta = \exp\left(\sum_{\beta \neq 0, \, g \geq 0} N_{g,\beta} \, u^{2g-2} v^\beta \right).$$

A version of the GW/DT correspondence with insertions for non Calabi–Yau 3-folds can be found in [69]. Various refinements, involving theories relative to a divisor, or equivariant with respect to a group action, are also conjectured. All of these conjectures have been proved for toric 3-folds in [70].

The GW/DT conjecture should be viewed as involving an analytic continuation and series expansion about two different points ($q = 0$ and $q = -1$, corresponding to $u = 0$). Therefore, the conjecture cannot be understood term by term[13] – to determine a single invariant on one side of the conjecture, knowledge of all of the invariants on the other side is necessary.

The overall shape of the conjecture is clear: the two different ways of counting curves in a fixed class β are entirely equivalent, with *integers* determining the Gromov–Witten invariants of 3-folds. By [82, Theorem 3.20], the integrality prediction of the GW/DT correspondence is entirely equivalent to the integrality prediction of the Gopakumar–Vafa formula (6.11).

Advantages

The integrality of the invariants is a significant advantage of using the Hilbert schemes $I_n(X, \beta)$ to define a counting theory. Also, the virtual counting of

---

[13] When combined with the Gopakumar–Vafa formula (6.11) and the relationship to the stable pairs discussed below, the GW/DT conjecture will become rather more comprehensible, see (6.29).

subschemes, at least in the Calabi–Yau 3-fold case, fits into the larger context of counting higher rank bundles, sheaves, and objects of the derived category of X. The many recent developments in wall-crossing [48, 26] apply to this more general setting. We will see an example in the next section.

Behrend's constructible function sometimes makes computations (in the Calabi–Yau case at least) more feasible – we can use cut and paste techniques to reduce to more local calculations. See for instance [4].

Drawbacks

The theory only works for nonsingular projective varieties of dimension at most 3. While the Hilbert scheme of curves is always well-defined, the deformation/obstruction theory fails to be 2-term in higher dimensions. By contrast, Gromov–Witten theory is well-defined in all dimensions and is proved to be a symplectic invariant. While we expect Donaldson–Thomas theory to have a fully symplectic approach, how to proceed is not known.

In Gromov–Witten theory, the genus expansion makes a connection to the moduli of curves (independent of X). The Euler characteristic n plays a parallel role in Donaldson–Thomas theory, but is much less useful. While there are very good low genus results in Gromov–Witten theory, there are few analogues for the Hilbert scheme.

Behrend's constructible function approach for the Calabi–Yau case is difficult to use. For example, the constructible functions even for toric Calabi–Yau 3-folds have not been determined.[14] So far, Behrend's theory has been useful mainly for formal properties related to motivic invariants and wall-crossing. For more concrete calculations involving Behrend's functions see [4, 5].

Serious difficulties

For the GW/DT correspondence, the division by $Z_0(q)$ confuses the geometric interpretation of the invariants. In fact, the subschemes of X with free points make the theory rather unpleasant to work with. This "compactification" of the space of embedded curves is much larger than the original space, adding enormous components with free points. In practice, the free points lead to constant technical headaches (which play little role in the main development of the invariants).

It is tempting to think of working with the closure of the "good components" of the Hilbert scheme instead, but such an approach would not

---

[14] Amazingly, we do not even know whether the constructible functions are non-constant in the toric Calabi–Yau case!

have a reasonable deformation theory nor a virtual class. However, a certain birational modification of the idea does work and will be discussed in the next section.

$4\frac{1}{2}$ Stable pairs

Limits revisited

Consider again the family of Figure 6.2. For $t \neq 0$, denote the disjoint union (6.14) by

$$C_t = C_t^1 \cup C_t^2 \,.$$

The ideal sheaf $\mathcal{I}_{C_t}$, central to the Hilbert scheme analysis, is just the kernel of the surjection

$$\mathcal{O}_X \to \mathcal{O}_{C_t^1} \oplus \mathcal{O}_{C_t^2} \,. \tag{6.21}$$

For the moduli of stable pairs, the map itself (not just the kernel) will be fundamental. We will take a natural limit of the map given by

$$\mathcal{O}_X \to \mathcal{O}_{C_0^1} \oplus \mathcal{O}_{C_0^2} \tag{6.22}$$

where the limits of the component curves are

$$C_0^1 = \{x = 0 = z\} \quad \text{and} \quad C_0^2 = \{y = 0 = z\} \,.$$

The result is a map which is *not* a surjection at the origin (where C_1 and C_2 intersect and the sheaf on the right has rank 2). In the limit, there is a nonzero cokernel, the structure sheaf of the origin $\mathcal{O}_0$, which accounts for the extra point lost in the intersection. Losing surjectivity replaces the embedded point arising in the limit of ideal sheaves (6.16).

The cokernels of the above maps (6.21) are *not* flat over $t = 0$ even though the sheaves $\mathcal{O}_{C_t^1} \oplus \mathcal{O}_{C_t^2}$ *are* flat. Similarly the kernels of the maps (6.21) are *not* flat over $t = 0$. In fact, at $t = 0$, we get the ideal (xy, z) of $C_0^1 \cup C_0^2$ which we already saw in (6.16) is not the flat limit of the ideal sheaves of C_t.

Moduli

The limit (6.22) is an example of a stable pair. The moduli of stable pairs provides a different sheaf-theoretic compactification of the space of embedded curves. The moduli space is intimately related to the Hilbert scheme, but is much more efficient.

Let X be a nonsingular projective 3-fold. A *stable pair* (F, s) is a coherent sheaf F with dimension 1 support in X and a section $s \in H^0(X, F)$ satisfying the following stability condition:

- F is *pure*, and
- the section s has 0-dimensional cokernel.

Let C be the scheme-theoretic support of F. Condition (i) means all the irreducible components of C are of dimension 1 (no 0-dimensional components). By [82, Lemma 1.6], C has no embedded points. A stable pair

$$\mathcal{O}_X \to F$$

therefore defines a Cohen–Macaulay curve C via the kernel $\mathscr{I}_C \subset \mathcal{O}_X$ and a 0-dimensional subscheme of C via the support of the cokernel.[15]

To a stable pair, we associate the Euler characteristic and the class of the support C of F,

$$\chi(F) = n \in \mathbb{Z} \quad \text{and} \quad [C] = \beta \in H_2(X, \mathbb{Z}) .$$

For fixed n and β, there is a projective moduli space of stable pairs $P_n(X, \beta)$ [82, Lemma 1.3] by work of Le Potier [59]. While the Hilbert scheme $I_n(X, \beta)$ is a moduli space of curves plus free and embedded points, $P_n(X, \beta)$ should be thought of as a moduli space of curves plus points *on the curve* only. Even though points still play a role (as the example (6.14) shows), the moduli of stable pairs is much smaller than $I_n(X, \beta)$.

Deformation theory

To define a flexible counting theory, a compactification of the family of curves in X should admit a 2-term deformation/obstruction theory and a virtual class. As in the case of $I_n(X, \beta)$, the most immediate obstruction theory of $P_n(X, \beta)$ does *not* admit such a structure. For $I_n(X, \beta)$, a solution was found by considering a subscheme C to be equivalent to a sheaf $\mathscr{I}_C$ with trivial determinant. For $P_n(X, \beta)$, we consider a stable pair to define an object of $D^b(X)$, the quasi-isomorphism equivalence class of the complex

$$I^\bullet = \{\mathcal{O}_X \xrightarrow{\ s\ } F\} . \tag{6.23}$$

For X of dimension 3, the object $I^\bullet$ determines the stable pair [82, Proposition 1.21], and the fixed-determinant deformations of $I^\bullet$ in $D^b(X)$ match those of

---

[15] When C is Gorenstein (for instance if C lies in a nonsingular surface), stable pairs supported on C are in bijection with 0-dimensional subschemes of C. More precise scheme theoretic isomorphisms of moduli spaces are proved in [83, Appendix B].

the pair (F, s) to all orders [82, Theorem 2.7]. The latter property shows the scheme $P_n(X, \beta)$ may be viewed as a moduli space of objects in the derived category.[16] We can then use the obstruction theory of the complex $I^\bullet$ in place of the obstruction theory of the pair.

The deformation/obstruction theory for complexes is governed at $[I^\bullet] \in P_n(X, \beta)$ by

$$\text{Ext}^1(I^\bullet, I^\bullet)_0 \quad \text{and} \quad \text{Ext}^2(I^\bullet, I^\bullet)_0. \tag{6.24}$$

Formally, the outcome is parallel to (6.17). The obstruction theory (6.24) has all the attractive properties of the Hilbert scheme case: 2 terms, a virtual class of dimension $\int_\beta c_1(X)$, and a description via the χ^B-weighted Euler characteristics in the Calabi–Yau case.

Invariants

After imposing incidence conditions (when $\int_\beta c_1(X)$ is positive) and integrating against the virtual class, we obtain stable pairs invariants for 3-folds X. In the Calabi–Yau case, the invariant is the length of the virtual class:

$$P_{n,\beta} = \int_{[P_n(X,\beta)]^{vir}} 1 = e(P_n(X, \beta), \chi^B).$$

For fixed curve class $\beta \in H_2(X, \mathbb{Z})$, the stable pairs partition function is

$$Z^P_\beta(q) = \sum_n P_{n,\beta} q^n.$$

Again, elementary arguments show the moduli spaces $P_n(X, \beta)$ are empty for sufficiently negative n, so Z^P_β is a Laurent series in q. Since the free points are now confined to the curve instead of roaming over X, we do not have to form a reduced series as in (6.20). In fact, we conjecture [82, Conjecture 3.3] the partition function Z^P_β to be precisely the reduced theory of Section $3\frac{1}{2}$.

DT/Pairs Conjecture: $Z^P_\beta(q) = Z^{red}_\beta(q).$

The DT/Pairs correspondence is expected for all 3-folds X with the incidence conditions playing no significant role [69]. Using the definition $Z^{red}_\beta = Z^{DT}_\beta / Z^{DT}_0$, we find

$$\sum_m P_{n-m,\beta} \cdot I_{m,0} = I_{n,\beta}. \tag{6.25}$$

[16] Studying the moduli of objects in the derived category is a young subject. Usually, such constructions lead to Artin stacks. The space $P_n(X, \beta)$ is a rare example where a component of the moduli of objects in the derived category is a scheme (uniformly for all 3-folds X).

Relation (6.25) should be interpreted as a wall-crossing formula for counting invariants in the derived category of coherent sheaves $D^b(X)$ under a change of stability condition.

For invariants of Calabi–Yau 3-folds, wall-crossing has been studied intensively in recent years, and we give only the briefest of descriptions. Ideal sheaves parameterised by $I_n(X, \beta)$ are Gieseker stable. We can imagine changing the stability condition[17] to destabilise the ideal shaves with free and embedded points. If Z is a 1-dimensional subscheme, then Z has a maximal pure dimension 1 subscheme C defining a sequence

$$0 \to \mathscr{I}_Z \to \mathscr{I}_C \to Q \to 0,$$

where Q is the maximal 0-dimensional subsheaf of $\mathcal{O}_Z$. In $D^b(X)$, we equivalently have the exact triangle

$$Q[-1] \to \mathscr{I}_Z \to \mathscr{I}_C. \tag{6.26}$$

We can imagine the stability condition crossing a wall on which the phase (or slope) of $Q[-1]$ equals that of $\mathscr{I}_C$. On the other side of the wall, $\mathscr{I}_Z$ will be destabilised by (6.26). Meanwhile, extensions E in the opposite direction

$$\mathscr{I}_C \to E \to Q[-1] \tag{6.27}$$

will become stable. But stable pairs are just such extensions! The exact sequence

$$0 \to \mathscr{I}_C \to \mathcal{O}_X \xrightarrow{s} F \to Q \to 0$$

yields the exact triangle

$$\mathscr{I}_C \to I^\bullet \to Q[-1].$$

The moduli space of pairs $P_n(X, \beta)$ should give precisely the space of stable objects for the new stability condition.

The formula (6.25) for $I_{n,\beta} - P_{n,\beta}$ should follow from the more general wall-crossing formulae of [48, 26]. The mth term in (6.25) is the correction from subschemes Z whose maximal 0-dimensional subscheme (or total number of free and embedded points) is of length m. It involves both the space $\mathbb{P}(\mathrm{Ext}^1(\mathscr{I}_C, Q[-1]))$ of extensions (6.26) and the space $\mathbb{P}(\mathrm{Ext}^1(Q[-1], \mathscr{I}_C))$ of extensions (6.27). Though both are hard to control, they contribute to the wall-crossing formula through the difference in their Euler characteristics,[18] which is the topological number

---

[17] Ideally, we would work with Bridgeland stability conditions [9], but that is not currently possible. The above discussion can be made precise using the limiting stability conditions of [1, 98], or even Geometric Invariant Theory [94].

[18] Really, we need to weight by the restriction of the Behrend function χ^B. To make the above analysis work then requires χ^B to satisfy the identities of [48, 26]. In fact, the automorphisms of Q make the matter much more complicated than we have suggested.

$$\chi(\mathscr{I}_C, Q) = \text{length}(Q) = m.$$

The above sketch has now been carried out at the level of (unweighted) Euler characteristics [99, 94] and for χ^B-weighted Euler characteristics in [10] in the Calabi–Yau case. The upshot is the DT/Pairs conjecture is now proved for Calabi–Yau 3-folds. The rationality of $Z_\beta^{\text{red}}(q)$ and the symmetry under $q \leftrightarrow q^{-1}$ is also proved [10].

Example

Via the Behrend weighted Euler characteristic approach to the invariants of a Calabi–Yau 3-fold, we can talk about the contribution of a single curve $C \subset X$ to the stable pairs generating function $Z_\beta^P(q)$. No such discussion is possible in Gromov–Witten theory.

If C is nonsingular of genus g, then the stable pairs supported on C with $\chi = 1 - g + n$ are in bijection with $\text{Sym}^n C$ via the map taking a stable pair to the support of the cokernel Q. Therefore, C contributes[19]

$$Z_C^P(q) = c \sum_n (-1)^{n-g} e(\text{Sym}^n C) q^{1-g+n} = c(-1)^g q^{1-g}(1+q)^{2g-2}.$$

(6.28)

The rational function on the right is invariant under $q \leftrightarrow q^{-1}$. We view the symmetry as a manifestation of Serre duality (discussed below). Control of the free points in stable pair theory makes the geometry more transparent. The same calculation for $Z_C^{\text{red}}(q)$ based on the Hilbert scheme is much less enlightening. The above calculation is closely related to the BPS conjecture for stable pairs.

Stable pairs and BPS invariants

By a formal argument [82, Section 3.4], the stable pairs partition function can be written uniquely in the following special way:

$$Z^P(q,v) := 1 + \sum_{\beta \neq 0} Z_\beta^P(q) v^\beta$$

$$= \exp\left(\sum_r \sum_{\gamma \neq 0} \sum_{d \geq 1} \tilde{n}_{r,\gamma} \frac{(-1)^{(1-r)}}{d} (-q)^{d(1-r)}(1-(-q)^d)^{2r-2} v^{d\gamma} \right),$$

where the $\tilde{n}_{r,\gamma}$ are integers and vanish for fixed γ and r sufficiently large.

[19] The Behrend function restricted to $\text{Sym}^n C$ can be shown [83, Lemma 3.4] to be the constant $(-1)^{n-g}c$, where $c = \chi^B(\mathcal{O}_C)$ is the Behrend function of the moduli space of torsion sheaves evaluated at $\mathcal{O}_C$.

We can compose the various conjectures to link the BPS counts of Gopakumar and Vafa to the stable pairs invariants. The form we get from the conjectures is almost exactly as above:

$$Z^P(q, v) =$$
$$\exp\left(\sum_{r\geq 0}\sum_{\gamma\neq 0}\sum_{d\geq 1} n_{r,\gamma}\frac{(-1)^{(1-r)}}{d}(-q)^{d(1-r)}(1-(-q)^d)^{2r-2}v^{d\gamma}\right).$$

The only difference is the restriction on the r summation. Hence, we can *define* the BPS state counts by stable pairs invariants via the $\tilde{n}_{r,\gamma}$!

BPS conjecture II. *For the $\tilde{n}_{r,\beta}$ defined via stable pairs theory, the vanishing*

$$\tilde{n}_{r<0,\beta} = 0$$

holds for $r < 0$.

By its construction, the approach to defining the BPS states counts via stable pairs satisfies the full integrality condition and half of the finiteness of BPS conjecture I. We therefore regard the stable pairs perspective as better than the path via Gromov–Witten theory. Still, a direct construction of the BPS invariants along the lines discussed in Section $2\frac{1}{2}$ would be best of all.[20]

For irreducible classes,[21] the BPS formula for the stable pairs invariants can be written as

$$Z^P_\beta(q) = \sum_{r\geq 0}^{g} n_{r,\beta}\, q^{1-r}(1+q)^{2r-2}, \tag{6.29}$$

with $n_{r,\beta} = 0$ for all sufficiently large r. There is a beautiful interpretation of (6.29) in the light of (6.28): to the stable pairs invariants, the curves in class β look like a disjoint union of a finite number $n_{r,\beta}$ of nonsingular curves of genus r.

We can prove directly that the partition function Z^P_β can be written in the form (6.29). For $r\geq 1$, the functions $q^{1-r}(1+q)^{2r-2}$,

$$1, \qquad q^{-1}+2+q, \qquad q^{-2}+4q^{-1}+6+4q+q^2, \qquad q^{-3}+\ldots$$

form a natural $\mathbb{Z}$-basis for the Laurent polynomials invariant under $q \leftrightarrow q^{-1}$. For $r = 0$, the coefficients of the Laurent series do not satisfy the same symmetry,

---

[20] For curves with only reduced plane curve singularities, both constructions of BPS numbers have been shown to coincide after making the $\chi^B = (-1)^{\dim}$ approximation to the virtual class [71, 72].

[21] A class $\beta \in H_2(X, \mathbb{Z})$ is irreducible if it cannot be written as a sum $\alpha + \gamma$ of nonzero classes containing algebraic curves.

$$q(1+q)^{-2} = q - 2q^2 + 3q^3 - 4q^4 + \ldots = \sum_{n \geq 1} (-1)^{n-1} n q^n.$$

To prove (6.29), it is therefore equivalent to show the coefficients $P_{n,\beta}$ of the partition function satisfy not the $q \leftrightarrow q^{-1}$ symmetry but

$$P_{n,\beta} = P_{-n,\beta} + c(-1)^{n-1} n \qquad (6.30)$$

for some constant c.

Relation (6.30) is a simple consequence of Serre duality for the fibres of the Abel–Jacobi map. By forgetting the section, we obtain a map from stable pairs to stable sheaves,[22]

$$P_n(X, \beta) \quad \longrightarrow \quad \mathcal{M}_n(X, \beta),$$

$$(F, s) \quad \longmapsto \quad F.$$

The fibre of the map is $\mathbb{P}(H^0(F))$ with weighted Euler characteristic[23] $(-1)^{n-1} c \cdot h^0(F)$. There is an isomorphism

$$\mathcal{M}_n(X, \beta) \quad \longrightarrow \quad \mathcal{M}_{-n}(X, \beta),$$

$$F \quad \longmapsto \quad F^\vee,$$

where $F^\vee = \mathcal{E}xt^2(F, K_X)$. If F is the push-forward of a line bundle L from a nonsingular curve C, then $F^\vee$ is the push-forward of $L^* \otimes \omega_C$, see [83] for details. The fibre $\mathbb{P}(H^0(F^\vee))$ over $F^\vee$ is $\mathbb{P}(H^1(F)^*)$ by Serre duality, with weighted Euler characteristic $(-1)^{-n-1} c \cdot h^1(F)$.

To prove relation (6.30), we calculate the difference between the two above contributions:

$$(-1)^{n-1} c(h^0(F) - h^1(F)) = (-1)^{n-1} c \chi(F) = (-1)^{n-1} cn.$$

Summation over the space of stable sheaves (in the sense of Euler characteristics) yields the relation

$$P_{n,\beta} - P_{-n,\beta} = (-1)^{n-1} n \, e(\mathcal{M}_n(X, \beta), c). \qquad (6.31)$$

The weighted Euler characteristics

$$e(\mathcal{M}_n(X, \beta), c) = e(\mathcal{M}_{n+1}(X, \beta), c) \qquad (6.32)$$

---

[22] The irreducibility of β implies the sheaves with arise are stable since F has rank 1 on its irreducible support.

[23] As proved in [83, Theorem 4], the Behrend function is constant on $\mathbb{P}(H^0(F))$ with value $(-1)^{n-1} c$ where $c = \chi^B(\mathcal{O}_C)$. On a first reading, the Behrend function can be safely ignored here.

are independent of n: tensoring with a degree 1 line bundle relates sheaves supported on C with $\chi = n$ to those with $\chi = n + 1$. We have proved the relation (6.30).

The above argument shows the coefficient $n_{0,\beta}$ of $q(1 + q)^{-2}$ is the χ^B-weighted Euler characteristic of $\mathcal{M}_n(X, \beta)$. In fact, Katz [49] had previously proposed the DT invariant of $\mathcal{M}_1(X, \beta)$ as a good definition of $n_{0,\beta}$ for *any* class β, not necessarily irreducible. Naively, Katz's definition sees only the rational curves because for a curve of higher genus the action of the Jacobian on the moduli space of sheaves forces the (weighted) Euler characteristic of the latter to be zero. Katz's proposal can be viewed as a weak analogue of the genus by genus methods in Gromov–Witten theory.

Identity (6.31) is easily seen to be another wall-crossing formula [1, 83, 98]. In [100], Toda has extended the above analysis to all curve classes by extending the methods of Joyce [48] and the ideas of Kontsevich and Soibelman [58] on BPS formulations of general sheaf counting. His main result reduces BPS conjecture II to an analogue of identity (6.32) for DT invariants for dimension 1 sheaves for *all* classes β.[24]

Advantages

The stable pair theory has the advantages of the ideal sheaf theory – integer invariants conjecturally equivalent to the rational Gromov–Witten

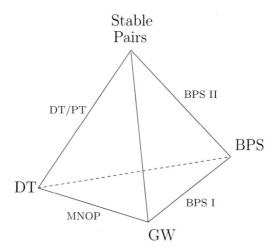

Figure 6.3. Conjectures connecting curve counting theories

---

[24] When β is not irreducible, sheaf stability issues change the definition of the DT invariant, see [100, Conjecture 6.3] for details.

invariants – but with the bonus of eliminating the free points on X. The geometry of the BPS conjectures is more clearly explained by stable pairs than any of the other approaches.

If descendent insertions (coming from higher Chern classes of the tautological bundles) are considered, the theory of stable pairs behaves much better than the parallel constructions for the moduli of stable maps or the Hilbert scheme. For example, the descendent partition functions for stable pairs are rational in q. See [80] for proofs in toric cases and further discussion.

At least for 3-folds, stable pairs appears to be the best counting theory to consider at the moment. The main hope for a better approach lies in the direct geometric construction of the BPS counts.

Drawbacks

Just as for the Donaldson–Thomas theory of ideal sheaves, the stable pairs invariants have only been constructed on nonsingular projective varieties of dimension at most 3. While we expect a parallel theory for symplectic invariants, how to proceed is not clear.

Serious difficulties

In the theory of stable pairs, free points are allowed to move along the support curve C. The free points are necessary to probe the geometry of the curve (and the associated BPS contributions in all genera) but in a rather roundabout way. An alternative opened up by the Behrend function might be to work with open moduli spaces (on which the arithmetic genus does not jump), but deformation invariance then becomes problematic.

As we have said repeatedly, a rigorous and sheaf theoretic approach to BPS invariants (at least for 3-folds and possibly in higher dimensions as well) would be highly desirable.

$5\frac{1}{2}$ Stable unramified maps

Singularities of maps

A difficulty which arises in Gromov–Witten theory is the abundance of collapsed components. In the moduli space of higher genus stable maps to $\mathbb{P}^1$ of degree 1, the entire complexity comes from such collapsed components attached to a degree 1 map of a genus 0 curve to $\mathbb{P}^1$. Collapsed contributions have to be removed to arrive at the integer counts underlying Gromov–Witten theory.

A map f from a nodal curve C to a nonsingular variety X is *unramified* at a nonsingular point $p \in C$ if the differential

$$df : T_{C,p} \to T_{X,f(p)}$$

is injective. If f is unramified at p, the component of C on which p lies cannot be collapsed.

The idea of stable unramified maps, introduced by Kim, Kresch, and Oh [51], is to control both the domain (allowing only nodal curves) *and* the singularities of the maps (essentially unramified and with no collapsed components). The price for these properties is paid in the complexity of the target space X. The target cannot remain inert, but must be allowed to degenerate.

Degenerations

Let X be a nonsingular projective variety of dimension n. The Fulton–MacPherson [31] configuration space $X[k]$ compactifies the moduli of k distinct labelled points on X. The Fulton–MacPherson compactification may be viewed as a higher dimensional analogue of the geometry of marked points on stable curves – when the points attempt to collide, the space X degenerates and the colliding points are separated in a bubble.

The possible degenerations of X which occur are easy to describe. We start with the trivial family

$$\pi : X \times \triangle_0 \to \triangle_0 ,$$

with fibre X over the disk $\triangle_0$ with base point 0. Next, we allow an iterated sequence of finitely many blow-ups of the total space $X \times \triangle_0$ at points which, at each stage,

(i) lie over $0 \in \triangle_0$ and
(ii) lie in the smooth locus of the morphism to $\triangle_0$.

After the sequence of blow-ups is complete, we take the fibre $\widetilde{X}$ of the resulting total space $\widetilde{X \times \triangle_0}$ over $0 \in \triangle_0$. The space $\widetilde{X}$, a *Fulton–MacPherson degeneration* of X, is a normal crossings divisor in the total space.

The Fulton–MacPherson degeneration $\widetilde{X}$ contains a distinguished component X_+ which is a blow-up of the original X at distinct points. The other components of $\widetilde{X}$ are simply blow-ups of $\mathbb{P}^n$. Of the latter, there are two special types

(i) *ruled* components ($\mathbb{P}^n$ blown-up at 1 point),
(ii) *end* components ($\mathbb{P}^n$ blown-up at 0 points).

The singularities of $\widetilde{X}$ occur only in the intersections of the components.

By construction, there is a canonical morphism

$$\rho : \widetilde{X} \to X$$

which blows-down X_+ and contracts the other components of $\widetilde{X}$. The automorphisms of $\widetilde{X}$ which commute with ρ can only be nontrivial on the components of type (i) and (ii).

For the moduli of stable unramified maps, the target X is allowed to degenerate to any Fulton–MacPherson degeneration $\widetilde{X}$.

Moduli

Let X be a nonsingular projective variety of dimension n. The moduli space $\mathcal{M}_g(X, \beta)$ of *stable unramified maps* to X parameterises the data

$$C \xrightarrow{f} \widetilde{X} \xrightarrow{\rho} X$$

satisfying the following conditions:

(i) C is a connected nodal curve of arithmetic genus g,

(ii) $\widetilde{X}$ is a Fulton–MacPherson degeneration of X with canonical contraction ρ,

(iii) $\rho_* f_*[C] = \beta \in H_2(X, \mathbb{Z})$,

(iv) the nonsingular locus of $\widetilde{X}$ pulls-back to exactly the nonsingular locus of C,

$$f^{-1}(\widetilde{X}^{ns}) = C^{ns} ,$$

(v) f is *unramified* on C^{ns},

(vi) at each node $q \in C$, the two incident branches $B_1, B_2 \subset C$ map to two *different* components $Y_1, Y_2 \subset \widetilde{X}$ and meet the intersection divisor at q with equal multiplicities,

$$\left[B_1 \cdot Y_1 \cap Y_2 \right]_{Y_1, q} = \left[B_2 \cdot Y_1 \cap Y_2 \right]_{Y_2, q} ,$$

(vii) for each ruled component $R \subset \widetilde{X}$, there is a component of C which is mapped by f to R with image *not* equal to a fibre of the ruling,

(viii) for each end component $E \subset \widetilde{X}$, there is a component of C which is mapped by f to E with image *not* equal to a straight line.

By (v), the map f is unramified everywhere except possibly at the nodes of C (which must map to the singular locus of $\widetilde{X}$). Constraint (vi) is the standard *admissibility* condition for infinitesimal smoothing which arises in relative Gromov–Witten theory [45, 61, 62]. Conditions (vii) and (viii) serve to stabilize the components of $\widetilde{X}$ with automorphisms over ρ.

The moduli space $\mathcal{M}_g(X, \beta)$ of unramified maps is a proper Deligne–Mumford stack. The unramified map limits of our two simple examples of degenerations (6.12) and (6.14) are easily described. For (6.12), the stable map limit is a double cover which is ramified over two branch points. In the unramified limit, we take the Fulton–MacPherson degeneration which blows up these points in X and adds projective space components. The proper transform of the double cover is then attached to nonsingular plane conics in the two added projective spaces. The conics are tangent to the intersection divisors at the points hit by the double cover. For (6.14), the limit is the same as in Gromov–Witten theory: the normalisation of the image (6.15) in the trivial Fulton–MacPherson degeneration of X.

A central result of [51] is the identification of the deformation/obstruction theory of an unramified map mixing the (unobstructed) deformation theory of Fulton–MacPherson degenerations with the usual deformation theory of maps to X. The deformation/obstruction theory is 2-term, and a virtual class is constructed on $\mathcal{M}_g(X, \beta)$ of dimension

$$\int_\beta c_1(X) + (\dim_{\mathbb{C}} X - 3)(1 - g)$$

as in Gromov–Witten theory.

There is no difficulty to include marked points in the definition of unramified maps [51]. Via incidence conditions imposed at the markings, a full set of unramified invariants can be constructed for any X.

Connections to BPS counts: CY case

How do the unramified invariants relate to all the other counting theories we have discussed? Since unramified invariants have been introduced very recently, not many calculations have been done. In the case of Calabi–Yau 3-folds X, an attempt [91] at finding the analogue of the Aspinwall–Morrison formula for multiple covers of an embedded $\mathbb{P}^1 \subset X$ with normal bundle $\mathcal{O}_{\mathbb{P}^1}(-1) \oplus \mathcal{O}_{\mathbb{P}^1}(-1)$ showed the invariant was different for double covers.

A full transformation relating the unramified theory to the other Calabi–Yau counts has not yet been proposed. Surely such a transformation exists and has an interesting form.

Question: What is the relationship between unramified invariants and Gromov–Witten theory for the Calabi–Yau 3-folds?

Connections to BPS counts: positive case

Let X be a nonsingular projective 3-fold and let $\beta \in H_2(X, \mathbb{Z})$ be a curve class satisfying

$$\int_\beta c_1(X) > 0. \tag{6.33}$$

Let $\gamma_1, \ldots, \gamma_n \in H^*(X, \mathbb{Z})$ be integral cohomology classes Poincaré dual to cycles in X defining incidence conditions for curves. We require the dimension constraint

$$n + \int_\beta c_1(X) = \sum_{i=1}^n \operatorname{codim}_{\mathbb{C}}(\gamma_i)$$

to be satisfied. Let

$$N^{\mathrm{UR}}_{g,\beta}(\gamma_1, \ldots, \gamma_n) \in \mathbb{Q}$$

be the corresponding genus g unramified invariant.

The BPS state counts of Gopakumar and Vafa were generalized from the Calabi–Yau to the positive case in [77, 76]. The BPS invariants $n_{g,\beta}(\gamma_1, \ldots, \gamma_n)$ are *defined* via Gromov–Witten theory by:

$$\sum_{g \geq 0} N^{\mathrm{GW}}_{g,\beta}(\gamma_1, \ldots, \gamma_n) \, u^{2g-2} =$$

$$\sum_{g \geq 0} n_{g,\beta}(\gamma_1, \cdots, \gamma_n) \, u^{2g-2} \left(\frac{\sin(u/2)}{u/2} \right)^{2g-2+\int_\beta c_1(X)}. \tag{6.34}$$

Zinger [108] proved the above definition yields *integers* $n_{g,\beta}(\gamma_1, \ldots, \gamma_n)$ which vanish for sufficiently high g (depending upon β) when the positivity (6.33) is satisfied. The following conjecture[25] connects the unramified theory to BPS counts.

BPS conjecture III: $N^{\mathrm{UR}}_{g,\beta}(\gamma_1, \ldots, \gamma_n) = n_{g,\beta}(\gamma_1, \ldots, \gamma_n).$

The above simple BPS relation should be true because the moduli space of unramified maps avoids all collapsed contributions. If proved, unramified maps may be viewed as providing a direct construction of the BPS counts in the positive case.

---

[25] BPS conjecture III for unramified invariants was made by R.P. and appears in Section 5.2 of [51].

Advantages

The main advantage of the unramified theory is the simple form of the singularities of the maps. In particular, avoiding collapsed components leads to (the expectations of) much better behaviour than Gromov–Witten theory.

The theory also enjoys many of the advantages of Gromov–Witten theory: definition in all dimensions, relationship to the moduli of curves, and connection with naive enumerative geometry for $\mathbb{P}^2$ and $\mathbb{P}^1 \times \mathbb{P}^1$.

Drawbacks

The Fulton–MacPherson degenerations add a great deal of complexity to calculations in the unramified theory. Even in modest geometries, a large number of components in the degenerations are necessary. In localization formulas, Hodge integrals on various Hurwitz/admissible cover moduli spaces occur (analogous to the standard Hodge integrals on the moduli space of curves appearing in Gromov–Witten theory). While the latter have been studied for a long time, the structure of the former has not been so carefully understood.

Unramified maps remove the degenerate contributions of Gromov–Witten theory, but keep the multiple covers. For Calabi–Yau 3-folds, the invariants are rational numbers. The BPS invariants are expected to underlie the theory, but how is not yet understood.

The unramified theory is expected to be symplectic, but the details have not been worked out yet.

Serious difficulties

The theory has been studied for only a short time. Whether the complexity of the degenerating target is too difficult to handle remains to be seen.

$6\frac{1}{2}$ Stable quotients

Sheaves on curves

We have seen compactifications of the family of curves on X via maps of nodal curves to X and via sheaves on X. The counting theory obtained from the moduli space of stable quotients [67], involving sheaves on nodal curves, takes a hybrid approach. The stable quotients invariants are directly connected to Gromov–Witten theory in many basic cases. However, the main application of stable quotients to date has been to the geometry of the moduli space of curves.

Moduli

Let $(C, p_1, \ldots, p_n)$ be a connected nodal curve with nonsingular marked points. Let q be a quotient of the rank N trivial bundle C,

$$\mathbb{C}^N \otimes \mathcal{O}_C \xrightarrow{q} Q \to 0 .$$

If the quotient sheaf Q is locally free at the nodes of C, then q is a *quasi-stable quotient*. Quasi-stability of q implies the associated kernel,

$$0 \to S \to \mathbb{C}^N \otimes \mathcal{O}_C \xrightarrow{q} Q \to 0 ,$$

is a locally free sheaf on C. Let r denote the rank of S.

Let C be a curve equipped with a quasi-stable quotient q. The data (C, q) determine a *stable quotient* if the $\mathbb{Q}$-line bundle

$$\omega_C(p_1 + \ldots + p_n) \otimes (\wedge^r S^*)^{\otimes \epsilon} \tag{6.35}$$

is ample on C for every strictly positive $\epsilon \in \mathbb{Q}$. Quotient stability implies $2g - 2 + n \geq 0$.

Viewed in concrete terms, no amount of positivity of S^* can stabilize a genus 0 component

$$\mathbb{P}^1 \stackrel{\sim}{=} P \subset C$$

unless P contains at least two nodes or markings. If P contains exactly two nodes or markings, then S^* *must* have positive degree.

Isomorphism

Two quasi-stable quotients on a fixed curve C

$$\mathbb{C}^N \otimes \mathcal{O}_C \xrightarrow{q} Q \to 0, \qquad \mathbb{C}^N \otimes \mathcal{O}_C \xrightarrow{q'} Q' \to 0 \tag{6.36}$$

are *strongly isomorphic* if the associated kernels

$$S, S' \subset \mathbb{C}^N \otimes \mathcal{O}_C$$

are equal.

An *isomorphism* of quasi-stable quotients

$$\phi : (C, q) \to (C', q)$$

is an isomorphism of curves

$$\phi : C \overset{\sim}{\to} C'$$

with respect to which the quotients q and $\phi^*(q')$ are strongly isomorphic. Quasi-stable quotients (6.36) on the same curve C may be isomorphic without being strongly isomorphic.

The moduli space of stable quotients $\overline{Q}_g(\mathbb{G}(r, N), d)$ parameterising the data

$$(C, \ 0 \to S \to \mathbb{C}^N \otimes \mathcal{O}_C \overset{q}{\to} Q \to 0),$$

with $rank(S) = r$ and $deg(S) = -d$, is a proper Deligne–Mumford stack of finite type over $\mathbb{C}$. A proof, by Quot scheme methods, is given in [67].

Every stable quotient (C, q) yields a rational map from the underlying curve C to the Grassmannian $\mathbb{G}(r, N)$. If the quotient sheaf Q is locally free on all of C, then the stable quotient yields a regular map from C to the Grassmannian. Hence, we may view stable quotients as compactifying the space of maps of genus g curves to Grassmannians of class d times a line.

Deformation theory

The moduli of stable quotients maps to the Artin stack of pointed domain curves

$$\nu^A : \overline{Q}_g(\mathbb{G}(r, N), d) \to \mathfrak{M}_{g,n} .$$

The moduli of stable quotients with fixed underlying curve

$$[C] \in \mathfrak{M}_{g,n}$$

is simply an open set of the Quot scheme of C. The deformation theory of the Quot scheme determines a two-term obstruction theory on $\overline{Q}_g(\mathbb{G}(r, N), d)$ relative to ν^A given by $(R \operatorname{Hom}(S, Q))^\vee$.

More concretely, for the stable quotient,

$$0 \to S \to \mathbb{C}^N \otimes \mathcal{O}_C \overset{q}{\to} Q \to 0,$$

the deformation and obstruction spaces relative to ν^A are $\operatorname{Hom}(S, Q)$ and $\operatorname{Ext}^1(S, Q)$ respectively. Since S is locally free and C is a curve, the higher obstructions

$$\operatorname{Ext}^k(S, Q) = H^k(C, S^* \otimes Q) = 0, \quad k > 1$$

vanish.

A quick calculation shows the virtual dimension of the moduli of stable quotients equals the virtual dimension of the moduli of stable maps to $\mathbb{G}(r, N)$.

Invariants

There is no difficultly in adding marked points to the moduli of stable quotients, see [67]. Therefore, we can define a theory of stable quotients invariants for Grassmannians. Similar targets such as flag varieties for $\mathbf{SL}_n$ admit a parallel development. An enumerative theory of stable quotients was sketched in [67] for complete intersections in such spaces. Hence, there is a stable quotients theory for the Calabi–Yau quintic in $\mathbb{P}^4$.

Since [67], the construction of stable quotient invariants has been extended to toric varieties [20] and appropriate GIT quotients [21]. The associated counting theories (well-defined with 2-term deformation/obstruction theories) should be regarded as depending not only on the target space, but *also* on the quotient presentation. The direction is related to the young subject of gauged Gromov–Witten theory (and in particular to the rapidly developing study of theories of Landau–Ginzburg type [17, 24]).

Question: What is the relationship between stable quotient invariants and Gromov–Witten theory for varieties?

For all flag varieties for $\mathbf{SL}_n$, the above question has a simple answer: the counting by stable quotients and Gromov–Witten theory agree exactly [67]. Perhaps exact agreement also holds for Fano toric varieties, see the conjectures in [20]. But in the non-Fano cases, and certainly for the Calabi–Yau quintic, the stable quotient theory is very different. There should be a wall-crossing understanding [101] of the transformations, but much work remains to be done.

Advantages

Stable quotients provide a more efficient compactification than Gromov–Witten theory. In the case of projective space, there is a blow-down morphism

$$\overline{\mathcal{M}}_g(\mathbb{P}^{N-1}, d) \to \overline{\mathcal{Q}}_g(\mathbb{G}(1, N), d)$$

which pushes-forward the virtual class of the moduli of stable curves to the virtual class of the moduli of stable quotients [67]. A principal use of the moduli of stable quotients has been to explore the tautological rings of the moduli of curves [81] – and in particular to prove the Faber–Zagier conjecture for relations among the κ classes on $\mathcal{M}_g$ [79]. The efficiency of the boundary plays a crucial role in the analysis.

The difference between stable maps and stable quotients can be seen already for elliptic curves in projective space. For stable maps, the associated moduli space is singular with multiple components. A desingularization, by *blowing-up*, is described in [102] and applied to calculate the genus 1 Gromov–Witten invariants of the quintic Calabi–Yau in [109]. On the other hand, the moduli of stable quotients related to such elliptic curves is a nonsingular *blow-down* of the stable maps space [67]. The stable quotients moduli here is a much smaller compactification.[26] A parallel application to the genus 1 stable quotients invariants of the quintic Calabi–Yau is a very natural direction to pursue.

Drawbacks

The stable quotients approach to the enumeration of curves, while valid for different dimensions, appears to require more structure on X (embedding, toric, or quotient presentations). The method is therefore not as flexible as Gromov–Witten theory.

Also, unlike Gromov–Witten theory, there is not yet a symplectic development. However, the connections with gauged Gromov–Witten theory may soon provide a fully symplectic path to stable quotients.

Serious difficulties

The theory has been studied for only a short time. The real obstacles, beyond those discussed above, remain to be encountered.

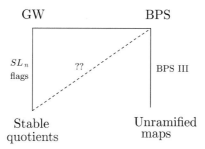

Figure 6.4. Conjectures relating curve counting theories

---

[26] A geometric investigation by Cooper of the stable quotients spaces in genus 1 for projective spaces can be found in [22].

Appendix: Virtual classes

Physical motivation

There are countless ways to compactify the spaces of curves in a projective variety X. What distinguishes the six main approaches we have described is the presence in each case of a virtual fundamental class.

Moduli spaces arising in physics should naturally carry virtual classes when cut out by a section (the derivative of an action functional) of a vector bundle over a nonsingular ambient space (the space of fields). While both the space and bundle are usually infinite dimensional, the derivative of the section is often Fredholm, so we can make sense of the difference in the dimensions. The difference is the *virtual dimension* of the moduli space – the number of equations minus the number of unknowns. The question, though, of what geometric objects to place in the boundary is often not so clearly specified in the physical theory.

As an example, the space of C^∞-maps from a Riemann surface C to X, modulo diffeomorphisms of C, is naturally an infinite dimensional orbifold away from the maps with infinite automorphisms. Taking $\bar\partial$ of such a map gives a Fredholm section of the infinite rank bundle with fibre $\Gamma(\Omega_C^{0,1}(f^*T_X))$ over the map f. The zeros of the section are the holomorphic maps

$$f : C \to X.$$

However, to arrive at the definition of a stable map requires further insights about nodal curves.

The Fredholm property allows us to take slices to reduce locally to the following finite dimensional model of the moduli problem.

Basic model

Consider a nonsingular ambient variety A of dimension n. Let E be a rank r bundle on A with section $s \in \Gamma(E)$ with zero locus $\mathcal{M}$:

$$\begin{array}{c} E \\ \downarrow{\scriptstyle s} \\ \mathcal{M} = Z(s) \subset A \end{array} \qquad (6.37)$$

Certainly, $\mathcal{M}$ has dimension $\geq n - r$. We define

$$\mathrm{vdim}(\mathcal{M}) = n - r$$

to be the *virtual dimension* of $\mathcal{M}$.

The easiest case to understand is when s takes values in a rank r' subbundle

$$E' \subset E$$

and is transverse to the zero section in E'. Then, $\mathcal{M}$ is nonsingular of dimension

$$n - r' = \mathrm{vdim}(M) + (r - r').$$

If E splits as $E = E' \oplus E/E'$, we can write $s = (s', 0)$. We can then perturb s to the section

$$s_\epsilon = (s', \epsilon)$$

with new zero locus given by

$$Z(\epsilon) \subset \mathcal{M}.$$

In particular, if ϵ can be chosen to be transverse to the zero section of E/E', we obtain a smooth moduli space $Z(\epsilon)$ of the "correct" dimension $\mathrm{vdim}(\mathcal{M})$ cut out by a transverse section s_ϵ of E. The fundamental class is

$$[Z(\epsilon)] = c_r(E)$$

in the (co)homology of A. If we work in the C^∞ category, we can always split E and pick such a transverse C^∞-section.

Even when E/E' has no algebraic sections (for instance if E/E' is negative), the fundamental class of $Z(\epsilon)$ is clearly $c_{r-r'}(E/E')$ in the (co)homology of $\mathcal{M}$. The "correct" moduli space, obtained when sufficiently generic perturbations of s exist or when we use C^∞ sections, has fundamental class given by the push-forward to A of the top Chern class of E/E'. The result is called the *virtual fundamental class*:

$$[\mathcal{M}]^{vir} = c_{r-r'}(E/E') \in A_{\mathrm{vdim}}(\mathcal{M}) \to H_{2\mathrm{vdim}}(\mathcal{M}).$$

Here, E/E', the cokernel of the derivative of the defining equations s, is called the *obstruction bundle* of the moduli space $\mathcal{M}$, for reasons we explain below.

More generally s need not be transverse to the zero section of any subbundle of E, and we must use the excess intersection theory of Fulton–MacPherson [30]. The limit as $t \to \infty$ of the graph of ts defines a cone

$$C_s \subset E|_{\mathcal{M}}.$$

We define the virtual class to be the refined intersection of C_s with the zero section $0_E : \mathcal{M} \hookrightarrow E$ inside the total space of E:

$$[\mathcal{M}]^{vir} = 0_E^![C_s] \in A_{\mathrm{vdim}}(\mathcal{M}) \to H_{2\mathrm{vdim}}(\mathcal{M}). \tag{6.38}$$

The result can also be expressed in terms of $c(E)s(C_s)$, where c is the total Chern class, and s is the Segre class.

In the easy split case with $s = (s', 0)$ discussed, C_s is precisely E'. We recover the top Chern class $c_{r-r'}(E/E')$ of the obstruction bundle for the virtual class.

Deformation theory

While the basic model (6.37) for $\mathcal{M}$ rarely exists in practice (except in infinite dimensions), an infinitesimal version can be found when the moduli space admits a 2-term deformation/obstruction theory. The excess intersection formula (6.38) uses data only on $\mathcal{M}$ (rather than a neighbourhood of $\mathcal{M} \subset A$) and can be used in the infinitesimal context.

At a point $p \in \mathcal{M}$, the basic model (6.37) yields the following exact sequence of Zariski tangent spaces

$$0 \to T_p\mathcal{M} \to T_pA \xrightarrow{ds} E_p \to \mathrm{Ob}_p \to 0. \tag{6.39}$$

So to first order, at the level of the Zariski tangent space, the moduli space looks like $\ker ds$ near $p \in \mathcal{M}$. Higher order neighbourhoods of $p \in \mathcal{M}$ are described by the implicit function theorem by the zeros of the nonlinear map $\pi(s)$, where π is the projection from E_p to Ob_p. The obstruction to prolonging a first order deformation of p inside $\mathcal{M}$ to higher order lies in Ob_p.[27]

The deformation and obstruction spaces, $T_p\mathcal{M}$ and Ob_p, have dimensions differing by the virtual dimension

$$\mathrm{vdim} = \dim A - \mathrm{rank}\, E$$

and are the cohomology of a complex of *vector bundles*[28]

$$B_0 \to B_1$$

over $\mathcal{M}$ restricted to p. The resolution of $T_p\mathcal{M}$ and Ob_p is the local infinitesimal method to express that $\mathcal{M}$ is cut out of a nonsingular ambient space by a section of a vector bundle.

Li and Tian [65] have developed an approach to handling deformation/obstruction theories over $\mathcal{M}$. If a global resolution $B_0 \to B_1$ exists, Li and Tian construct a cone $C_s \subset E_1$ and intersect with the zero cycle as in (6.38) to

---

[27] The obstruction space is not unique. Analogously, a choice of generators for the ideal of a subscheme $\mathcal{M} \subset A$ is not unique. For instance in our basic model we could have taken the obstruction bundle to be E/E' or zero. In each of the six approaches to curve counting, a natural *choice* of an obstruction theory is made.

[28] $B_0 = TA$ and $B_1 = E$ are vector bundles since A is smooth and E is a bundle.

define a virtual class on $\mathcal{M}$. Due to base change issues, the technique is difficult to state briefly, but the upshot is that if the deformation and obstruction spaces of a moduli problem have a difference in dimension which is *constant* over $\mathcal{M}$ we can (almost always) expect a virtual cycle of the expected dimension.

Behrend–Fantechi

We briefly describe a construction of the virtual class proposed by Behrend and Fantechi [6] which is equivalent and also more concise.

Dualising and globalising (6.39), we obtain the exact sequence of sheaves

$$E^*|_\mathcal{M} \xrightarrow{ds} \Omega_A|_\mathcal{M} \to \Omega_\mathcal{M} \to 0,$$

where the kernel of the leftmost map contains information about the obstructions. The sequence factors as

$$
\begin{array}{ccc}
E^*|_\mathcal{M} & \xrightarrow{ds} & \Omega_A|_\mathcal{M} \\
\downarrow s & & \| \\
I/I^2 & \xrightarrow{d} & \Omega_A|_\mathcal{M} \longrightarrow \Omega_\mathcal{M} \to 0,
\end{array}
$$

where I is the ideal of $\mathcal{M} \subset A$ and the bottom row is the associated exact sequence of Kähler differentials. We write $E^*|_\mathcal{M} \xrightarrow{ds} \Omega_A|_\mathcal{M}$ as

$$B^{-1} \to B^0,$$

a two-term complex of *vector bundles* because A is nonsingular and E is a bundle. The complex

$$\{I/I^2 \to \Omega_A|_\mathcal{M}\}$$

is (quasi-isomorphic to) the *truncated cotangent complex* $\mathbb{L}_\mathcal{M}$ of $\mathcal{M}$. Our data is what Behrend and Fantechi call a *perfect obstruction theory*: a morphism of complexes

$$B^\bullet \to \mathbb{L}_\mathcal{M}$$

which is an isomorphism on h^0 (the identity map $\Omega_\mathcal{M} \to \Omega_\mathcal{M}$) and a surjection on h^{-1} (because $E^* \to I/I^2$ is onto). The definition can also be interpreted in terms of classical deformation theory [6, Theorem 4.5].

Behrend and Fantechi show how a perfect obstruction theory leads to a cone in $B_1 = (B^{-1})^*$ which can be intersected with the zero section to give a virtual class of dimension

$$\text{vdim} = \text{rank } B^0 - \text{rank } B^{-1}.$$

The virtual class is the usual fundamental class when the moduli space has the correct dimension and is the top Chern class of the obstruction bundle when $\mathcal{M}$ is nonsingular. The virtual class is also deformation invariant in an appropriate sense that would take too long to describe here.

References

[1] A. Bayer, *Polynomial Bridgeland stability conditions and the large volume limit*, Geom. Topol. **13**, 2389–2425, 2009. arXiv:0712.1083.

[2] A. Beauville, *Counting rational curves on $K3$ surfaces*, Duke Math. J. **97**, 99–108, 1999. alg-geom/9701019.

[3] K. Behrend, *Donaldson-Thomas invariants via microlocal geometry,* Ann. of Math. **170**, 1307–1338, 2009. math.AG/0507523.

[4] K. Behrend and J. Bryan. *Super-rigid Donaldson-Thomas invariants*, Math. Res. Lett. **14**, 559–571, 2007. math.AG/0601203.

[5] K. Behrend, J. Bryan. and B. Szendrői. *Motivic degree zero Donaldson-Thomas invariants,* arXiv:0909.5088.

[6] K. Behrend and B. Fantechi, *The intrinsic normal cone*, Invent. Math., **128**, 45–88, 1997. alg-geom/9601010.

[7] K. Behrend and B. Fantechi. *Symmetric obstruction theories and Hilbert schemes of points on threefolds*, Alg. Numb. Theor. **2**, 313–345, 2008. math.AG/0512556.

[8] P. Belorousski and R. Pandharipande, *A descendent relation in genus 2*, Ann. Scuola Norm. Sup. Pisa **29**, 171–191, 2000. math.AG/9803072.

[9] T. Bridgeland. *Stability conditions on triangulated categories*, Ann. of Math. **166**, 317–345, 2007. math.AG/0212237.

[10] T. Bridgeland. *Hall algebras and curve-counting invariants*, J. AMS **24**, 969–998, 2011. arXiv:1002.4374.

[11] J. Bryan and C. Leung, *The enumerative geometry of K3 surfaces and modular forms*, J. AMS, **13**, 371–410, 2000. math.AG/0009025.

[12] J. Bryan and R. Pandharipande, *BPS states of curves in Calabi-Yau 3-folds*, Geom. Topol. **5**, 287–318, 2001. math.AG/0306316.

[13] J. Bryan and R. Pandharipande, *Curves in Calabi-Yau threefolds and TQFT*, Duke J. Math. **126**, 369–396, 2005.

[14] J. Bryan and R. Pandharipande, *On the rigidity of stable maps to Calabi-Yau threefolds*, in *The interaction of finite-type and Gromov-Witten invariants (BIRS 2003)*, Geom. Top. Monogr. **8**, 97–104, 2006. math.AG/0405204.

[15] P. Candelas, X. de la Ossa, P. Green, and L. Parks, *A pair of Calabi-Yau manifolds as an exactly soluble superconformal field theory*, Nucl. Phys. **B359** (1991), 21–74.

[16] L. Caporaso and J. Harris, *Counting plane curves of any genus*, Invent. Math. **131** (1998), 345–392. alg-geom/9608025.

[17] H.-L. Chang and J. Li, *Gromov-Witten invariants of stable maps with fields*, arXiv:1101.0914.

[18] X. Chen, *Rational curves on $K3$ surfaces*, J. Alg. Geom. **8**, 245–278, 1999. math/9804075.

[19] W.-Y. Chuang, D.-E. Diaconescu and G. Pan, *BPS states, Donaldson-Thomas invariants and Hitchin pairs*, preprint.

[20] I. Ciocan-Fontanine and B. Kim, *Moduli stacks of stable toric quasimaps*, Adv. in Math. **225**, 3022–3051, 2010. arXiv:0908.4446.

[21] I. Ciocan-Fontanine, B. Kim and D. Maulik, *Stable quasimaps to GIT quotients*, arXiv:1106.3724.

[22] Y. Cooper, *The geometry of stable quotients in genus one*, arXiv:1109.0331.

[23] D. Cox and S. Katz, *Mirror symmetry and algebraic geometry*. Mathematical Surveys and Monographs, **68**. AMS, Providence, RI, 1999.

[24] H.-J. Fan, T. J. Jarvis, and Y. Ruan, *The Witten equation and its virtual fundamental cycle*, arXiv:0712.4025.

[25] P. Di Francesco, C. Itzykson, *Quantum intersection rings*. in *The moduli space of curves*, R. Dijkgraaf, C. Faber, and G. van der Geer, eds., Birkhauser, 81–148, 1995.

[26] R. Dijkgraff, E. Verlinde, and H. Verlinde, *Topological strings in d < 1*, Nucl. Phys. **B352** 59–86, 1991.

[27] S. K. Donaldson and R. P. Thomas. *Gauge theory in higher dimensions*. In *The geometric universe (Oxford, 1996)*, 31–47. Oxford Univ. Press, Oxford, 1998.

[28] A. Douady and J.-L. Verdier, *Séminaire de géometrie analytique á ENS 1974/1975*, Astérisque **36-37**, 1976.

[29] C. Faber and R. Pandharipande. *Hodge integrals and Gromov-Witten theory*, Invent. Math., **139**, 173–199, 2000. math.AG/9810173.

[30] W. Fulton, *Intersection theory*. Springer-Verlag, Berlin, 1984.

[31] W. Fulton and R. MacPherson, *A compactification of configuration spaces*, Ann. of Math. **139**, 183–225, 1994.

[32] W. Fulton and R. Pandharipande, *Notes on stable maps and quantum cohomology*, Algebraic geometry (Santa Cruz 1995), 45–96, Proc. Sympos. Pure Math. 62, Part 2, Amer. Math. Soc., Providence, RI, 1997. alg-geom/9608011.

[33] E. Getzler, *Intersection theory on $\overline{M}_{1,4}$ and elliptic Gromov-Witten invariants*, Jour. AMS **10**, 973–998, 1997. alg-geom/9612004.

[34] A. Givental, *Equivariant Gromov-Witten invariants*, Internat. Math. Res. Notices **13**, 613–663, 1996. alg-geom/9603021.

[35] R. Gopakumar and C. Vafa, *M-theory and topological strings–I*, hep-th/9809187.

[36] R. Gopakumar and C. Vafa, *M-theory and topological strings–II*, hep-th/9812127.

[37] L. Göttsche, *A conjectural generating function for numbers of curves on surfaces*, Comm. Math. Phys. **196**, 523–533, 1998.

[38] L. Göttsche and R. Pandharipande, *The quantum cohomology of blow-ups of $\mathbb{P}^2$ and enumerative geometry*, J. Diff. Geom. **48**, 61–90, 1998. alg-geom/9611012.

[39] M. Gromov, *Pseudo holomorphic curves in symplectic manifolds*, Invent. Math. **82**, 307–347, 1985.

[40] M. Gross, *Tropical geometry and mirror symmetry*, AMS, 2011.

[41] M. Gross, R. Pandharipande, B. Siebert, *The tropical vertex*. Duke Math. J. **153**, 297–362, 2010. arXiv:0902.0779.

[42] M. Gross and B. Siebert, *From real affine geometry to complex geometry*, Ann. of Math. **174**, 13011428, 2011. math.AG/0703822.

[43] K. Hori, S. Katz, A. Klemm, R. Pandharipande, R. Thomas, C. Vafa, R. Vakil, and E. Zaslow, *Mirror Symmetry*, AMS: Providence, R.I., 2003.

[44] S. Hosono, M. Saito, and A. Takahashi, *Relative Lefschetz action and BPS state counting,* Internat. Math. Res. Notices, **15**, 783–816, 2001. math.AG/0105148.

[45] E.-N. Ionel and T. Parker, *Relative Gromov-Witten Invariants.* Ann. of Math. **157**, 45–96, 2003. math.SG/9907155.

[46] I. Itenberg, G. Mikhalkin, and E. Shustin, *Tropical algebraic geometry.* Oberwolfach Seminars, 35. Birkhuser Verlag, Basel, 2007.

[47] D. Joyce, *Kuranishi homology and Kuranishi cohomology.* arXiv:0707.3572.

[48] D. Joyce and Y. Song, *A theory of generalized Donaldson Thomas invariants,* to appear in Memoirs of the AMS, 2011. arXiv:0810.5645.

[49] S. Katz, *Genus zero Gopakumar-Vafa invariants of contractible curves,* J. Diff. Geom. **79**, 185–195, 2008. math.AG/0601193.

[50] S. Katz, A. Klemm, and C. Vafa, *M-theory, topological strings and spinning black holes*, Adv. Theor. Math. Phys., **3**, 1445–1537, 1999. hep-th/9910181.

[51] B. Kim, A. Kresch and Y-G. Oh, *A compactification of the space of maps from curves,* arXiv:1105.6143.

[52] A. Klemm, D. Maulik, R. Pandharipande and E. Scheidegger, *Noether-Leftschetz theory and the Yau-Zaslow conjecture* J. AMS **23**, 1013–1040, 2010. arXiv:0807.2477.

[53] A. Klemm and R. Pandharipande, *Enumerative geometry of Calabi-Yau 4-folds.* Comm. Math. Phys. **281**, 621–653, 2008. math.AG/0702189.

[54] J. Kollár, *Rational curves on algebraic varieties*, Springer-Verlag: Berlin, 1999.

[55] M. Kontsevich, *Intersection theory on the moduli space of curves and the matrix Airy function.* Comm. Math. Phys. **147**, 1–23, 1992.

[56] M. Kontsevich, *Enumeration of rational curves via torus actions*, in *The moduli space of curves*, R. Dijkgraaf, C. Faber, and G. van der Geer, eds., Birkhauser, 335–368, 1995. hep-th/9405035.

[57] M. Kontsevich and Yu. Manin, *Gromov-Witten classes, quantum cohomology, and enumerative geometry*, Comm. Math. Phys. **164**, 525–562, 1994. hep-th/9402147.

[58] M. Kontsevich and Y. Soibelman, *Stability structures, motivic Donaldson-Thomas invariants and cluster transformations*, arXiv:0811.2435.

[59] J. Le Potier, *Systèmes cohérents et structures de niveau,* Astérisque, **214**, 143, 1993.

[60] M. Levine and R. Pandharipande, *Algebraic cobordism revisited*, Invent. Math. **176**, 63–130, 2009. math.AG/0605196.

[61] A-M. Li, and Y. Ruan, *Symplectic surgery and Gromov-Witten invariants of Calabi-Yau 3-folds I*, Invent. Math. **145**, 151–218, 2001. math.AG/9803036.

[62] J. Li, *Stable morphisms to singular schemes and relative stable morphisms*, J. Diff. Geom. **57**, 509–578, 2001. math.AG/0009097.

[63] J. Li. *Zero dimensional Donaldson-Thomas invariants of threefolds*, Geom. Topol. **10**, 2117–2171, 2006. math.AG/0604490.

[64] J. Li. *Recent progress in GW-invariants of Calabi-Yau threefolds*, In *Current Developments in Mathematics, 2007*, International Press, 2009.

[65] J. Li and G. Tian, *Virtual moduli cycles and Gromov-Witten invariants of algebraic varieties,* Jour. AMS, **11**, 119–174, 1998. alg-geom/9602007.

[66] B. Lian, K. Liu, and S.-T. Yau, *Mirror principle I*, Asian J. Math. **4**, 729–763, 1997. alg-geom/9712011.

[67] A. Marian, D. Oprea and R. Pandharipande, *The moduli space of stable quotients*, Geom. Topol. **15**, 1651–1706, 2011. arXiv:0904.2992.

[68] D. Maulik, N. Nekrasov, A. Okounkov, and R. Pandharipande, *Gromov-Witten theory and Donaldson-Thomas theory. I,* Compos. Math., **142**, 1263–1285, 2006. math.AG/0312059.

[69] D. Maulik, N. Nekrasov, A. Okounkov, and R. Pandharipande. *Gromov-Witten theory and Donaldson-Thomas theory. II*, Compos. Math., **142**, 1286–1304, 2006. math.AG/0406092.

[70] D. Maulik, A. Oblomkov, A. Okounkov, and R. Pandharipande. *Gromov-Witten/Donaldson-Thomas correspondence for toric 3-folds*, Invent. Math. **186**, 435–479, 2011. arXiv:0809.3976.

[71] D. Maulik and Z. Yun, *Macdonald formula for curves with planar singularities.* arXiv:1107.2175

[72] L. Migliorini and V. Shende, *A support theorem for Hilbert schemes of planar curves.* arXiv:1107.2355.

[73] G. Mikhalkin, *Enumerative tropical algebraic geometry in* $\mathbb{R}^2$. J. AMS. **18**, 313–377, 2005. math/0312530.

[74] A. Okounkov and R. Pandharipande, *Gromov-Witten theory, Hurwitz numbers, and matrix models.* Algebraic geometry, Seattle 2005. Proc. Sympos. Pure Math., **80**, Part 1, 325–414. math.AG/0101147.

[75] A. Okounkov and R. Pandharipande, *Gromov-Witten theory, Hurwitz theory, and completed cycles.* Ann. of Math. **163**, 517–560, 2006. math.AG/0204305.

[76] R. Pandharipande, *Hodge integrals and degenerate contributions*, Comm. Math. Phys. **208**, 489–506, 1999. math/9811140.

[77] R. Pandharipande. *Three questions in Gromov-Witten theory.* In *Proceedings of the International Congress of Mathematicians, Vol. II (Beijing, 2002)* 503–512, Beijing, 2002. Higher Ed. Press. math.AG/0302077.

[78] R. Pandharipande, *Maps, sheaves, and K3 surfaces,* arXiv:0808.0253.

[79] R. Pandharipande, *The kappa ring of the moduli of curves of compact type*, to appear in Acta. Math. arXiv:0906.2657 and arXiv:0906.2658.

[80] R. Pandharipande and A. Pixton, *Descendent theory for stable pairs on toric 3-folds*, arXiv:1011.4054.

[81] R. Pandharipande and A. Pixton, *Relations in the tautological ring*, arXiv:1101.2236.

[82] R. Pandharipande and R. P. Thomas, *Curve counting via stable pairs in the derived category*, Invent. Math. **178**, 407–447, 2009. arXiv:0707.2348.

[83] R. Pandharipande and R. P. Thomas, *Stable pairs and BPS invariants*, J. AMS. **23**, 267–297, 2010. arXiv:0711.3899.

[84] R. Pandharipande and R. P. Thomas, *Almost closed 1-forms*, arXiv:1204.3958.

[85] R. Pandharipande and A. Zinger, *Enumerative geometry of Calabi-Yau 5-folds*, New Developments in Algebraic Geometry, Integrable Systems and Mirror Symmetry (RIMS, Kyoto, 2008), 239–288, Adv. Stud. Pure Math., **59**, Math. Soc. Japan, Tokyo, 2010. arXiv:0802.1640.

[86] R. Piene and M. Schlessinger *On the Hilbert scheme compactification of the space of twisted cubics*. Amer. J. Math. **107**, 761–774, 1985.

[87] Z. Ran, *The degree of a Severi Variety*, Bull. AMS, 125–128, 1997.

[88] Y. Ruan, *Topological sigma model and Donaldson type invariants in Gromov theory*, Duke Math. J. **83**, 461–500, 1996.

[89] Y. Ruan and G. Tian, *A mathematical theory of quantum cohomology*, Math. Res. Lett. **1**, 269–278, 1994.

[90] H. Schubert, *Kalkül der abzählenden Geometrie*, B. G. Teubner, Leipzig, 1879.

[91] I. Setayesh, *Multiple cover calculation for the unramified compactification of the moduli space of stable maps*, arXiv:1305.3404.

[92] V. Shende, *Hilbert schemes of points on a locally planar curve and the Severi strata of its versal deformation*, to appear in Compositio Math. arXiv:1009.0914.

[93] W. Stevens, *Thirteen ways of looking at a blackbird*, Ateliers Leblanc, 1997.

[94] J. Stoppa and R. P. Thomas, *Hilbert schemes and stable pairs: GIT and derived category wall crossings*. To appear in Bull. SMF., 2011. arXiv:0903.1444.

[95] C. H. Taubes, SW $\Rightarrow$ Gr: *from the Seiberg-Witten equations to pseudo-holomorphic curves*. Jour. AMS. **9**, 845–918, 1996.

[96] R. P. Thomas, *A holomorphic Casson invariant for Calabi-Yau 3-folds, and bundles on K3 fibrations*, J. Diff. Geom. **54**, 367–438, 2000. math.AG/9806111.

[97] Y. Toda. *Birational Calabi-Yau 3-folds and BPS state counting*, Comm. Numb. Theor. Phys. **2**, 63–112, 2008. math.AG/07071643.

[98] Y. Toda, *Limit stable objects on Calabi-Yau 3-folds*, Duke Math. J. **149**, 157–208, 2009. arXiv:0803.2356.

[99] Y. Toda, *Curve counting theories via stable objects I: DT/PT correspondence*. J. AMS., **23**, 1119–1157, 2010. arXiv:0902.4371.

[100] Y. Toda, *Stability conditions and curve counting invariants on Calabi-Yau 3-folds*, arXiv:1103.4229.

[101] Y. Toda, *Moduli spaces of stable quotients and wall-crossing phenomena*, Compositio Math. **147** 1479–1518, 2011. arXiv:1005.3743.

[102] R. Vakil and A. Zinger, *A desingularization of the main component of the moduli space of genus-one stable maps into $\mathbb{P}^n$*, Geom. Topol. **12**, 1–95, 2008. math.AG/0603353.

[103] E. Witten, *On the structure of the topological phase of two-dimensional gravity*, Nucl. Phys. **B340**, 281–332, 1990.

[104] E. Witten, *Two dimensional gravity and intersection theory on moduli space*, Surv. Diff. Geom. **1**, 243–310, 1991.

[105] E. Witten, *Chern-Simons gauge theory as a string theory*. In *The Floer memorial volume*, Progr. Math., **133**, 637–678. Birkhuser, Basel, 1995. hep-th/9207094.

[106] S.-T. Yau and E. Zaslow, *BPS states, string duality, and nodal curves on K3*, Nucl. Phys. **B457**, 484–512, 1995. hep-th/9512121.

[107] H. G. Zeuthen *Lehrbuch der abzählenden methoden der geometrie*, Teubner, Leipzig, 1914.

[108] A. Zinger, *A comparison theorem for Gromov Witten invariants in the symplectic category*, Adv. Math. **228**, 535–574, 2011. arXiv:0807.0805.

[109] A. Zinger, *The reduced genus-one Gromov-Witten invariants of Calabi-Yau hypersurfaces*, J. AMS **22**, 691–737, 2009. arXiv:0705.2397.

Printed in the United States
By Bookmasters